Sustainable Production, Life Cycle Engineering and Management

Series Editors

Prof. Christoph Herrmann
Institut für Werkzeugmaschinen und Fertigungstechnik
Technische Universität Braunschweig
Braunschweig
Germany
E-mail: C.Herrmann@tu-braunschweig.de

Prof. Sami Kara
School of Mechanical & Manufacturing Engineering
The University of New South Wales
Sydney
Australia
E-mail: S.Kara@unsw.edu.au

For further volumes:
http://www.springer.com/series/10615

Sustainable Production, Life Cycle Engineering and Management

Modern production enables a high standard of living worldwide through products and services. Global responsibility requires a comprehensive integration of sustainable development fostered by new paradigms, innovative technologies, methods and tools as well as business models. Minimizing material and energy usage, adapting material and energy flows to better fit natural process capacities, and changing consumption behaviour are important aspects of future production. A life cycle perspective and an integrated economic, ecological and social evaluation are essential requirements in management and engineering. This series will focus on the issues and latest developments towards sustainability in production based on life cycle thinking.

Axel Sommer

Managing Green Business Model Transformations

Springer

Author
Mr. Axel Sommer
Centre for Sustainability Management (CSM)
Leuphana University Lüneburg
Germany

Dissertation, Leuphana Universtät Lüneburg, 2011
ISSN 2194-0541 e-ISSN 2194-055X
ISBN 978-3-642-28847-0 e-ISBN 978-3-642-28848-7
DOI 10.1007/978-3-642-28848-7
Springer Heidelberg New York Dordrecht London

Library of Congress Control Number: 2012935656

Printed on acid-free paper

Springer is part of Springer Science+Business Media (www.springer.com)

Foreword

The term "green" as used in the present book aims at the environmental dimension of sustainability. Environmental sustainability requires extensive changes in technologies, processes, products and consumption patterns. Innovation plays a key role. They must contribute to ensure that natural resources are used efficiently and effectively, to close product and material cycles, and to minimize adverse environmental impacts. Examples of such innovations are known. But often it is more about evolutionary improvement or revolutionary individual cases. A holistic approach requires a perspective that goes beyond the consideration of product and process innovations. The comprehensive transformation of the business models has to be at the center. This is where this book provides deep insight as it shows us how new (green) business models can be designed, assessed and implemented. Axel Sommer develops a comprehensive framework to support companies in the transformation towards green business models. Such green business models on the one hand represent a significant improvement in overall environmental performance relating to its entire value chain system compared to that of conventional business models and on the other hand contribute to maintaining or increasing the competitiveness of companies.

Prof. Christoph Herrmann
Technische Universität Braunschweig

Prof. Sami Kara
The University of New South Wales

Foreword

The question how companies can transform their business models through more environmental sustainability is both, highly relevant for society and management, and poorly understood to date. As a difference to much of the environmental management literature, business models do not just address some specific issues of production, products or organization. Business models are at the core of shaping all corporate activities and therefore key to corporate sustainability.

This book addresses the unfolding megatrend of companies dealing with environmental sustainability in different and often ever more fundamental manners. Axel Sommer's research sheds light on this complex business challenge of high strategic importance. He managed to strike a balance between the breadth and depth of reviewing literature that is relevant for managing green business model transformations. The systems perspective is used to link the broad fields of research that build the foundation of this thesis. In addition to the achievement of synthesising such diverse research domains, Axel Sommer provides new theoretical directions, e.g. related to the conceptualisation of "green" business models, the formation of business model paradigms, and dynamic patterns of competition resulting from green business model transformations.

The reasoning for radical change is carefully examined. This includes ethical arguments; but the focus lies on economic motives as they may be crucial for most corporate decision-makers. For this purpose, a strategic management view is taken that rationally describes corporate opportunities and risks alike which may go along with green business model transformations.

The book provides a rich variety of practical case examples that illustrate and support the conclusions of this thesis. The use of the business model concept turns out to be particularly useful as it makes analyses tangible while still maintaining a broad strategic perspective. This aspect is of high importance as sustainability initiatives can fail because of diverse, often subtle gaps in the business rationale which too focussed management approaches may fail to uncover.

The final outcome of this piece of research – a framework for management practice – is both sophisticated and useful. It starts with challenging the mental models of managers, an often neglected but critical aspect for successful radical change. The author furthermore provides a rich set of management approaches - some newly developed for this purpose – which managers can use to break outdated mental models, to generate new ideas and to translate them into a worthwhile green business model transformation. The framework's inevitable limitations are highlighted which enables practitioners to apply it in a thoughtful way.

In conclusion, this book will be of interest both for researchers and practitioners, for sustainability advocates and sceptics. It presents a way forward to solving one of the key challenges that many companies face today: being both sustainable and competitive.

Prof. Dr. Stefan Schaltegger
Centre for Sustainability Management (CSM)
Leuphana University Lüneburg
Germany

Preface

About the Content of this Book

Environmental sustainability creates both tremendous business opportunities and formidable threats to established companies across virtually all industry sectors. Yet many companies tackle the issue in a superficial or passive way through increased environmental reporting, the use of "greenspeak" in their corporate communication activities or isolated efforts to create green products or reduce pollution. In contrast, there are a small but increasing number of firms that employ a holistic approach to sustainability and consider fundamental changes to their existing business models.

By ignoring the opportunities of Green Business Model Transformations, companies exclude themselves from a large variety of potential means to create economic value. In addition to ordinary product and process innovations, they can change "the rules of the game" within an industry towards environmental sustainability. This can facilitate

- the commercialisation of new green products that would not be competitive otherwise
- targeting new customer segments with previously unmet needs
- improved economics of value creation by developing a green architecture of the firm and its business network

Green Business Model Transformations, however, are challenging ventures in many respects: Firstly, it is difficult to develop and accurately assess the prospects of new, green business models due to uncertainties and inherent complex systemic characteristics. It can therefore be very challenging for individual managers of established companies to question the status quo and seize new, green opportunities. The implementation of Green Business Model Transformations therefore requires strong change capabilities: The stakes are high, many otherwise disconnected parts of the organisation are involved, and the subject matter is complex in nature.

As a result, there is a great need for guidance in management practice that current research does not address. This work aims to fill this gap with a general approach to *Managing Green Business Model Transformations* by defining a process to evaluate green business models and providing a methodology for their realisation.

The proposed management framework is divided into six phases:

1. **Embracing ambiguity:**

   As a first step, managers need reconsider their preset patterns of thought, disregard short-term pressures and instead consider the tectonic shifts that environmental sustainability may have set in motion within their industry and the business environment at large. Scenario planning is proposed as a management tool. The tailored Accelerated Green Scenario Planning approach is both effective and efficient in replacing outdated mental models. It focuses attention to environmental issues that are critical for the firm's future success.

2. **Rethinking old certainties:**

   Building on the newly gained openness, management can rethink old certainties. New ideas are created through Executive Ideation Events and may be further refined and complemented through large-scale Crowd Ideation Events.

3. **Translating into business impact:**

   A systematic approach to design new, green business models is then needed to commercialise the new business ideas. A newly developed Business Model Refinement Heuristic should greatly increase the chances of success compared to the non-systematic approaches prevalent in current management practice.

4. **Planning for action:**

   Based on the new green business model design, a detailed economic model can be developed. If necessary, the model is enhanced by simulation based on system dynamics. A strategic plan is produced that considers the competitive environment, determines strategic measures and targets, and that makes strategic risks and challenges explicit. Lastly, a change plan is produced that contains all implementation-related planning information.

5. **Making it happen:**

   Even the most ingenious green business model can fail if not properly implemented. Therefore, special attention is given to the usual characteristics of Green Business Model Transformations, including potential ambiguity and controversy regarding the business case, its thematic complexity, and the necessary involvement of many different parts of the organisation.

6. **Finding new signals:**

   The last phase is intended to ensure that changes in business model assumptions and the external environment are monitored and addressed adequately.

Unlike many simplistic management frameworks the proposed transformation methodology draws upon a broad range of theoretical works and considers practical experiences from a wide variety of industries. The theoretical foundation is based on the three domains of environmental sustainability in business, business models, and organisations, change and innovation. Relevant insights from other adjacent fields like strategic management, system dynamics, stakeholder theory, and project management are also incorporated. As a result, respective underlying theoretical assumptions and limitations of the approach are made explicit. The relatively new concept of business models is thereby central to deal holistically with the complex economic effects of environmental sustainability – a task that traditional management concepts fail to fulfil satisfactorily. Moreover, each phase is complemented by reflections on specific change management challenges that can easily be overlooked or underestimated.

The practical experience is based on own project work and numerous interviews. In total, nearly 40 individuals have been interviewed, including senior management consultants and corporate managers from various industries in different functions. Companies that have been covered with interviews correspond to about three million employees and revenues of over US$ 1 trillion.

Ethical as well as economic justifications for Green Business Model Transformations are discussed. However, the approach taken aims to appeal to a mainstream management audience, not just environmentalist managers who are likely to engage the sustainability issue already anyway – albeit maybe not in a systematic manner. Consequently, the transformation methodology does not presuppose a green mindset of the managers applying it. Rather it builds on the business case for sustainability to determine whether or not a Green Business Model Transformation should be undertaken. As a result, this work does not suggest that Green Business Model Transformations are imperative for all companies, or the most attractive option irrespective of the individual circumstances. On the other hand, the research for this thesis also led me to believe that green opportunities are overlooked to an extent that should be unacceptable both to the business world and society at large. A bold but conscious move towards a more sustainable economy is considered desirable by many academics, managers and ordinary citizens for the collective good of future generations. It seems that market forces allow that transition to happen not at the expense, but to the benefit of economic prosperity of this generation: through well-managed Green Business Model Transformations.

Acknowledgements

The deeper I looked into the subject addressed in this work, the more I realised how incredibly complex it is, and hence I am indebted to all the people who helped me to understand its facets a little bit better. This includes the interviewees I was able to speak with from a large variety of firms, many of which have been incredibly open and helpful. Furthermore, I would like to thank my colleagues from BCG that took the time to share their insights with me.

I would like to thank my supervisor Professor Dr. Stefan Schaltegger, who forced me to scrutinise my thinking and approach from the beginning, thereby presumably saving me a lot of time and sorrow later on. At the same time, I was glad to have a supervisor that gave me the freedom to develop my own train of thought, without imposing his own views on my work. Moreover, both my first and my second supervisor, Professor Dr. Uwe Schneidewind, demonstrated great flexibility to overcome scheduling difficulties.

I am also indebted to all my friends for supporting me in numerous ways during the writing of my thesis. This includes content-related discussions I had with Ina Horn, Love Edquist, Tim Kunde, Ulrike Arnhold, as well as – maybe even more importantly – the many pleasant distractions that helped me enjoying life during my PhD time. As a non-native speaker I was very lucky to have Laura Kilian and Richa Menke help me with the English language, even at short notice.

Lastly, I would like to thank my parents Hans and Inge, my sister-in-law Jana, and in particular my brother Steffen for their exceptional emotional and intellectual support.

Axel Sommer

Contents

Part IV: Conclusion and Outlook

List of Figures

List of Tables

List of Abbreviations

3P	Pollution Prevention Pays
AFDC	Alternative Fuels and Advanced Vehicles Data Center
ATAG	Air Transport Action Group
BEV	Battery electric vehicle
BM	Business model
BMU	Bundesumweltministerium (Federal Ministry for the Environment, Nature Conservation and Nuclear Safety)
BoP	Base (or bottom) of the pyramid
CCS	Carbon capture and storage
CEO	Chief Executive Officer
CFRP	Carbon Fibre Re-Inforced Plastic
CLD	Causal loop diagram
CO_2	Carbon dioxide
CR	Corporate responsibility
CSR	Corporate social responsibility
CVP	Customer value proposition
DCF	Discounted cash flow
DCTI	Deutsches CleanTech Institut
DJSI	Dow Jones Sustainability Index
ESG	Environmental, social and governance
ETP	European Technology Platform
EU ETS	European Union Emissions Trading Scheme
EV	Electric vehicle
GDP	Gross domestic product
GHG	Greenhouse gas
GNP	Gross national product
GNF	Global Nature Fund
Gt	Gigatonne
HEV	Hybrid electric vehicle
HR	Human resources
HSE	Health, safety and environment
IATA	International Air Transport Association
ICE	Internal combustion engine
ICT	Information and communication technologies
IGCC	Integrated gasification combined cycle
IMO	International Maritime Organization
IPCC	Intergovernmental Panel on Climate Change
IRR	Internal rate of return

ISO	International Organization for Standardization
IT	Information technology
KPI	Key performance indicator
LCA	Life cycle assessment
M&A	Mergers and acquisitions
MA	Millennium Ecosystem Assessment
MSC	Marine Stewardship Council
Mt	Megatonne
NGO	Non-governmental organisation
NPV	Net present value
OECD	Organisation for Economic Co-operation and Development
OEM	Original equipment manufacturer
OGC	Office of Government Commerce
PHEV	Plug-in hybrid electric vehicles
PMI	Project Management Institute
PPP	Public private partnership
R&D	Research and development
RBV	Resource-based view
ROA	Return on assets
ROE	Return on equity
ROS	Return on sales
SESA	Sustainable Energy Solutions for Africa
SRI	Socially responsible investment
SWOT	Strengths, weaknesses, opportunities, and threats
TCE	Transaction cost economics
TCO	Total cost of ownership
TEU	Twenty-foot equivalent unit
UN	United Nations
UNEP	United Nations Environment Programme
WBCSD	World Business Council for Sustainable Development
WCED	World Commission on Environment and Development
WFP	World Food Programme
WHO	World Health Organization
WRI	World Resources Institute

Part I
Introduction

Chapter 1
The Emergence of Green Business Models

1.1 The Green (r)evolution

Climate change and its potential impact has had an enormous increase in attention in the media, in politics and in the public at large. Consequently, the topics *climate change* and *sustainability* have more and more become a priority for top management of corporations (Economist Intelligence Unit 2008b). This trend is also reflected in increased communication about sustainability: According to a survey of the professional services firm KPMG, almost 80 percent of the largest 250 corporations has published a corporate responsibility report by 2008, up from 52 percent in 2005 (KPMG 2008, 14). Another trend is that sustainability is not considered solely a threat anymore that has to be addressed via defensive strategies in order to mitigate reputation or litigation risks. Moreover, sustainability is no longer considered a question of philanthropy. Instead, more and more companies view sustainability as an opportunity to gain a competitive edge and create tangible, economic value (Esty & Winston 2009).

Nevertheless, the debate on the core issue – namely whether sustainability practices as such create value for the company or not – is far from over. In this respect, a distinction has to be made between *sustainability markets* like clean energy, and *sustainability management* that aims to integrate sustainability aspects into the regular operations of a company. The former can undoubtedly be very attractive: global new investment in clean energy alone has more than quadrupled since 2004 to US$ 155 billion (circa € 106 billion) in 2008 (New Energy Finance 2009, 4). The latter refers to whether a company operates in a sustainable way or not, i.e. minimises its social and environmental impact within the constraints of the technological possibilities and market mechanics of the industry it belongs to. Clarification of the relation between sustainability performance and economic value creation is difficult: The link between sustainability performance and economic value creation is often indirect and hard to quantify (Steger 2006).

Whether a company should be expected to act in a sustainable ("ethical") manner – even if it contradicted its economic interests – remains controversial in academic circles (Schaltegger & Burritt 2005). It seems sensible to claim that a company should at least seek to improve its sustainability performance as long as this creates value for the company in the long run. In practice, it is unrealistic to expect

A. Sommer: Managing Green Business Model Transformations, SPLCEM, pp. 3–6.
springerlink.com

companies to improve their sustainability performance beyond that point at the expense of economic value. Overall progress in corporate sustainability has thus been gradual rather than revolutionary so far - most companies have refrained from challenging their business models fundamentally with regard to sustainability. However, recent history clearly demonstrates that such a passive approach may not be expedient in the wake of a new "megatrend" unfolding.

1.2 The Origins of the Business Model Concept

A few years before climate change became a major concern in public perception, another topic was omnipresent in the media: e-business. In 1999, technology stocks skyrocketed and the era of the so-called "old" economy seemed to have come to an end (Brynjolfsson & Urban 2001). Valuations of some dot-com firms reached heights that appeared to be completely out of proportion to their revenue and earnings – or, not uncommon, losses (de Figueiredo 2000). The brick and mortar counterparts of these dot-coms were suddenly dwarfed in terms of market capitalisation. The perhaps most intriguing example in this context is the merger of AOL and Time Warner which was announced in January 2000: At the time, the internet service provider AOL was worth twice as much, but generated only about a quarter of Time Warner's revenue (BBC News 2000; AOL Time Warner Inc. 2001).

What was this phenomenon based on? – It was based on the firm belief that these highly valued companies would demonstrate fantastic growth and profits in the future – created by their superior *business models*. The term business model may preliminarily be defined as *a blueprint of the value proposition offered to the customer, the way the business creates and delivers that value and extracts profits from it* (a detailed discussion on the concept is provided in chapter 4.2). Although the roots of the business model concept in academia date back a few decades, it was not widely used until the e-business boom started in the late 1990ies (Osterwalder 2004, 23). At this time, entrepreneurs, managers and investors realised how useful business models were as a medium to propagate their dot-com fantasies that seemed to overturn conventional rules of doing business. After the bubble burst, the term *business model* obtained a negative connotation due to the abrupt end of many ill-conceived ventures that consumed large amounts of capital (Magretta 2002). Consequently, the merger of AOL and Time Warner is now considered one of the worst in history: Shareholders had lost more than US$ 200 billion (circa € 160 billion) by the end of 2003 (Bodie 2006). Not only were initial expectations unreasonable, but competition soon started to make AOL's original core business model obsolete and dramatically eroded its base of subscribers.

Nevertheless, Lai et al. (2006) point out that the term business model continues to be considered useful by business people to explain the strategies and plans for value creation of firms – including both those from the new and old economy. Apart from these practical applications in communication, the business model concept is also adjuvant for researchers trying to analyse the logic behind economic value creation (Osterwalder 2004). And despite its roots, the concept's usefulness is certainly not restricted to dot-coms (Linder & Cantrell 2001).

How is the e-business boom and crash relevant to the research of this work? Firstly, it originated the business model concept as a suitable unit of analysis in strategic management. Maybe even more importantly, it illustrated how entire industries can be revolutionised by introducing innovative new business models. Just like information technology, the challenge of global sustainability represents a "major discontinuity with the power to radically transform many industries" (Hart & Milstein 1999, 25).

However, the digital revolution also demonstrated spectacularly how hype effects combined with superstitious use of the business model concept can produce large-scale failure (which, in turn, can provide most valuable insights for learning). Thus, the business model concept can be utilised to distinguish real sustainability opportunities from thin air, and help seizing these opportunities while avoiding the usual pitfalls of business model transformations. In other words, it can guide companies to become sustainable *and* more competitive.

1.3 Green Business Models as a Means towards Competitive Sustainability

There is no doubt: the sustainability megatrend will change the business world. In fact, the economy has already started to respond, for example in the form of green marketing efforts (most apparent in consumer goods), changes in product portfolios (e.g., low carbon cars) or capital flows (e.g., renewable energy stocks). Some scholars even proclaim the emergence of a new form of capitalism (Porter & Kramer 2011). Similarly to the dot.com bubble, certain signs of hype can also be observed. But even if some activities are part of a sustainability fad, sustainability will continue to lead to fundamental changes in the business world (Esty & Winston 2009). Naturally, the resulting dynamic business environment brings about immense opportunities and challenges that can shake the competitive landscape of industries to the core (e.g., see Lash & Wellington 2007; Porter et al. 2007). Thus, hesitant managers should be asking themselves: "How do sustainability issues influence the future success of our current business models?" And: "How can we adapt them to best mitigate the risks and take advantage of opportunities arising from sustainability issues?"

Sustainability issues are similar to digital technologies in that they also have the capability to fundamentally change the logic of existing market mechanisms (Senge et al. 2008). On a detailed level, however, there are major differences. Among others, green business models often require significant investments of capital and other resources, and are intertwined with the existing business environment in complex ways. Green business models therefore tend to conflict with traditional business practices and structures. Many researches pin their hopes on start-up firms with their entrepreneurial spirit to introduce revolutionary green business models and associate large, incumbent firms mainly with gradual greening of their operations (Hockerts & Wüstenhagen 2010). Yet, despite the momentum in recent years, progress in sustainable development has been slower than hoped for (Baumgartner & Korhonen 2010). The reach of green start-ups is limited, and the changes made by most established firms are indeed incremental in nature. This suggests that radical changes from large incumbent firms are possible and necessary to transform the economy at a sufficient pace.

Yet, so far, established companies frequently neglect disruptive opportunities for too long – even if they can build on the experiences of pioneering small firms and possess superior resources to seize the opportunities. All too often managers are misled trying to protect previous investments in technologies or markets, and fail to question their beliefs that were valid for years or even decades (Hart & Milstein 1999; Hockerts & Wüstenhagen 2010). This work aims to help rectify this situation and assist them in *Managing Green Business Model Transformations*, thereby accelerating the transition towards an environmentally more sustainable economy.

Chapter 2
Purpose of this Work and Research Approach

2.1 Current State of Research

2.1.1 Overview of Relevant Research Domains

There are three relevant broad research domains that provide the theoretical foundation for this work (see figure 2.1): The first domain, *environmental sustainability in business*, subsumes the tectonic forces that currently reshape the business world and also constitutes the underlying motivation for the writing of this monograph. The second domain is the concept of *business models* that was identified to be especially suited as a unit of analysis for the intended research work. The third domain is the broad field of research related to *organisations, change and innovation*. These topics are central to the transition from old, conventional business models to new, (more) sustainable ones.

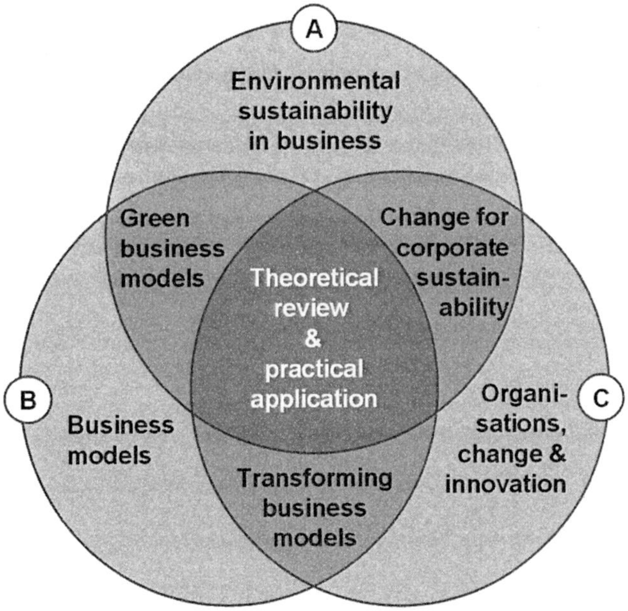

Fig. 2.1 Central research domains of this work

A. Sommer: Managing Green Business Model Transformations, SPLCEM, pp. 7–20.
springerlink.com

The three research domains, at whose intersection this publication aims to contribute, have very different histories and theoretical foundations that are briefly outlined in the following.

A: Environmental sustainability in business

There is a large body of research dealing with different aspects of environmental sustainability in business (research domain A, see figure 2.1). Generally, three dimensions of sustainability can be distinguished: the financial, the social, and the environmental aspect (Schaltegger & Wagner 2006a). Individual topics and the general view on the issue have shifted significantly over time: The "traditionalist", sceptic view of sustainability assumes that a higher sustainability performance corresponds to a lower economic performance of a company or industry. This view is reflected in Milton Friedman's famous essay with the telling title "The social responsibility of business is to increase its profits" (Friedman 1970). Consistent with this view, management and measurement of financial performance is the oldest and most advanced of the three dimensions (Schaltegger & Wagner 2006a). Social and environmental sustainability beyond legal compliance were mainly associated with voluntary, philanthropic activities at the periphery of business. However, social and environmental aspects are now more and more considered relevant for economic success in management literature. This led researchers to explore the so-called *business case for sustainability* (e.g., Schaltegger & Wagner 2006b). The shift towards expected tangible benefits from corporate sustainability efforts is easier achieved with respect to environmental issues (e.g., Amory et al. 1999; Hart 1995) compared to the harder to measure impact of social sustainability (Goddard 2006). Consequently, this work focuses on *environmental sustainability.*

Empirical evidence on whether environmental sustainability pays or not remains inconclusive to-date. Only few quantitative studies find a significant negative correlation. Although there is some evidence for a positive relationship, many studies find no significant relationship at all (Ambec & Lanoie 2007; Molina-Azorin et al. 2009). Disregarding the methodological issues of respective studies (see Ambec & Lanoie 2008), one conclusion that can be drawn from several decades of research in this field is that there is no simple correlation, let alone a simple causal link (Salzmann et al. 2005). Sustainability performance is difficult to operationalise (Porter & Kramer 2006) and the link is dependent on many factors, some of them company-specific (Lankoski 2000).

Case studies offer in-depth – albeit not necessarily generalisable – analyses for the factors that make environmental sustainability efforts profitable (Schaltegger & Wagner 2006a). Therefore, interviews with business managers from a broad range of industry sectors have been conducted to fill the knowledge gap from existing research.

B: Business models

As mentioned earlier, the business model concept became popular with managers, the media, and academia during the dot-com boom (Osterwalder 2004, 23). However, there is no theory of business models available yet; instead, there are "a myriad of concepts, ontologies and frameworks of business models all of which have merit, but none of which have been universally accepted" (Lambert 2006, 2). Although business models possess no long-established theoretical roots in economics or business research (Teece 2010), the concept is compatible with a number of theoretical frameworks in strategic management that are relevant to economic value creation (Amit & Zott 2001). The business model unites the resource-based view of the firm (e.g., Penrose 1959; Wernerfelt 1984; Barney 1991a) with that of the positioning perspective of the industrial organisation (e.g., Porter 1985; 1980). Moreover, the business model concept provides enough detail to relate to specific problems that may arise during large corporate transformations; yet it is not too detailed to obscure big picture issues. The business model concept has seen a revival in practical application, this time based on more structured foundations (e.g., Osterwalder & Pigneur 2010; Johnson 2010). Apart from analytical reasons, the great practical importance for company success of business model innovation (compared to isolated product or process-based innovations) is brought forward in favour of the concept (e.g., Johnson et al. 2008; Lindgardt et al. 2009). This means that instead of applying a narrow product- or process-based view, customer needs and functions that can satisfy them are (re-)considered broadly. Compatibility with strategic management concepts, analytical capabilities, and high practical relevance make business models an ideal unit of analysis for the purpose of this work.

One field of research in particular enables business models to serve as a valuable tool for analysis of complex management issues: system dynamics. System dynamics has been pioneered by Jay W. Forrester more than fifty years ago (see Forrester 2003) and is driven by the insight that some problems cannot be solved satisfactorily by traditional means of analysis, i.e. by dividing the problem into smaller pieces until each individual piece can be solved. Instead, dynamic interactions and feedback loops between the relevant system elements need to be considered holistically (Forrester 1958). In this context, *systems thinking* has been developed as a powerful analytical tool and is thus introduced in conjunction with business models. Addressable problems include many sustainability issues (Senge et al. 2008), as well as organisational challenges (Senge 1990). Hence, systems thinking principles will be used in various ways to facilitate successful Green Business Model Transformations.

C: Organisations, change, and innovation

Organisation research is a very large and heterogeneous field of research (Scherer 2002). However, respective organisation theories form the basis of the transformation aspect examined in this work and are therefore considered in broad terms. Various systems perspectives of organisations exist - rational, natural, open systems (Scott 1992). The open systems view is particularly instructive as it emphasises the importance of interactions between organisations and their (changing) environments. Thus, both exogenous and endogenous change is relevant for this study. The former determines the setting within which corporations do business. This type of change is external and usually cannot be controlled by the company – rather it imposes change on it. Driving forces include changes of the natural environment and scientific understanding (e.g., climate science), as well as political and social forces (e.g., societal change, political movements, or consumer behaviour). The above-mentioned forces are considered a critical input to determine the economic justification of Green Business Model Transformations. However, what is at the centre of consideration of this publication is another aspect of change: the deliberate transformation of an organisation – also referred to as *organisational change. Change management* thereby refers to the managerial process of implementing organisational change. There are a multitude of change management models that may be applied. An early one is the "Unfreeze-Change-Refreeze" approach by Lewin (1951). Academics and practitioners have suggested a wide array of other approaches since then. Yet, owing to the great complexity and dynamism of organisations and their relationships with the external environment (Daft & Lewin 1990), prescriptive research on the topic is still far from being concluded.

When confronted with challenges and opportunities from inside or outside the organisation, companies can try to innovate and improve their competitive position. Innovation (pioneered by Schumpeter 1912) is considered both in the context of organisational factors and the external environment. Innovation is thereby not limited to product innovation like many executives seem to think (Linder et al. 2003, 44f.). For example, Sawhney et al. (2007) have identified 12 ways for companies to innovate, with new product or service offerings being only one of them. Furthermore, new technology is not a sufficient (or even necessary) condition for successful innovation. Rather, effective business models are key to successful innovation as they allow companies to extract economic value from technology (Chesbrough & Rosenbloom 2002). Thus, innovation related to business models (e.g., Johnson 2010) obviously receives special attention as well within this work. The same is true for innovation based on or leading to environmental sustainability (e.g., Fussler 1996; Fichter et al. 2006).

As illustrated in figure 2.1, the three research domains are overlapping. These intersections are adjacent to the topic of this publication and have attracted varying levels of attention by researchers.

A/B: Green business models

Literature that explicitly discusses business models for sustainability is still very rare (Stubbs & Cocklin 2008; Lüdeke-Freund 2009). However, there is research that considers selected elements of business models in relation to sustainability issues (green products, consumers, processes, supply chains, etc.). Although these works inevitably lack the integrated view that a business model approach offers, they provide valuable insights for classifying and designing various forms of green business models.

In the absence of an existing taxonomy for green business models, it needs to be developed within this work. One requirement therefore is to provide a conceptualisation that provides the basis for the categorisation of real-world business models.

B/C: Transforming business models

A considerable number of researchers have dealt with the topic of switching from *old economy* to *new economy* business models; and many failures of such attempts are also well-documented (e.g., Markus & Benjamin 1997; Pinker et al. 2002; Weill & Broadbent 1998). Linder & Cantrell (2000) describe various types of business model transformations distinguished by the degree of change to the core logic. Furthermore, some literature on innovation relates to business models, although often focused on products and services only (e.g., Chesbrough & Rosenbloom 2002; Linder et al. 2003; Bjelland & Wood 2008). Authors have rarely addressed fundamental changes to business models as such; exceptions include the works of Voelpel et al. (2005), Johnson et al. (2008), and Lindgardt et al. (2009). Only recently, the two works of Osterwalder & Pigneur (2010; building on Osterwalder 2004) and Johnson (2010) address the practical challenge of transforming business models in a comprehensive way.

The business model concept does not solely focus on the organisation, but also considers external parties that participate in or benefit from the company's value creation activities. These external parties are not limited to suppliers or customers but also include various partners that need to be considered for any transformation effort. Consequently, literature on strategic networks (e.g., Doz & Hamel 1998; Gulati et al. 2000) is relevant, too. Moreover, other external parties that do not directly participate in the value creation process (e.g. competitors, NGOs, the public at large) cannot be ignored – at least not generally. Stakeholder management (e.g., Freeman 1984; Donaldson & Preston 1995) thus plays a vital role for many business model transformations. For didactic reasons, the relevant literature of stakeholder theory is reviewed in the context of sustainability in business.

A/C: Change towards corporate sustainability

Natural scientists and environmentalists have long warned that humanity's exploitation of natural resources and consumption patterns are not sustainable and need to be changed fundamentally (e.g., Meadows et al. 1972; WCED 1987; von Weizsäcker 1988; Daly & Cobb 1989; Hawken et al. 1999). Many of today's business leaders have joined early environmentalists in their demand for far-reaching change to more sustainable business practices. Accordingly, the topic has gained popularity among researchers of strategic management, too (e.g., Hart 1995; Hart & Milstein 1999; Porter & van der Linde 1995b). In contrast, organisation theorists have been slower to address sustainability (Shrivastava 1994).

The move toward the mainstream had the positive effect that academia now deals with the topic in a more differentiated way, thus increasing the impact of the results: Early advocates for a change for sustainability often alienated practitioners by delivering one-sided success stories. Today, many works are much more balanced, emphasising also the risks and pitfalls of transforming business practices to be more sustainable (Esty & Winston 2009).

2.1.2 Research Gap

So far, there is a big void at the combined intersection of environmental sustainability in business, business models, and organisations, change and innovation. Current sustainability performance management techniques are well suited to minimise company risks and realise incremental improvements that are linked to economic value creation. However, these approaches only address isolated parts of business models. What is missing is how companies can systematically manage fundamental transformations of their business models to make them green – and profitable. These transformations may be implemented by visionary and environmentally proactive companies. But they may also be dictated to formerly passive companies by competition and changed market conditions. In both cases, such transformations are usually unprecedented events for these companies. Either the company is the first in its industry, and no company has done something similar before. Or, the company is following, but is still confronted with challenges unique to the company, including the question whether it can defy potential first mover advantages. Consequently, business model transformations are usually based on more or less substantiated beliefs of top management, and – due to a lack of appropriate frameworks – are treated as one-time initiatives that rely largely on ad-hoc management approaches.

This publication aims to fill that gap, both in terms of the lack of theoretical foundation as well as by providing a management framework that can be applied in practice. To give a first idea of the process of a Green Business Model Transformation, a simplified, schematic overview is presented in figure 2.2:

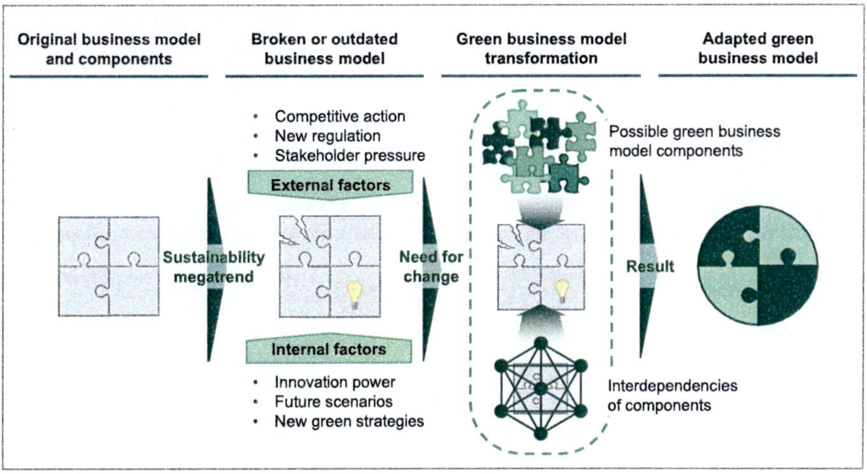

Fig. 2.2 Schematic overview of Green Business Model Transformations

2.2 The Central Research Question

The central research question is formulated as follows:

> ***"How can established companies successfully manage a fundamental transformation of their business model(s) based on green value propositions and value creation, thereby improving or sustaining economic performance?"***

The central research question contains a number of specifications that need clarification: Firstly, the motivation for the transformation is based on – or at least compatible with – economic interests, i.e. it does not serve philanthropic purposes. Hence, the new, green business model has to offer competitive financial long-term prospects compared to the previous, or alternative, conventional business model. In order to achieve this, the green characteristics of the new business model can relate to its value proposition, to the value creation, or often both.

Furthermore, this work focuses on established companies as opposed to start-ups, making the transformation aspect a key focus. This is further emphasised by the circumstance that fundamental transformations – not incremental adaptations – are examined.

The central research question leads to a number of related theoretical and practical sub-questions that need to be answered in order to deliver a comprehensive work on the issue:

Issues that relate the *theoretical basis* for Green Business Model Transformations:

- What is the general link between corporate environmental sustainability and the economic success of established companies?

- What are relevant environmental issues and how can they be addressed?
- What specific value creation levers does corporate environmental sustainability offer?
- How can business models be conceptualised and used systematically as a management tool?
- What constitutes a green business model?
- What are useful perspectives on organisational change that help to understand and manage Green Business Model Transformations in established companies effectively?
- How does (radical) innovation work in established companies?

Challenges that are to be addressed by an effective *management framework*:

- How do Green Business Model Transformations take place in practice?
- What managerial implications can be derived from practical experiences?
- How can companies evaluate the (future) fitness of their business models regarding sustainability issues and learn to appreciate opportunities offered by green business models?
- How can companies unleash the creativity of their employees or other parties in order to address environmental sustainability in a fresh, innovative way?
- How can ideas be translated systematically into viable green business models, despite complex first- and second-order effects on internal operations, on the market, on stakeholders?
- What tools, methods, and approaches for preparation and implementation are appropriate?
- What change management challenges are to be mastered during various stages of Green Business Model Transformations?
- How can be ensured that the new green business model – once implemented successfully – remains compatible with the company's strategy and sustains high fitness with respect to the competitive environment?

The questions above will all be addressed in the chapters that follow. However, due to the breath and heterogeneity of the topics related to the central research question, I will first define the scope of this work and clarify the used terminology.

2.3 Definition of Scope

2.3.1 Definition of the Term "Green" in this Work

Organisational changes concerning corporate environmental sustainability are often referred to as "turning green". The term "green" is used (and misused) in many ways in this context. It can even include social elements. For example, "green cosmetics" may refer to products that are produced free of chemicals (environmental, narrow sense of "green"), or under the rules of fair trade (broad sense of "green"). Because the terms "sustainable business model" and "business model for sustainability" are too broad in scope, the still somewhat fuzzy but very intuitive term "green business model" is used instead. The study will focus on the narrow, environmental definition of "green" and consider other aspects of sustainability only if they are part of an inseparable, collective green value proposition. Non-profit or philanthropy-based activities that are unrelated to the core business are not in scope.

2.3.2 Types of Corporations in Scope

Corporations that are *established* players in their respective markets are in focus. Schaltegger (2002) categorises companies according to their environmental positioning in a matrix with the following two dimensions:

- Priority of environmental issues as a business goal ("core" vs. "supplementary" vs. "trustee duty")
- Market effect of business ("alternative scene" vs. "eco-niche" vs. "mass market")

Hence, companies under consideration are active in the mass market and move (or have moved) towards "green" being a core element of their business model. This relates to the position of companies labelled "Ecopreneurs", a portmanteau word combining "ecological" and "entrepreneur". However, only companies that are already established in the mass market are addressed in this work; out of scope are newly founded companies that aim to reach widespread dissemination of their products and services in the future (as their journeys tend to be very different in nature). Moreover, the scope is not limited to companies with products whose function is to protect or restore the environment (e.g., pollution control products). Even companies from environmental "vice" industries (like oil firms) are relevant as long as they transform their business models in a way that is fundamentally different from their less green peers. No focus in terms of industry sector or similar is set as many sustainability issues span across industry borders. Restricting industry scope would also reduce the sample of relevant study objects too much. However, observed systematic differences between certain types of companies (e.g., consumer vs. industrial goods companies) will be considered as necessary.

2.3.3 Green versus Conventional Business Models

The distinction between green and non-green business models can be difficult in practice: Some companies already apply advanced business practices to minimise their environmental impact, but do not call themselves "green". Other firms, by and large, continue with their previous practices that do not represent an especially strong environmental performance level, but now put a green label on them as part of corporate communications activities (see Economist Intelligence Unit 2008b). Even if adequate criteria were available to determine and compare the environmental performance of companies objectively, it would still often be a matter of degree of "greenness". More fundamentally, the question touches upon the very foundation of how sustainability is understood (Ehrenfeld 2005). Depending on the underlying understanding of sustainability the business models that qualify as green differ greatly. Hence, an important task of this work will be to clarify the concept of (environmental) sustainability and subsequently develop a taxonomy of green vs. non-green business models upon it (see chapter 5).

There is a big difference between "greening" a conventional business model and transforming a business model into a green one. The former will bring about incremental change and make existing operations converge to a local environmental optimum (Hart & Milstein 1999). The latter form of change seeks a radical new way of doing business.

A simplified interim classification based on the degree of radicalness of change can be used to clarify the scope of this work until the formal definition of green business models is presented in chapter 5. Four cases are distinguished for this purpose (see figure 2.3): (1) *Green Evolution* (small routine improvements), (2) *Isolated Green Adaptation*, (3) *Staged Green Transformation*, and (4) *Green Revolution* ("big bang" approach).

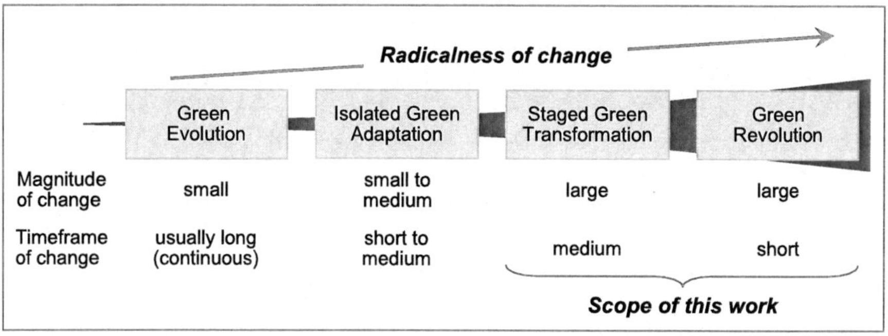

Fig. 2.3 Different degrees of change distinguished by radicalness

Only Staged Green Transformations and Green Revolutions are in scope – the former two cases would just be considered as part of a larger programme that fundamentally transforms major business model elements. The following two criteria may be used in order to identify transformations that are in scope:

- The magnitude of (green) change: the scope of change affects a considerable part of the company (e.g. in terms of share of revenue of the affected parts of the company), and the depth of change is substantial, i.e. significant non-green parts of the business model are transformed to become green.
- The timeframe of change: major change tends to happen rapidly in the considered cases – within months or a couple of years for large-scale green revolutions (from planning to realisation), or possibly a few years for staged green transformations.

A Green Business Model Transformation does not necessarily have to happen organically. Ilinitch & Schaltegger (1995) propose a framework for an ecologically-oriented portfolio analysis that is based on an adapted BCG portfolio matrix, complemented by a third, environmental impact dimension. As a result of such an analysis, companies may decide to sell "dirty cash cows" or buy companies with green business models to bring the overall portfolio more in line with the company's new green strategy. However, this approach is only of interest for the purpose of this work if respective M&A activity is part of a larger strategic programme.

Lastly, one complication that also needs to be dealt with in the context of green business models is the possible systemic effects that they may have beyond the direct sphere of influence of the company. For instance, a principally green business model may – ceteris paribus – lead to a sharp increase of consumption of green products that offsets all relative environmental performance gains of the business model.

2.3.4 Analytical Focus Regarding the Transformation Process

A Green Business Model Transformation can be divided into various distinct phases conceptually, although the process will never follow a pure sequential pattern in reality. In the context of the business case for sustainability, for example, Steger (2006, 440) distinguishes four main steps that are analogous to that of the transformation process under examination:

1. "Identifying issues" (opportunities, need for change)
2. "Building the business case" (justification and planning of the proposed transformation)
3. "Implementing the business case" (execution of the transformation)
4. "Monitoring and controlling"

Due to the involved magnitude of change, all four phases are especially delicate compared to "ordinary" environmental initiatives and therefore need to be thoroughly considered by companies. However, implementation entails similar characteristics to any large-scale business transformation and monitoring and controlling are similar to common financial and environmental performance management practices. Therefore, stronger emphasis will be put on earlier phases of Green Business Model Transformations in this work. As will be laid out in chapter 9, a more detailed framework with respect to these tasks is appropriate compared to the example above. As a result, a six-phase approach is proposed.

2.4 Structure of this Work

Figure 2.4 summarises the structure of this book. It is divided into four parts: After the introduction, the second part constitutes the theoretical foundation. Relevant fields of research are reviewed and related to the various aspects of the central research question. The third part draws upon short case studies ("vignettes") of Green Business Model Transformations and extends the theoretical part towards a framework for management practice. Finally, a short conclusion and future outlook is provided.

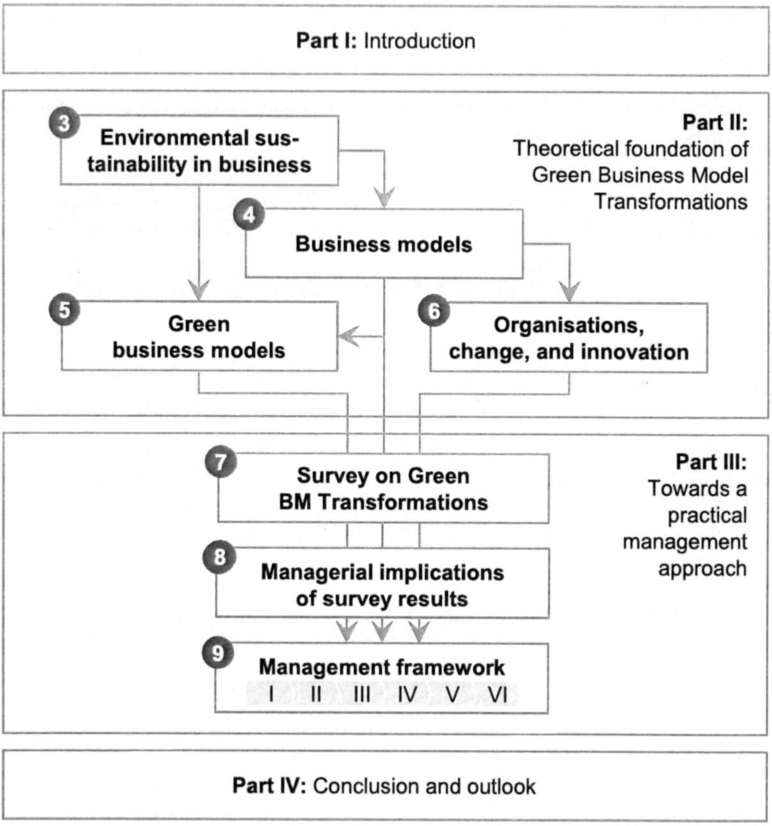

Fig. 2.4 Structure of this work

Due to the broad and heterogeneous theoretical foundation of this work, integrating the respective fields of research is a formidable challenge. One of the few commonalities across all research domains is the applicability of the systems perspective: Organisations can be viewed as systems, sustainability can be considered a systemic phenomenon, and the business model concept explicitly appreciates systemic characteristics in the business context. Hence, the systems perspective will be a recurring theme throughout the theoretical review of existing literature within chapters 3 through 6 (**Part II**).

Chapter 3 outlines the relevance of environmental sustainability in business, both as a management task, and with respect to its link to the economic success of the company. Important environmental issues are placed into the context of corporations and their stakeholders. The chapter closes with the appreciation of possible corporate environmental strategies.

Chapter 4 introduces and defines the concept of business models. Business models are delineated from business strategy and operations, their generic elements are discussed. Furthermore, the systems perspective and its importance for business models are explained. A brief discussion of change methodologies and evaluation measures for business models follows.

Chapter 5 presents a taxonomy for green business models. Drawing upon existing taxonomies for business models in general, four types of business models are proposed based on their aggregated environmental impact. Finally, practical application of the taxonomy and a number of Green Business Model Prototypes are presented.

Chapter 6 addresses organisation theory. Theoretical perspectives on organisational change are reviewed and put into context with the challenge of Green Business Model Transformations. Change management strategies and existing prescriptive approaches for management practice are examined. The last part of the chapter deals with relevant aspects of innovation theory.

Part III starts with **chapter 7**, which complements the synthesised theoretical perspectives from the previous chapters with seven case vignettes, i.e. brief, instructive analyses of business situations. Each vignette highlights different aspects of Green Business Model Transformations. The question why some companies or industries have not (yet) considered such transformations is also investigated. Taken together, the vignettes provide a rich survey on Green Business Model Transformations. The chapter closes with an evaluation of the environmental performance of the described green business models.

Chapter 8 explains the managerial implications based on the gathered industry examples and related theoretical work. Both static and dynamic implications are discussed on different levels (industry, business model, individual company). Furthermore, five ideal-type corporate journeys towards green business models are distilled from the survey.

Chapter 9 contains a major contribution of this publication: the framework for management practice. It is divided into six phases: (I) Embracing ambiguity; (II) Rethinking old certainties; (III) Translating into business impact; (IV) Planning for action; (V) Making it happen; (VI) Finding new signals. For each phase, the problem context is explained and change management challenges are discussed. Moreover, specific management tools and advice for suitable courses of action are provided.

Part IV concludes with a recapitulation of the contributions of this work, discusses its limitations and possibilities for future research.

2.5 Research Approach for the Practical Part

The case vignettes in chapter 7 are used to validate and illustrate the conceptual work throughout the book. A wide variety of companies from different industries have been examined based on publically available documents such as newspaper and journal articles, company websites and reports. Most importantly, dozens of company representatives have been interviewed in order to extract the key learnings from the cases. Moreover, industry and functional experts from The Boston Consulting Group have been consulted. Not only did this sharpen the focus on key issues for the company interviews, but discussions with these consultants also helped to increase the validity of information and opinions gathered with respect to biases of interviewees.

Investigated firms include recognised sustainability leaders as well as firms from industries that have not yet produced serious attempts of Green Business Model Transformations. Study subjects were selected based on the proposition that possible forms of green business models will vary by industry sector, and depend on characteristics of the company (e.g., ownership structure, culture, organisational capabilities). Preferably, at least two firms per industry sector have been studied in order to avoid false generalisations about industry-specific characteristics. Industries that are covered in the seven vignettes include food and packaged consumer goods, lighting, building services, chemicals, airlines, shipping, automobiles, and industrial conglomerates. For a full list of interviewees see Appendix 1: List of interviews. Moreover, transcripts of interviews conducted on sustainability in business by the MIT Sloan Management Review and BCG during 2009 (see Berns et al. 2009a, 34ff.) have been considered, albeit not explicitly.

The possibility of conducting a management survey was dismissed because of its very limited potential contribution to the central research question. First, some recent surveys (e.g., Economist Intelligence Unit 2008b; Berns et al. 2009b; Haanœs et al. 2011) with similar foci are already available. Second, and more importantly, the most critical questions not covered in previous surveys are difficult to pose in survey format due to the complex causality of sustainability in business that often depends on context and requires follow-up questions.

Hence, semi-structured interviews have been conducted instead. They were loosely based on a case study protocol (e.g., Yin 2003, 67ff.) (see Appendix 2: Case study protocol excerpt). The interviews have not been recorded and transcribed for the following two reasons: First, due to the high sensitivity of the required information (e.g., strategic plans and decisions, opinions about customers, competitors, or certain parts of the own organisation) many interviewees would not have accepted recording. Second, even if an interviewee had agreed to record the interview, an important share of information would likely have been held back in this case. In fact, on several occasions, interviewees insisted that some statements must not show in the text of this publication. Although these parts of the interviews have obviously not been included – at least not explicitly – they were often crucial for understanding certain important aspects of business models and motives for management decisions.

Part II

Theoretical Foundation of Green Business Model Transformations

Chapter 3
Environmental Sustainability in Business

"I think the world has reached a tipping point now. We're beyond the debates over whether [addressing sustainability] is something that needs to be done or not – it's now mostly about how we do it. [...] it's not about altruism, it's about creating value."

Steve Fludder, Vice President, Ecomagination, General Electric (Berns et al. 2009b, 7)

In recent years, some very large, established companies like General Electric have completely overturned their previous negligence of sustainability and now consider it a central aspect of their business strategy. However, many other companies still regard sustainability to be a side issue, and few act decisively in order to derive real value from it. The divide becomes evident in a recent global survey of business leaders[1]: 68 percent of those who considered themselves to be experts with respect to sustainability said that their investments in sustainability aimed at financial returns, as opposed to only 32 percent among self-identified novices. The difference is even greater – 50 versus 10 percent – with respect to whether or not the company has developed "a compelling business case for sustainability" (Berns et al. 2009b, 8f.). Overall, improved company or brand image still represents the most important perceived benefit (Berns et al. 2009a, 58). This is somewhat conflicting with the results from an earlier survey with managers from US-based firms[2]: the top three answers for the primary motivation of the firm for corporate citizenship were revenue growth (16%), increasing profit (16%), and cost savings (13%) (Economist Intelligence Unit 2008a, 24). The discrepancy may stem from the different demographics of the respondents, yet both surveys highlight the fact that sustainability in business is no longer a matter of philanthropy.

---

[1] 1,560 business leaders from for-profit organisations replied to an electronic survey during March/April 2009. One third of respondents are from the executive suite, fifty percent were senior managers. All regions are represented, the strongest being North America (28%), Europe (14%), Asia Pacific (13%), but also a large number (27%) of global companies (Berns et al. 2009a; Berns et al. 2009b).

[2] The Economist Intelligence Unit (2008a, 2) electronically surveyed 566 managers from US-based companies during September 2008 (39 percent of the respondents were on the level of vice president or above).

A. Sommer: Managing Green Business Model Transformations, SPLCEM, pp. 23–47.
springerlink.com

It is the goal of this chapter to substantiate the assertion that addressing sustainability, more precisely environmental sustainability, makes economic sense, too.

One of the first and most persistent challenges with respect to sustainability in business both to academics and practitioners is to define it (Schaltegger & Burritt 2005, 186ff.).

3.1 Sustainability Defined

For decades, the definition of sustainability has remained a contested issue despite – or maybe because of – the far-reaching consequences of how the term is understood.

A widely used definition among academics and sustainability experts is the so-called Brundtland Commission definition:

> *"Humanity has the ability to make development sustainable – to ensure that it meets the needs of the present without compromising the ability of future generations to meet their own needs."* (WCED 1987, 8)

Sustainable development thereby represents a process that aims at reaching sustainability as its end state (Schaltegger & Burritt 2005, 185).

Another popular interpretation of sustainability by sustainability pioneer John Elkington has become known as the *triple bottom line* (Elkington 1994; 1999). According to this view, companies should complement their attention to the financial bottom line with consideration of the social and environmental bottom lines. It is argued that there are many triple wins such that the financial bottom line need not to suffer from an expanded, integrated management focus on sustainability.

In contrast, among business leaders who consider themselves novices in the sustainability domain, sustainability is often understood as "maintaining the viability of our business" (Berns et al. 2009a, 37). This definition obviously is a lot less strict than that of the Brundtland Commission and much vaguer than the triple bottom line. In fact, one might actually make a case against environmental protection in general based on the assumption that it often increases cost today and offers limited returns tomorrow. In order to examine environmental sustainability on a microeconomic level, it is necessary to establish a broad understanding of its macroeconomic context first as it has profound influence on future policies and the business environment in general.

3.1.1 The Macroeconomic Perspective: Weak versus Strong Sustainability

Many major controversies on sustainability relate to the rivalling perspectives of weak versus strong sustainability (Schaltegger & Burritt 2005, 187f.). The *weak sustainability* paradigm (tracing back to works of Solow 1974; Hartwick 1977) is rooted in neoclassical economics and postulates that sustainability only requires that the aggregated stock of all forms of capital[3] – including natural and man-made capital – remains at least constant over time (Droste-Franke 2004, 40ff.). Weak sustainability thereby implies that natural capital is either "super-abundant", or that it can be substituted with man-made capital – both as an input factor and to provide direct utility (Neumayer 2003, 22), or that "technical progress can overcome any resource constraint" (Neumayer 2003, 23). As a result, economic growth can justify the depletion of resources and pollution of the environment as long as the growth in man-made capital is stronger than the damage done to the environment.

In reference to Rawls' *theory of justice* (Rawls 1999, first published 1971) and based on the belief that future generations will be wealthier anyway, one may question current efforts on environmental protection and deduce that it is justified for the current generation to maximise its own welfare instead. Along these lines, Lomborg (2001) demands that available resources should be invested in issues with relatively near-term returns like economic growth or education in poor countries – as opposed to the prevention of (in his view overstated) environmental issues like climate change that mainly concern future generations. In addition, Lomborg argues, future technological advances will make it much more efficient to deal with long-term environmental issues in the future than it would be today. Nordhaus (2008) provides an example of a well-established – but not undisputed (Krugman 2010) – neoclassical cost-benefit analysis of climate change. It is implicitly based on the central weak sustainability assumption of substitutability, and suggests only modest (initial) interventions for combating climate change. It should be noted that these interventions are the result of a utilitarian approach – weak sustainability principally does not warrant any active climate policy given that total welfare is assumed to increase even without it (Neumayer 2003, 31f.).

Environmental optimism as described above is strongly contested. Counterarguments range from rather technical arguments like the use of high discount rates that make future returns of environmental protection look small compared to its cost today (e.g., Weitzman 1998; Portney & Weyant 1999) to very fundamental arguments that relate to the relationship between human welfare and material prosperity (for a comprehensive discussion, see Neumayer 2003). The growth imperative has been contested noticeably at least since the early 1970ies (e.g., see Nordhaus & Tobin 1972), with the most prominent call for reconsidering the current path of development issued by the Club of Rome report "The Limits to Growth" (Meadows et al. 1972). Modern societies' fixation on consumption is considered a dead end by many academics that are active in the sustainability domain today (e.g., see Ehrenfeld 2005; 2008; Jackson 2009; Fedrigo & Hontelez

[3] Various forms of capital can be distinguished: Classical economics distinguishes land, labour and human-made capital (Ekins et al. 2003, 166). In the context of sustainability, other disaggregations are common. For instance, Ekins (1992) differentiates manufactured (machines, buildings, infrastructure, etc.), human (knowledge, skills, etc.), social/organisational (networks, organisations, etc.), and natural (natural resources, pollution sinks, etc.) capital. For the sake of simplicity, I will distinguish only natural capital and man-made capital.

2010). Yet, economic growth as measured by the disputed gross domestic product (GDP) is still at the top of the political agenda of developed and developing countries alike (for a discussion of alternatives to GDP see Stiglitz et al. 2009).

Advocates of *strong sustainability* in the tradition of Herman Daly (e.g., Daly 1992, first published in 1977), by contrast, would argue that the majority of current human economic activity – that involves burning fossil fuels, producing waste, emitting greenhouse gases, etc. – is non-sustainable business practice. One of their main arguments against the weak sustainability paradigm is directed against the underlying assumption that natural capital is substitutable (Neumayer 2003). One interpretation of strong sustainability therefore demands that physical stocks of so-called *critical natural capital* must not be used beyond their regenerative capacity and cannot be substituted with each other or other forms of capital (Droste-Franke 2004, 42). Indispensable ecosystem services like climate regulation or fresh water obviously belong to this category. However, a definitive conceptualisation of criticality of natural capital is still pending (Brand 2009).

Ekins et al. (2003, 168f.) present various theoretical and practical reasons why they think that scientists should prefer strong sustainability as the a priori position and revert to weak sustainability only in cases where it has been shown to be appropriate. Maybe the most intuitive among them is that the loss of natural capital can be irreversible, which is very rarely the case for manufactured capital.

Based on earlier works by Endres & Radke (1998) and Pezzey (1994), Droste-Franke attempts to integrate the concepts of weak and strong sustainability and suggests the following three priorities for sustainable development (Droste-Franke 2004, 63): First, all stocks of capital considered relevant to society must be kept above critical levels. Second, appropriately valued changes to all relevant forms of capital must always be non-declining in total. Third, the present value of intertemporal utility needs to be maximised under the constraints of priorities 1 and 2.

The first priority may be interpreted as a safety margin against imminent unacceptable environmental damage which must not be violated even if efficiency considerations suggest otherwise. For example, Nordhaus considers such non-economic constraints in his analysis (Nordhaus 2008, 17): He argues that governments may want to limit the maximum temperature rise from global warming to 2.5° C if climate scientists conclude that a stronger rise could trigger fatal feedback processes. Priority 2 is satisfied as long as a continuous increase in total welfare is projected. Priority 3 combines the concept of intergenerational equity with welfare maximisation. Interestingly, Nordhaus found that most of the tested climate-constraint cases are still close to the economic optimum (Nordhaus 2008, 15).

In summary, few scientists are extreme proponents of either weak or strong sustainability in that they claim natural capital is either completely substitutable or not at all, respectively. Moreover, as Neumayer (2003, 89) remarks, the controversy between the two rivalling paradigms can neither be resolved theoretically nor empirically – at least not in the near-term future.

3.1.2 *Corporate Sustainability*

The lack on consensus in defining sustainability is problematical as companies are left to interpret sustainability as they choose. If companies follow the weak sustainability paradigm – which most companies active in the field today do – sustainability is usually equated with *eco-efficiency* (e.g., Schmidheiny 1992), i.e. "creating more goods and services with ever less use of resources, waste and pollution" (WBCSD 2000, 1). Most corporate environmental initiatives are designed in the spirit of eco-efficiency. Of course, the problem with eco-efficiency is that absolute environmental degradation still grows if efficiency gains are fully consumed by even stronger increases in economic output. In fact, within the paradigm of weak sustainability such developments are desirable as long as the aggregated capital increases. The cumulative effects of such a development may still be devastating for the environment in the long run.

One proposal for solution are so-called "factor X" concepts that demand radical improvements in resource productivity. For example, reaching factor four allows doubling prosperity while requiring half the natural capital (von Weizsäcker et al. 1995). By decoupling economic growth from environmental pollution and resource use, its proponents are hoping to avoid the painful trade-off between a healthy environment and a high (material) standard of living.

However, critics argue that reaching factor X is a necessary, but not a sufficient condition for saving the planet for future generations. The requirements from strong sustainability go further and may be translated into management rules for individual companies that include the following (see Neumayer 2003, 25; based on Daly 1992): A company must not harvest more than the highest sustainable yield, and pollute the environment only to an extent that does not harm its absorptive capacity.

Some large companies have responded to the demands from strong sustainability advocates and set highly ambitious long-term targets. For instance, Wal-Mart has formulated the goal of creating zero waste, using only renewable energy, and "to sell products that sustain our natural resources and the environment" (Wal-Mart 2007, 41). However, it has to be acknowledged that most of the few companies that have formulated respective goals are still far away from living up to their ambitious aspirations.

In this work, the chosen focus lies on the field of environmental sustainability (and not social or governance aspects per se). As explained earlier, this is not to suggest that the other dimensions can be ignored (to the contrary). It is merely a reflection of the fact that environmental sustainability generally represents the largest opportunity for companies, or at least, it is the easiest to identify and quantify for the purpose of business model transformations (the reputational and litigation risks from social and governance issues can be huge). Moreover, the focus on environmental sustainability should not hide the fact that companies who wish to take sustainability seriously need to adopt an integrated, holistic perspective on economic, social, and environmental aspects of sustainability (see figure 3.1). Often, these three dimensions are indivisible. Independent of the respective organisational structure for sustainability management and the company-specific focus on individual sustainability issues, advanced companies share the understanding that there are links between all three sustainability dimensions that need to be actively managed (Schaltegger & Wagner 2006a). Integrating all sustainability dimensions is thereby a greater challenge than optimising the effectiveness of each aspect individually (Schaltegger & Burritt 2005). This work focuses on the opportunities arising from ecological issues and their interfaces with the social and the economic dimension in particular (eco-efficiency).

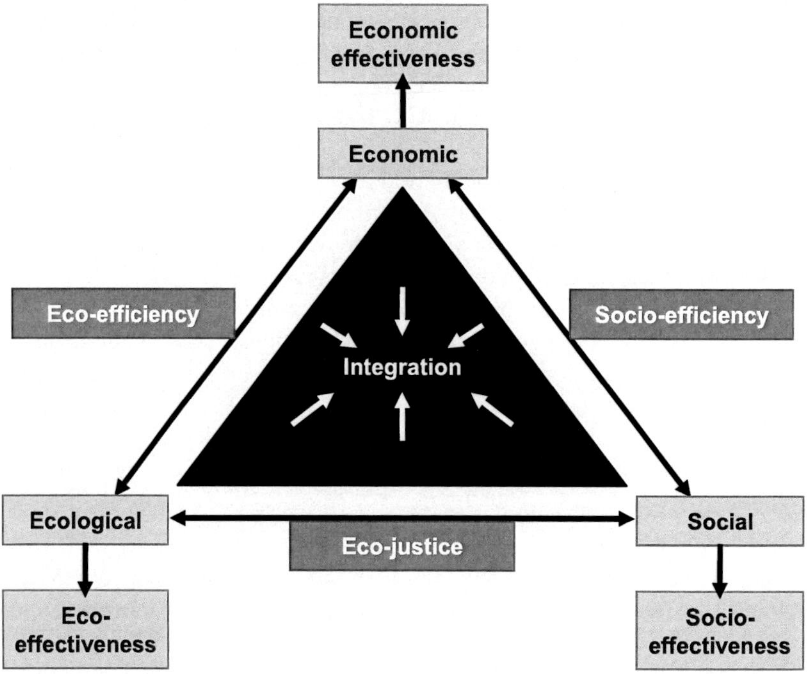

Fig. 3.1 Corporate Sustainability aspects and their interrelations (Schaltegger & Burritt 2005, 189)

Although this integrated, holistic understanding has started to take hold in the academic and business community (Forstmoser 2006) it does not mean that there is a consensus on sustainability language and terms yet. In corporate settings, sustainability is often referred to as *Corporate Social Responsibility* (CSR). In some manufacturing companies, so-called *Health, Safety and Environment* (HSE) departments cover much of what is usually attributed to corporate sustainability. In the financial industry, the term *ESG* (Environmental, Social and Governance) is commonly used. All these terms may mean the same or not, depending on who uses them and in which context. In the following, the term *corporate sustainability* will be used, thereby adopting the definition below:

> *"[Corporate sustainability management] deals with both the analysis and management of the effects of environmental and social activities on the competitiveness and economic success of a company, as well as with the analysis and management of the social and environmental effects of business activities."* (Schaltegger & Wagner 2006a, 4)

Despite the presented theoretical arguments and empirical evidence that the corporate mindset has shifted in favour of corporate sustainability, one objection must not be lightly dismissed: that it may be just a passing management fad.

3.2 The Relevance of Corporate Sustainability for Economic Success

> *"All truth passes through three stages. First, it is ridiculed. Second, it is violently opposed. Third, it is accepted as being self-evident."*
> Attributed to Arthur Schopenhauer (1788-1860)[4]

Corporate sustainability advocates are rarely ridiculed these days. Many people now find claims of its importance for a company's success self-evident, at least on a generic level (Lee 2008). Although much has changed since the famous criticism of Milton Friedman (Friedman 1970), not all business managers have been converted. From a practical point of view, it seems unlikely that corporate sustainability will turn out to be a management fad – the fact that companies adhere to their commitment to corporate sustainability even during the downturn and with scarce financial resources suggests otherwise: 60 percent of respondents in the already quoted management survey indicated unchanged or increasing commitment for their companies (Berns et al. 2009b, 8). Rather, managers' increased interest in corporate sustainability has led to an urgent need for new strategic frameworks and approaches to manage it (Berns et al. 2009a, 73) – a condition that this publication seeks to improve.

Nevertheless, sceptics of corporate sustainability express doubts with respect to the postulated economic benefits of corporate sustainability.

3.2.1 Critical Views on Corporate Sustainability

Historically, most environmentalists have vigorously fought for their cause. Based on the understanding that our planet is seriously at risk, many have deduced that helping the environment must be good business – because "it *ought* to be" (Reinhardt 1999, 150). Countless examples of win-win situations for both the environment and the financial bottom line have been collected. Yet, corporate

---

[4] According to Wikiquote, the quotation is disputed (Wikiquote 2009).

sustainability has all too often been presented one-sidedly without acknowledging the difficulties and risks (Walley & Whitehead 1994; Esty & Winston 2009, xxf.). Ironically, these overzealous efforts to promote corporate sustainability may in fact have harmed the sustainability movement in the long run. Every manager knows that business reality is not quite as straightforward – it is full of ambiguity and difficult trade-offs (Reinhardt 1999). Many managers therefore consider corporate sustainability a shallow term; sustainability issues are believed to be non-substantial. These sceptics may further argue that all the talk is merely a form of political correctness (Berns et al. 2009b, 11). On this note, sustainability activities that are vital for the competitiveness of a firm may be argued to have just been relabelled and reassigned. Worse even, conventional wisdom holds that environmental protection in general does not improve competitiveness but instead entails additional cost (Ambec & Lanoie 2008). This line of thought is flanked by the belief that corporate sustainability is not only bogus in economic terms; that it is also the role of the government – and not business – to enforce sustainable business practices (Friedman 1970). Providing public goods like clean air beyond legal compliance is doubtful to be rewarded according to this view; market pressures would instead dictate maximising profits (Berchicci & King 2007, 5f.).

In a softened form of criticism, sceptics may argue that opportunities and threats stemming from corporate sustainability are real but insignificant. This fits with the view that sustainability is a matter of public relations and possibly product marketing – but that it is largely irrelevant to the core of the business. This view seems to become less widespread, but is still prevalent, especially among managers that are relatively inexperienced in the sustainability domain (Berns et al. 2009b).

The controversy around corporate sustainability also includes aspects of rivalling ideologies and contrasting ethical perspectives (Schaltegger & Wagner 2006b, 2) as explained above. However, these aspects go far beyond the scope of this work. I believe that ethical arguments will not persuade a sufficient fraction of corporate leaders to consider Green Business Model Transformations, at least not in the case of public companies. Consequently, and because I believe that a case for Green Business Model Transformation can often be made without explicitly relying on ethical arguments, I will concentrate my analysis on economic considerations that may be accepted both within the neoclassical and the (moderate) environmentalist camp (for a systematic comparison see Gladwin et al. 1995).

In conclusion, my claim is that the economic opportunities stemming from corporate sustainability are big and largely untapped; that green business models in particular can be used to create value, sometimes on a grand scale. In the following section I will present theoretical and empirical evidence to back these seemingly bold statements.

3.2.2 Evidence for the Value Creation Potential of Corporate Sustainability

3.2.2.1 Corporations' Blind Spot on Green Opportunities (and Risks)

The claim that corporate sustainability can create previously neglected value immediately points to the economist's famous metaphorical "$10 bills on the ground": According to traditional economists these bills can never be found, "because someone would have already picked them up" (Porter & van der Linde 1995b, 98). Thus, if someone spotted a bill nevertheless, it would have to be counterfeit.

However, such reasoning is based on the unrealistic assumption that information in the market is perfect and that all large opportunities from environmental sustainability have already been identified (Porter & van der Linde 1995b, 98f.). By contrast, Porter & van der Linde argue that – in order to recognise the opportunities to improve resource productivity, innovativeness, and competitiveness – it may even be beneficial to enact stricter (but flexible) environmental regulation. This way, so they argue, the widespread lack of attention and professionalism regarding environmental matters can be reduced (Porter & van der Linde 1995b, 99f.). The so-called *Porter Hypothesis* has been discussed controversially (e.g., Jaffe et al. 1995; Walley & Whitehead 1994; see also Ambec & Lanoie 2008). Nevertheless, it has to be acknowledged that general awareness and reporting capabilities have greatly increased since (KPMG 2008). However, some of the other impediments that Porter and van der Linde recognise are still very relevant today: The business environment is still complex and fast changing (Berns et al. 2009b, 14) – probably even more so than 15 years ago. Aragón-Correa & Sharma (2003) argue that while perceived unpredictability of the general business environment actually facilitates adopting proactive environmental strategies, uncertainty about the impact of external changes on the organisation and about the effects of individual decisions as well as complexity are considerable barriers. Combined with organisational inertia (Shrivastava 1995b, 942) the forces opposing a resolute pursuit of green opportunities can become very strong. This is especially true for established companies in mature, low growth industries in which substantial risk-taking is less rewarding (Russo & Fouts 1997). It seems a lot safer to stick to compliance and from time to time adopt established "best practices" from peers, rather than undertaking major novel green initiatives. However, this kind of passive, uncreative approach just reinforces the preconception that corporate sustainability offers only limited upsides to the company (if any) and may thus create a self-fulfilling prophecy.

While environmental reporting capabilities have improved significantly over the last years, there is still a lack of strategic management tools needed to move beyond incremental improvement of environmental performance (Berns et al. 2009a, 73). A cost-based mindset towards environmental management (Porter & van der Linde 1995a) as well as control problems to adequately align incentives

(Porter & van der Linde 1995b, 99) may further impede corporations from picking up the $10 bills – or seeing them in the first place. This includes the still widespread fixation on quantifiable short-term benefits (Christensen et al. 2008, 103f.) and coordination issues across organisational units (Johnson 2010, 155ff.). However, at least one condition has probably changed in favour of attempting bold, green initiatives: stakeholder pressure. The relationship between stakeholders and corporate sustainability will be explained in section 3.3.2. However, while the increased pressure has raised the topic on the corporate agenda, most activities are still to be considered passive and non-substantial in nature (Berns et al. 2009b, 12). Thus, the question whether sustainability pioneers are rewarded as a general rule has yet to be answered (Steger 2006, 442).

The considerations above explain why green opportunities may still be untapped. However, three issues remain open: How significant are they? What do they consist of? How can they be seized?

The following includes a short summary of the literature regarding these questions. Various practical examples of green value creation will be provided in part III, chapter 7.

3.2.2.2 Empirical Studies on the Link between Environmental and Economic Performance

An important insight that overenthusiastic environmentalists frequently fail to mention is the fact that while corporate sustainability *can* create value, it does not under every circumstance, nor does it happen automatically without adequate sustainability management (Schaltegger & Synnestvedt 2002).

A growing number of quantitative empirical studies try to clarify the link between environmental sustainability and economic/financial performance. Different variables are commonly used for this purpose (Molina-Azorin et al. 2009, 1093): Frequently used financial performance measures include ROA, ROS, ROE, stock market returns, stock price, and profits. Resource consumption, emissions, toxic waste, oil and chemical spills, and recovered, treated or recycled substances are commonly used measures of environmental performance. Some studies measure environmental management instead of performance, e.g. in terms of environmental strategy and practices, types of undertaken initiatives, environmental management systems, or ISO 14001 certification. This variety of measures already indicates the problem of methodological diversity in this field of research.

Three broad types of quantitative empirical studies have been used to investigate the link between environmental and economic performance: portfolio analyses, event studies, and studies measuring long-term effects based on regression analysis (Ambec & Lanoie 2007, 16ff.).

Portfolio analyses examine how companies with highly rated environmental performance perform at the stock market compared to the investment universe. In their meta-study Ambec & Lanoie (2007) examined 16 studies, five of which found a positive link, but 11 of which found no statistically significant differences. Another meta-study examining 20 studies also showed mixed results (UNEP & Mercer 2007). Overall, sustainable responsible investment (SRI) seems to offer comparable risk-adjusted returns – which is a success in itself as restricted portfolios are intrinsically disadvantaged due to their lower diversification.

Event studies consider certain environmental events and their effects on the stock price. Especially bad news has been found to have a noticeable impact: For the 14 studies considered by Ambec & Lanoie (2007; 2008) they report an average abnormal daily loss of 2.22%. In a way, this result is not surprising as unforeseen hits to the bottom line should result in adjustments by the stock market. However, there is less evidence that positive events are equally rewarded. For example, the announcement of the inclusion in the well-known Dow Jones Sustainability Index (DJSI) family was found to have only very small effects (< 0.1%) on the stock price (Consolandi et al. 2008).

A third type of studies analyses long-term effects of the environmental performance of selected companies by using regression analysis. Of the 12 studies in this category considered by Ambec & Lanoie (2007), nine show a positive, one a negative relationship, and two studies found no impact.

All three types of studies discussed above suffer from methodological problems which are not enlarged upon here in detail (for a discussion of issues see Ambec & Lanoie 2008). However, one challenge of particular relevance is to reveal the underlying causal chains that are responsible for the described statistical findings.

3.2.2.3 Examination of Causality behind Statistical Findings

Four basic causal relationships are applicable for the link under examination (Ambec & Lanoie 2008, 58f.): First, financial performance may influence environmental performance. A positive relationship is plausible based on the premise that companies that do well can afford to spend money on environmental protection. This possibility is relevant because a significant positive correlation between environmental and economic performance may be misinterpreted to be caused by environmental performance in cases where philanthropy is the true source. Empirical data from Japan supports this hypothesis (Nakao et al. 2007), while Wagner et al. (2002) found no respective evidence of significance. A second possibility is that there is a third factor that influences both environmental and economic performance. However, most studies of the subject (like the ones mentioned above) are led by the hypothesis that the causal relationship goes the other way round, i.e. environmental performance and environmental management influences economic performance. Lastly, there may be no significant linkage at all, a condition that could be true for firms operating in a business environment where ecological issues are of subordinate importance (Ambec & Lanoie 2008, 57f.; Molina-Azorin et al. 2009, 1095). In addition, even if there is a strong causal relationship, cause and effect may be separated by a delay of several years and are thus difficult to isolate (Hart & Ahuja 1996; King & Lenox 2002). The fact that meta-analyses of empirical studies (Ambec & Lanoie 2007; Molina-Azorin et al. 2009) found that results regarding the direction of the correlation are not univocal indicates that the underlying causal chains are non-trivial.

None of the four causal relationships can be ruled out yet. In fact, all may be present under certain circumstances and further research is needed to gain a deeper understanding of the topic (Ambec & Lanoie 2008). Especially the question which specific chains of cause and effect the relationships are based on would be of high interest. The use of case studies is essential to meet this challenge. Although case studies tend to lack representativeness due to limited sample sizes and company-specific circumstances, they are very well suited to unraveling complex causal relationships (Schaltegger & Wagner 2006a, 7). Hence, case studies represent a major source of insights on how to successfully manage Green Business Model Transformations (see chapters 7 and 8).

Schaltegger & Burritt (2005) aim to reconcile the seemingly contradicting views on the impact of environmental efforts on economic success by suggesting that these views reflect different management approaches and performance levels, not mere perceptions (see figure 3.2). For example, companies that employ reactive, end-of-pipe solutions can indeed expect increasing costs from any voluntary effort towards environmental protection (represented by the lower curve ES_0-EP_0). In contrast, well-managed pollution prevention can save resources and thus improve the cost base (curve ES_0-A-B-EP_1). Sometimes these savings can be substantial as the case of 3M's Pollution Prevention Pays (3P) programme demonstrates – the company saved US$ 500 million over 15 years (Shrivastava 1995a). The natural-resource-based view (Hart 1995) provides a theoretical framework to link competitiveness and environmental performance. Along with pollution prevention, Hart argues that product stewardship and sustainable development[5] can improve competitiveness. Analogous to the classic resource-based view (e.g., Wernerfelt 1984; Barney 1991b; Peteraf 1993), he argues that superior investment in environmental resources bestow competitive advantage to firms vis-à-vis competitors due to factor immobility and barriers to competition.

[5] In his paper, Hart understands sustainable development as a strategy to "sever the negative links between environment and economic activity in the developing countries" (Hart 1995, 996).

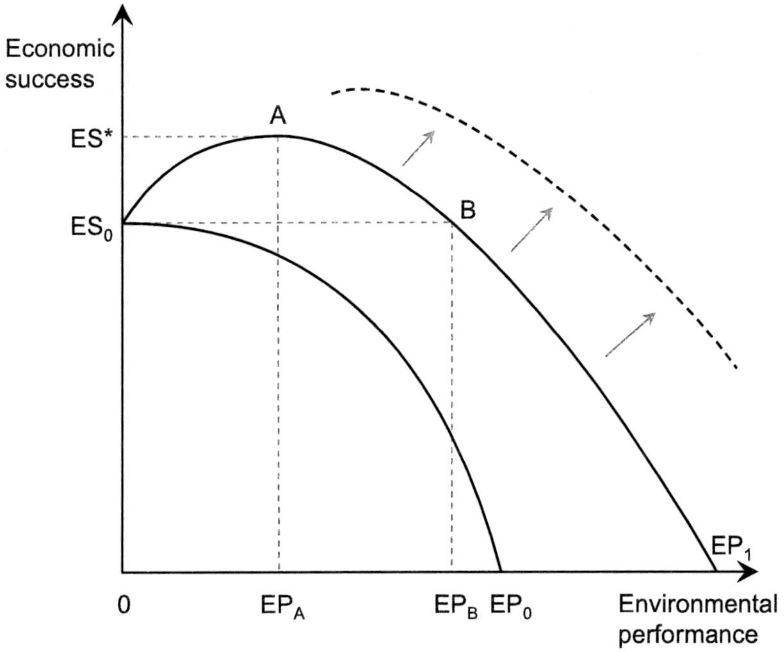

Fig. 3.2 Possible relations between voluntary corporate environmental performance and economic success (similar to Schaltegger & Synnestvedt 2002, 341; Schaltegger & Burritt 2005, 197; Schaltegger & Wagner 2006a, 11)

At some point, however, the marginal gain from environmental protection declines and becomes economically disadvantageous relative to mere compliance (point B). Hence, profit-maximising management should seek to find the "sweet spot" (point A) that offers maximum economic returns and yet encompasses superior environmental performance (EP_A) compared to legal minimum standards (0).

Exceptional environmental management may even push the curve further outwards (dashed curve), thus relaxing the trade-off and facilitating even better performances in both dimensions. Green business models may allow particularly large shifts, e.g. by enabling innovative new, green value propositions that expand existing or create new markets. Such discontinuities also suggest that the link may not be a smooth curve as depicted in figure 3.2, but involve "steps" (Schaltegger & Burritt 2005, 199).

In an effort to set in motion large-scale win-win dynamics as described above, General Electric has launched the company-wide "Ecomagination" initiative in 2005. GE's high expectations are reflected in the ambitious growth targets for products covered by the initiative from US$ 6 billion in 2004 to US$ 25 billion in 2010 (General Electric 2007). Although some of the reported revenue growth stems from products that were in the pipeline or already existed but were added to the portfolio later, the growth is still remarkable (General Electric 2010a).

In conclusion, differences in empirical findings may reflect differences in subsamples and their relative position between the two curves on the graph of figure 3.2, as well as the varying look of the curves themselves, depending on the industry, country and considered year of the study (Schaltegger & Burritt 2005, 200). It has been demonstrated that the relationship between environmental and economic performance can be positive based on costs as well as revenues. However, it has also become clear that a progressive mindset and sound management practices are needed – especially if the aspired results are in the magnitude of that of GE and 3M.

The following section provides a systematic overview of the business case for (environmental) sustainability. Later in chapters 7 and 8, these (mostly generic) considerations are complemented by exemplified but detailed analyses of how levers for green value creation can be used in practice.

3.2.3 The Business Case for Environmental Sustainability

The term *business case for sustainability* refers to the question how "the competitiveness and business success of a company [can] be improved with voluntarily created outstanding environmental and social performance" (Schaltegger & Wagner 2006a, 1). Ultimately, every business case can be reduced to a limited number of general levers for value creation (Schaltegger & Hasenmüller 2005, 8ff.; Berns et al. 2009b, 15; Esty & Winston 2009, 101ff.). Although environmental and social dimensions are often intertwined, the following summary focuses on environmental performance only.

3.2.3.1 Levers for Value Creation through Environmental Sustainability

The business case for sustainability can be evaluated with metrics on three different levels (Peloza & Yachnin 2008): *End state outcome metrics* (e.g., share price), *intermediate outcome metrics* (e.g., profits), and *mediating metrics* (e.g. energy consumption). For a causal analysis, understanding the mediating process is essential. Value creation levers are grouped into the following three categories that they most directly affect, and which ultimately determine the economic success of the firm:

- Profits, i.e. revenue (price or volume) and cost
- Tangible and intangible assets
- Risk

Corporate sustainability can either affect the categories above directly in financial terms, or affect value creation more indirectly through complex causal relationships. Moreover, value creation levers never produce isolated effects, which is one of the reasons why it is so difficult to prove the positive link between environmental and economic performance.

Value creation levers for generating profits

There are countless ways that superior environmental performance can improve both the top and the bottom line of companies (e.g., see Lankoski 2000, 26f. for a list). Key levers mentioned in the literature are specified below, but are not exhaustive.

Although not always the case, some green brands and products are capable of commanding a price premium (Manget et al. 2009, 14ff.), boost sales, or increase customer loyalty (Esty & Winston 2009, 127ff.). Moreover, some companies are able to capitalise on their sustainability efforts by entering completely new markets, e.g. hitherto underserved, developing markets (Hart & Milstein 1999, 29ff.; WEF 2009), or environmental goods and services markets (OECD & Eurostat 1999).

Apart from green product differentiation and addressing new markets, overall environmental performance of a company may also positively affect its access to certain markets (Ambec & Lanoie 2008, 47ff.), e.g. by acquiring a preferred supplier status with eco-minded customers or as a result of reputational spillover effects in business relationships (also see Lankoski 2000, 128f.). However, this lever is much harder to quantify in a business case than green product sales.

On the cost side, many companies realise substantial savings by using energy and other input factors more efficiently, by recycling waste and using environmentally-friendly substitutes, or by reducing cost related to emissions like taxes, treatment cost or pollution rights (Lankoski 2006, 34f.).

Like 3M's 3P programme, the use of respective equipment and operating methods to avoid quality defects and prevent pollution can create substantial cost savings in production (Shrivastava 1995a). In order to take full advantage of sustainability-related cost saving opportunities, however, companies need to consider business operations holistically. That is, product development should incorporate environmental aspects already in the design phase and take into account the whole life cycle (Dechant et al. 1994, 12f.); process improvements should be considered along the complete value chain (Nidumolu et al. 2009, 59ff.). For instance, Wal-Mart expects significant logistics savings from packaging reductions of its suppliers (Wal-Mart 2007, 58).

Despite the many ways to cut internal cost, companies should not necessarily limit their search for competitive advantage to their own cost base – sustainability may also facilitate *increasing* the cost of competitors, thus improving the company's *relative* cost base (Reinhardt 1999, 152ff.). A firm with strong environmental credentials can lobby for stricter regulation (McWilliams et al. 2002), thereby putting less sustainable rivals under pressure. It may also benefit from green subsidies or public investment (Schneidewind 1995, 40). In fact, Russo & Fouts (1997, 540) argue that the potential benefits go beyond simple cost advantages and revenue potential, that an environmental leader's "ability to influence public policies in ways that confer competitive advantage" should be considered a valuable intangible asset.

Value creation levers for creating tangible and intangible assets

Like political acumen, most benefits attributed to superior environmental performance are intangible assets. Reputation thereby represents a key item. Miles & Covin (2000) argue that environmental marketing performance is a major factor for reputation, and financial performance in general. Brand value is not easy to quantify, but estimates of successful green branding efforts claim gains in the billions of dollars in some instances (Interbrand 2008, 12ff.). Brand is long known to create demand and positively affect future sales. Pioneering brands (see Schmalensee 1982) – for which sustainability (still) offers lots of opportunities – can even realise demand advantages that go beyond classic advertising effects.

Another important potential benefit relates to human resources: Environmental (and social) sustainability can be used to increase productivity by boosting employee morale, and help recruiting and retaining high-quality talent (Steger 2006, 431; Siegel 2009, 14). On a more general note, CSR can arguably improve relationships and thus reduce transactions cost with any targeted stakeholder group (Jones 1995; Barnett 2007).

Moreover, in contrast to compliance-focused firms, environmentally proactive firms may be able to derive competitive advantages from deploying new technology and accumulating internal routines and know-how that is difficult to imitate (Russo & Fouts 1997, 537f.). In some cases these may qualify for legal protection as intellectual property. For instance, legislation for zero emission cars in California created a surge in respective U.S. patent filings in the 1990ies, especially by Japanese car manufacturers (Bedsworth & Taylor 2007, 9ff.). Proactive Toyota in particular was subsequently able to capitalise on its early investments (Maynard 2007; Avadikyan & Llerena 2010) (see also chapter 7.8).

Managing risk

The recent *Deepwater Horizon* oil spill that is estimated to cost BP dozens of billions of dollars (The Economist 2010) is a dramatic reminder how important managing environmental risk can be. Historic examples show that severe neglect of the sustainability dimension can even entail companies or entire industries losing their social (or legal) licence to operate. Instructive examples include the prolonged misconduct of the once widely admired American company Johns-Manville with respect to asbestos health risks; it culminated in 1982 in the largest bankruptcy in American history until then (Sells 1994). A more recent example is the scandal around tainted Chinese milk in 2008 that made the entire industry implode (Branigan 2008; Ramzy 2008).

The magnitude and type of environmentally driven business risks differ widely by industry sector (Steger 2006). Firms can systematically identify and mitigate these business risks, even if they are subtle or improbable to materialise (Esty & Winston 2009, 114ff.). Yet risk management does not eliminate the possibility of being attacked by an activist group. Companies can try unilaterally to create goodwill by going beyond compliance in environmental protection to avoid such attacks. At the same time, however, signalling responsiveness to stakeholder demands can make the company a more attractive target (Baron & Diermeier 2007). In many instances, seeking collaborative relationships with certain NGOs rather than taking a purely defensive approach will be the best option (Esty & Winston 2009, 69ff.).

Naturally, managing environmental risks is mainly about reducing potential downsides like lawsuits, consumer boycotts, or brand damage. This makes it difficult to demonstrate that a chosen level of effort is adequate (Reinhardt 1999, 155f.). However, there are also some benefits that are directly visible: First, sound environmental risk management can reduce the cost of capital (Sharfman & Fernando 2008) and insurance premia (Lankoski 2000, 27). It will also help firms to score high in sustainability rankings like that of the DJSI (SAM & PwC 2010). The direct effect of inclusion in respective indices on the stock price is still very small, but will probably grow in importance (Consolandi et al. 2008).

3.3 Building the Business Case for Sustainability

If environmental sustainability is to be recognised within the firm as a serious and significant business opportunity, building a robust business case is essential (Berns et al. 2009b, 20). This involves a systematic assessment of the value creation levers mentioned above (and potentially additional ones).

There are numerous known tools to identify, build, implement, and monitor the business case for sustainability (e.g., see Steger 2006, 440f.; Schaltegger et al. 2007). Yet, building a business case for sustainability can be challenging. Firstly, the costs of environmental initiatives are often underestimated (Walley & Whitehead 1994), and it is not clear if they will actually create value before a detailed analysis has been conducted. Especially for more radical endeavours like a Green Business Model Transformation, managers shy away from these business cases because of the associated risks, inherent uncertainty, and longer-term payback (Steger 2006, 432). As explained above, many of the major value creation levers are also hard to quantify in financial terms. On the one hand, this makes the proposed transformation harder to justify to sceptical managers. On the other hand, there is a risk of "massaging" the numbers, and that less obvious practical obstacles get ignored. For instance, sales targets of a new green product may be set unrealistically high if the overall value proposition is not sufficiently attractive to a broad target audience – relying primarily on green aspects will usually not work for the mainstream. Moreover, if the corporate brand or the company's business practices are not compatible with selling green products, even attractive value propositions may fail to deliver expected results. Lastly, even if environmental sustainability can be used successfully to boost sales as projected, unexpected cost increases may offset the benefits. Many firms will thus first need to acquire "green competencies" (Marcus & Fremeth 2009, 22f.) in order to operate a successful green business model.

In conclusion, opportunities are plentiful, as are the associated challenges. These intertwined issues will be elaborated upon from chapter 7 onwards based on practical examples. Next, the basics of environmental issues and practical means of dealing with them are outlined.

3.3.1 Overview of Environmental Sustainability as a Management Issue

Environmental problems are diverse (Millennium Ecosystem Assessment 2005): Their scope can be global (e.g. climate change), regional (e.g., water scarcity), or local (e.g., waste). Some issues build up over a long time (e.g., ozone depletion), other problems have short-term characteristics (e.g., local spills of toxic material). Moreover, while some environmental problems can be corrected (e.g., acid rain), others are irreversible (e.g., loss in biodiversity). There is also a complex system of direct and indirect drivers that reinforce environmental degradation. Some of the most fundamental sources of concern like population growth and changes in lifestyle lie well beyond the sphere of influence of individual corporations.

Obviously, different industries face different environmental issues, or do so with different intensities (Steger 2006, 420ff.; SAM & PwC 2010). Moreover, significant portions of the environmental impact of firms can occur upstream in the supply chain (e.g., for the food industry) or downstream at the customer (e.g., for the automotive industry).

The diversity and complexity of issues makes it difficult for companies to take a holistic approach towards environmental sustainability and find lasting solutions to the most pressing problems. Moreover, companies undertaking Green Business Model Transformations need to consider not only the objective characteristics of environmental issues, but also the perception that relevant groups (customers, politicians, etc.) have of them. "Objective characteristics" is an idealised term that refers to what science claims to know about an issue, knowing that conflicting claims may arise and research findings may be proven wrong in some cases.

3.3.2 Current Perceptions on Environmental Issues

There is a large temptation for companies to jump on the bandwagon and focus only on "hype issues" like climate change. Concentrating on issues based on current public sentiment can backfire as it has been seen historically that the perceived importance of environmental issues can shift dramatically over time. For example, while 30 percent of Europeans considered climate change to be the most serious problem for the world as a whole in 2008, only 18 percent did so less than a year later in the wake of the economic crisis (TNS Opinion & Social 2009). The Economist found similar fluctuations in the US. The survey also revealed a widespread lack of basic knowledge on the issue which may contribute to the strong volatility of opinions as well (The Economist 2009, 9). Apart from short-term fluctuation, long-term shifts can be observed as well: resource depletion, acid rain, the fear of nuclear accidents, ozone depletion, climate change, and now again the risks of nuclear power have all been prominently featured in the media over some period of time. At the same time, the number of industries that are "in the firing line" has increased dramatically since the 1960ies (Elkington 1994, 95). An overview of the ten environmental issues that most concerned European Union citizens end of 2007 is presented in table 3.1[6]:

Table 3.1 Top 10 environmental concerns in the European Union 2007 (TNS Opinion & Social 2008, 6)

1	Climate change (57%)
2	Water pollution of seas, rivers, lakes and underground sources (42%)
3	Air pollution (40%)
4	Man-made disasters, i.e. major oil spills or industrial accidents, etc. (39%)
5	Natural disasters, i.e. earthquakes, floods, etc. (32%)
6	The impact on our health of chemicals used in everyday products (32%)
7	Depletion of natural resources (26%)
8	Growing waste (24%)
9	Loss in biodiversity, i.e. the extinction of species, loss of wildlife and habitats (23%)
10	Agricultural pollution, i.e. use of pesticides, fertilizers, etc. (23%)

Esty & Winston (2009, 34) present a slightly different top-10 list based on a managerial perspective. Climate change is number one as well, but other items differ in their rank order, and disasters are omitted from the list. Instead "energy", a major controllable input factor for many firms and a primary source for climate change, has been added as second. However, such lists can only provide rough guidelines in any case. Industries and business models have very different impact profiles, and public perceptions of environmental issues change over time, too. Moreover, there are significant differences between countries and regions. Not all of these differences can be explained based on how much affected people are by a specific problem. For example, 41 percent of the Swedish population believe that the impact on health of chemicals used in everyday products is one of the five

[6] In the survey respondents were asked to name the five main issues they are worried about. The study was conducted from November until December 2007 with 26,730 citizens of the 27 member states. Lower rated anwers were: The use of genetically modified organisms in farming (20%); urban problems, i.e. traffic jams, pollution, lack of green spaces, etc. (15%); impact of current transport modes, i.e. more cars, more motorways, more air traffic, etc. (12%); our consumption habits (11%); noise pollution (8%); none of these (spontaneous) (1%); don't know (1%) (TNS Opinion & Social 2008).

most severe environmental issues, as opposed to only 26 percent of the Finnish population (TNS Opinion & Social 2008, 10). Not only does the perceived relative importance of environmental issues differ, the importance of environmental protection overall can vary substantially, too: While 89 percent of Swedes find it very important personally, only 47 percent of Finnish citizens share that view (TNS Opinion & Social 2008, 11). Such differences from country to country can become a severe problem for multinationals: Green value propositions may be compelling in one country, but not in another – even if they are geographically close and demographically (but not culturally) similar like Sweden and Finland. Similarly, the company's business practices may be considered acceptable in one country, but criticised harshly elsewhere – without an objective difference in the environmental impact caused.

Finally, even domestically the perceived urgency and gravity of environmental issues differs between stakeholder groups. Stakeholder groups are increasingly argued to be important not only in shaping the environmental agenda for businesses (Murillo-Luna et al. 2008), but also as a factor that influences the competitiveness of firms (Porter & Kramer 2006; Berchicci & King 2007; Harrison et al. 2010). Hence, it is worthwhile to insert a brief examination of the topic of stakeholder management (e.g., Freeman 1984).

3.3.3 Stakeholders Relevance for Environmental Management and Competitiveness

Stakeholders are "any group or individual who can affect or is affected by the achievement of the organization's objectives" (Freeman 1984, 46). Various categorisations of stakeholder groups exist (e.g., Henriques & Sadorsky 1999; Buysse & Verbeke 2003; Esty & Winston 2009, 65ff.). For example, Henriques & Sadorsky (1999, 89f.) distinguish between four broad groups of stakeholders: (1) *regulatory stakeholders* (governments, trade associations, informal networks, and competitors that shape the industry); (2) *organisational stakeholders* (customers, suppliers, employees, and shareholders); (3) *community stakeholders* (community groups, environmental organisations, and other potential lobbies); (4) *the media*.

The relative importance of each group depends on the firm's level of commitment: Henriques & Sadorsky (1999, 90) argue that the more proactive firms are, the more important they consider organisational and community stakeholders. The former are of course at the very core of business model considerations. Regulatory stakeholders will be of greatest importance to firms with medium environmental commitment, and the media are the main focus of reactive firms. The public at large (and thus also the mass media) tends to be especially influenced by negative incidents like environmental disasters.

In general, the public is swayed by emotional messages, like Al Gore's touching film (Guggenheim 2006) about climate change and himself combating the same has illustrated: It created enormous awareness among American citizens that scientists had largely failed to do so up to that point (Nielsen 2007). Since the advent of the blogosphere and other internet-based phenomena, the media increasingly loses its delimitation (Esty & Winston 2009, 80). For example, NGOs increasingly exploit the potential of the internet to advance their mission. Yet, NGOs are quite heterogeneous as a group (Friedman et al. 2002). Depending on the focus and characteristics of the particular NGO it may either pick topics primarily based on public attention or scientific grounds, act hostile and ideologically-driven or be constructive and engage with corporations in a cooperative way to improve their environmental business practices (Dahan et al. 2010). Conversely, firms can take an antagonistic stance on NGOs or seek to increase competitiveness *and* environmental performance through cooperative approaches. The same is true regarding regulatory stakeholders; leading firms may be able to shape their regulatory environment rather than react to it (Buysse & Verbeke 2003).

Managing Green Business Model Transformations thus also means managing stakeholders. An obvious problem in managing stakeholders is that stakeholder groups have competing interests (Eesley & Lenox 2006). Mitchell et al. (1997) suggest analysing stakeholders based on three criteria for stakeholder salience: *power*, *legitimacy*, and *urgency*. Stakeholders that possess all three attributes are *definite* stakeholders that must not be ignored. A company needs to carefully consider its approach to *expectant* (two of three attributes) and *latent* stakeholders (one attribute) – especially as stakeholders may acquire additional attributes over time.

Esty & Winston (2009, 265f.) recommend checking current stakeholder prioritisation by allocating stakeholder groups to a matrix based on the current level of focus and their influence today and in the future.

Even the considerations described above, however, do not provide a sufficient basis for effective stakeholder management if (heterogeneous) stakeholder groups are defined too broadly. For instance, customers may need to be assessed in a more fine-grained manner in order to draw meaningful conclusions (Harrison & Freeman 1999, 484) – much in line with the business model concept as will be explained in chapter 4.

The merits of stakeholder theory are still subject to heated debates among academics (e.g., Freeman et al. 2004; Sundaram & Inkpen 2004). However, many scholars agree that stakeholder orientation in general terms is helpful to achieve corporate objectives. Yet, there is some causal ambiguity surrounding specific stakeholder treatments and the derived competitive benefits (Harrison et al. 2010, 69). The authors suggest that limiting factors to successful stakeholder management – or "managing for stakeholders", as they call it – includes "over-allocating value to stakeholders, an inability to measure value created using accounting measures, a lack of ability to translate knowledge into value-creating opportunities, measuring value creation over a too short time frame, and the potential for opportunism" (Harrison et al. 2010, 69). Firms can try to overcome these barriers, but it seems unlikely that detailed prescriptions of general validity for stakeholder management are possible; most firms will need to find their own, tailored approach. Ultimately, the justification of stakeholder management may have to remain normative rather than instrumental in nature (Donaldson & Preston 1995).

3.3.4 Possible Courses of Corporate Action to Tackle Environmental Issues

A company may be tempted to adopt the agendas of its most relevant stakeholders, trying to appease their demands for environmental measures. However, it is advisable to first develop a proprietary perspective on what should be done (Piet 1994, 43), and only then try balancing the various internal and external interests.

A useful starting point may be to consider the environment in terms of the environmental services it provides, i.e. the natural value it creates. The company can then take actions to conserve, substitute, or restore certain environmental services it affects. The World Resources Institute (WRI) distinguishes four types of ecosystem services (Millennium Ecosystem Assessment 2005): (1) *Provisioning* (of food, fibre, genetic resources, biochemicals, natural medicine, pharmaceuticals, and fresh water); (2) *Regulating* (of air quality, climate, water, erosion, diseases and pests, pollination, natural hazards, water purification, and waste treatment), (3) *Cultural* (aesthetic, spiritual and religious values, recreation and ecotourism), and (4) *Supporting* ecosystem services (e.g., primary production, soil formation, and nutrient cycling).

Gladwin et al. (1995) have categorised the potential range of sustainable behaviours into eight sustainability and operational principles and provide more than 40 related sample techniques that companies may consider (see table 3.2).

Table 3.2 Principles and sample techniques for sustainable development (Gladwin et al. 1995, 892)

Sustainability principles (operational principles)	**Sample techniques**	**Sustainability principles** (operational principles)	**Sample techniques**
1) Assimilation Waste emissions ≤ Natural assimilative capacity	• Pollution prevention • Natural products • Detoxification • Biodegradability • Low input agriculture • Synthetic reduction	**5) Conservation** Energy-matter throughput per unit of output (time 2) ≤ Energy-matter throughput per unit of output (time 1)	• Fuel efficiency • Mass transit • Cogeneration • Computer controls • Demand side mgmt • Smart buildings
2) Regeneration Renewable harvest rate ≤ Natural regeneration rate	• Sustained yield mgmt • Safe minimum standards • Harvest certification • Access restriction • Exclusive harvest zones • Resource right systems	**6) Dissipation** Energy-matter throughput (time 2) ≤ Energy-matter throughput (time 1)	• Depackaging • Durable design • Repair/reconditioning • Telecommuting • Bioregional sourcing • Dematerialization
3) Diversification Biodiversity loss ≤ Biodiversity preservation	• Biosphere reserves • Extractive reserves • Buffer zones • Polyculture farming • Ecotourism • Debt for nature swaps	**7) Perpetuation** Nonrenewable resource depletion ≤ Renewable resource substitution	• Solar energy • Wind power • Hydrogen fuel • Bioenergy • Hydropower • Geothermal energy
4) Restoration Ecosystem damage ≤ Ecosystem rehabilitation	• Reforestation • Mine reclamation • Site decontamination • Bioremediation • Species reintroduction • Habitat restoration	**8) Circulation** Virgin / recycled material use (time2) ≤ Virgin / recycled material use (time1)	• Closed-loop manufact. • Industrial ecosystems • Internal recycling • Waste recovery • Design for disassembly • Water recirculation

The listed principles and techniques may be used as a pool of ideas for continuous improvement or more radical redesigns of current business practices.

In addition, a multitude of decision support tools can help to operationalise environmentally sustainable business practices. These tools, in their entirety sometimes referred to as *sustainability accounting*, are diverse as they need to satisfy various information needs (e.g., physical vs. monetary, different timeframes, scope, and so on) (Burritt & Schaltegger 2010; Schaltegger & Burritt 2010). Some tools like *life cycle assessment* (LCA) (see Schaltegger 1997; ISO 2010) or *eco-efficiency analysis* (see Verfaillie & Bidwell 2000) are well-known, others less so. However, a general review of available decision support tools goes well beyond the scope of this work. Instead, specific management tools for Managing Green Business Model Transformations will be described, enhanced, or newly developed in chapter 9.

The size of the opportunity to protect, restore, or make more productive use of ecosystem services is potentially gigantic: Costanza et al. (1998) calculated a total annual value of US$ 16 to 54 trillion for these services – the world's GNP was US$ 18 trillion at the time. Of course, most of the underlying ecosystem services are not part of conventional markets. Therefore, calculated values contain large uncertainties and methodological issues. However, as the discussion about putting a price tag on CO_2 shows, at least some of these services that represent externalities today may soon become internalised cost in the future. Moreover, even without formal internalisation, if the efforts to protect certain ecosystem services are embedded in a convincing business model, certain customers and other relevant stakeholders can be expected to endorse it nevertheless. Chapter 7 will provide some examples of companies who managed to do so.

3.4 Corporate Environmental Strategies

More and more companies embrace active environmental business strategies (Esty & Winston 2009). Irrespective of the specific content, having an environmental strategy distinguishes firms from those that consider environmental management a purely operational (or even rhetorical) task (Baumgartner & Ebner 2010).

Researchers have suggested numerous models to classify companies in terms of their environmental management approaches (see Kolk 2002 for an overview of publications until 2000). Most proposed classifications describe a continuum and range from being non-compliant or reactive to proactive and/or visionary (Lee & Ball 2006, 381ff.). Yet, the image of linear progression towards environmental excellence has been criticised. Ghobadian et al. (1998) argue that the communicated attitude towards the environment does not necessarily match the firm's actual course of action. A company with *restrained commitment* may thus be viewed externally as advanced, but in reality restricts its efforts to mere compliance (lack of "walk the talk"). Other firms that show *speculative commitment* try to cut a corner in pursuit of a major opportunity and leapfrog competitors to environmental leadership. However, this engagement is opportunistic, and the firm will revert to old practices if the presumed opportunity does not fulfil expectations (Ghobadian et al. 1998, 19ff.). GE may be argued to match this type, although there is no evidence to suggest that GE's initial commitment was not sincere (GE Ecomagination turned out successful, after all). Lastly, Ghobadian et al. (1998, 21) describe a company type with *conditional commitment* that applies more or less advanced environmental practices in different countries – depending on its respective interests.

Baumgartner & Ebner (2010) propose an even more fine-grained typology based on 21 sustainability aspects and four respective maturity levels (from poor to sophisticated). Identified sustainability strategies are argued to have distinct maturity level profiles, i.e. they require various minimum levels of maturity for each aspect in order to be implemented successfully.

Apart from maturity-based typologies, firms can also be distinguished from others with respect to the relationship of sustainability activities to the core business, their target, and the expected benefits. Halme & Laurila (2009) propose three corresponding corporate responsibility (CR) types: *philanthropy*, *CR integration*, and *CR innovation*. Obviously, the last strategy is the most likely to produce new green business models. However, even philanthropic activities may eventually generate green business models if they are somehow related to the firm's capabilities (e.g., compare Google 2010a). CR integration, i.e. efforts to make existing business operations more sustainable, can originate green business models if changes are radical rather than incremental in nature.

For the purpose of this work, the distinction between operational, incremental and strategic, radical approaches (e.g., Steger 2006, 427) is thus of particular interest. To illustrate the difference the case of Wal-Mart can be used: The decision of Wal-Mart to design future stores to be 25 to 30 percent more energy efficient and retrofit existing ones (Wal-Mart 2009b, 20) can be assumed to have been based on a sound business case in the wake of high and rising energy prices. But despite the considerable absolute investment required, this change is not fundamental to its business model. The success of this measure is not pivotal to the success of Wal-Mart as a whole. In contrast, its decision to "green" its supply chain may be just that. For example, Wal-Mart now requires its suppliers' to report on waste, water and energy use. It also introduced an initiative in 2007 to reduce suppliers' packing by 5 percent by 2013, claiming that it could save Wal-Mart's supply chain US$ 3.4 billion (circa € 2.5 billion) until 2013 (Wal-Mart 2007). Furthermore, Wal-Mart aims to make the sustainability performance of its (increasingly green) products transparent to consumers through a "sustainability index" (Wal-Mart 2009a). If a success, Wal-Mart's changes could save it billions and boost its revenues. If not, Wal-Mart may permanently hurt its own cost base and severely threaten its low-price value proposition. Each of these supply chain measures may not seem revolutionary at first glance, and do not yet qualify Wal-Mart's business model as being "green" (see chapters 5 and 7.10). Nevertheless, the changes do alter the existing business model logic, and even affect the entire retail industry (Rosenbloom 2010). In this context, it is interesting to note that the slogan "Always Low Prices." has meanwhile been replaced by "Save money. Live better." – after previously remaining unchanged for 19 years (Mui & Rosenwald 2007).

As mentioned above, management is generally more focused on operational, incremental sustainability management – despite the much higher impact of radical transformations. In addition to the described difficulties in building the business case, such transformations require special expertise, mindsets and decisions beyond business-as-usual (Steger 2006, 432). Furthermore, the potential collateral damage of large transformations to organisations can be significant. In an interview with the Harvard Business Review, Nestlé CEO at the time Peter Brabeck-Letmathe states: "You cannot underestimate the traumatic impact of abrupt change, the distraction it causes in running the business, the fear it provokes in people, the demands it makes on management's time." (Wetlaufer 2001, 113).

Green Business Model Transformations are, by necessity, radical in nature. As a result, they will rarely rise from reactive or defensive environmental strategies. It is also hard to imagine that a company can sustain a Green Business Model Transformation and at the same time uphold a strategy of restrained commitment. Even if the company managed to keep up green appearances and keep external pressure low, cynicism will likely spread within the company.

It is therefore crucial that the environmental strategy is well-aligned with the general business strategy (Butner & Gregory 2009), and not to consider the transformation an independent undertaking. Hence, Green Business Model Transformations may need to coincide with a strategic reorientation towards environmental sustainability.

Florida (1996) found that, within a manufacturing context, the ability to reap financial benefits from a proactive environmental strategy is path dependent and intertwined with a broader set of capabilities (related to innovation, productivity, etc.). Moreover, environmental strategies can be hard to reverse if they encompass high complexity and significant tacit knowledge (Rugman & Verbeke 1998, 368f.). Many Green Business Model Transformations possess these characteristics. Hence, not only are some companies better positioned initially than others to pursue such a transformation, but it can be considered what Rugman and Verbeke call a "green gamble" (Rugman & Verbeke 1998, 368).

As the industry survey in chapters 7 and 8 will demonstrate, however, the range of potential green business models is large enough to provide viable opportunities for many different kinds of companies. Moreover, business models as a concept for management science help to increase the odds that the "green gamble" turns out to be a "green winner".

Chapter 4
The Business Model Concept as a Unit of Analysis for Management Science

The last chapter demonstrated that corporate sustainability can indeed facilitate economic success in many ways. However, the link between the two is complex, and there is a risk that companies will engage in many disconnected environmental initiatives that fail to tap the full economic potential which environmental sustainability offers. At the same time, companies may find it difficult to translate abstract environmental strategies into viable business concepts. I propose to use the business model concept to overcome these problems. In order to use the concept as intended with the management framework of chapter 9, it is necessary to clarify the relationship between business models and strategy, and to define and explain the generic elements of business models and how they interact. Ways to identify and deal with systemic patterns and growth dynamics complete the theoretical foundation that is needed to design successful green business models.

4.1 Strategy and its Relation to Business Models

Business has often been compared to warfare. Some ancient books on war like Sun Tzu's "The Art of War" (Sun Tzu 1963) or Carl von Clausewitz' "Vom Kriege" (von Clausewitz 1832) have become highly appreciated works for many business strategists. War and business are fundamentally different in their motivation and outcome (von Oetinger et al. 2003, 5f.). However, they have many conceptual commonalities, and warfare can therefore be used as an allegory to illustrate the relationship between strategy, business models, and the operational layer as it is understood in this work.

If one compared a modern cooperation to a 14th century army, one may consider the army's artillery to be one of its strategic business units. Following this example, the *business model* would correspond to a blueprint that comprises the logic of how the artillery is supposed to contribute to the success of the army. This blueprint would be based on the artillery's armament – medieval catapults in this case – and depict the catapults' capabilities (e.g., impact, firing range, speed), principal means of use (e.g., combat procedures), the required resources (e.g. operating personnel, type of ammunition), and all other relevant information that describes the concept behind the artillery's function.

A. Sommer: Managing Green Business Model Transformations, SPLCEM, pp. 49–91.
springerlink.com

However, the business model needs to be distinguished from *strategy* (Magretta 2002). The term *strategy* derives from Greek *strategos*, an elected general in ancient Athens and refers to the "coordinated application of all the forces [...] to achieve a goal. [...] strategy's components include a long-range view, the preparation of resources, and planning for the use of those resources before, during, and after an action" (Britannica Online Encyclopædia 2009).

Therefore, *business strategy* plans for the future success of a business in a dynamic, competitive environment (Porter 2008b). In contrast, the *business model* can be viewed as its translation into a logical framework for economic value creation (Osterwalder 2004). Hence, *business strategy* and *business model* are interlinked (Zott & Amit 2006). If it becomes evident that the business model will no longer support the strategy effectively in the foreseeable future, companies should either adapt their business model or revisit their strategy (or both). None of the two principally precedes the other, just like one leg needs to follow the other when walking (Umbeck 2009, 55; based on Mintzberg et al. 2009, 37, who originally referred to strategy and structure). There is a risk, however, that an established company instead "loses connection" to its strategy in order to retain the business model that has worked well in the past and been optimised considerably over time. Or, managers neglect strategy – in a continuous effort to mimic successful competitors – and blindly chase every new hyped technology, much afraid of missing the next revolution (Porter 1996, 75ff.). To phrase it in the context of the artillery example: Even the army with the best developed catapults needs to realise at some point that gunpowder and cannons make catapults obsolete. The trick is to recognise when that point will be reached – because early, immature versions of cannons were still inferior and ineffective compared to catapults. In fact, catapults and cannons co-existed for a quite some time until the catapult concept was finally abandoned (Wikipedia 2009a).

One piece that determines the competitiveness of a firm, however, is still missing: the operational layer. In order to execute the business model efficiently, business organisation and related business processes need to be aligned. While the *architecture* of key processes that directly relate to the core logic of value creation belongs to the business model layer, the actual implementation of business processes does not (Osterwalder 2004, 14). Tools like benchmarking and best practice sharing often address primarily the operational layer (Porter 1996). On a similar note, Casadesus-Masanell & Ricart (2010) further distinguish *tactics* from strategy and business models. Strategy would then correspond to higher order choices including the selection and design of business models and strategic contingencies. Tactics therefore map to the operational layer and include the remaining choices given the employed business model of the firm. However, because the term tactics implies conscious decisions, the operational layer is more comprehensive as it also includes historic characteristics, or ones that emerge in a seemingly random fashion (but may eventually still be relevant for the success of the firm). Translated to the military example, strategy would include the selection of the artillery as a branch of service and its role within the campaign, the business model represents the blueprint and area of application of the artillery, and tactics would include choices made by soldiers on the battlefield.

The relationship between the business strategy, business model and the operational layer as it is understood in this work is summarised in figure 4.1:

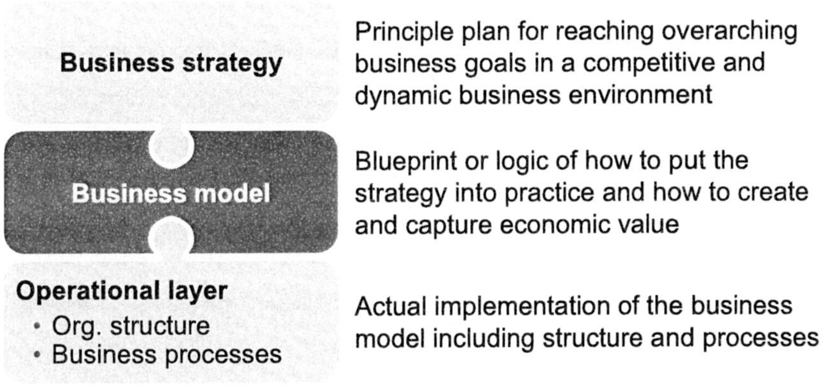

Fig. 4.1 Relationship between business strategy, business model, and the operational layer

A sound strategy that is poorly executed cannot create value. Conversely, operational efficiency without a distinct strategy will not deliver sustained success either – even if some individual elements of the business model are improved substantially. As Porter points out, operational effectiveness is necessary but not sufficient since "[competitive advantage] grows out of the *entire system* of activities" (Porter 1996, 73). With the business environment becoming more and more dynamic, competitive, and afflicted with uncertainty in many ways, the gap between these two layers has become wider than it used to be, and more difficult to bridge (Al-Debei et al. 2008). Al-Debei et al. (2008) argue that the digital revolution has not only produced a more challenging business environment, it has also essentially created the need for the business model concept to translate business strategy into business processes. It could be added that other (to a certain degree related) phenomena like globalisation or the sustainability movement have further contributed to that need. Therefore, the business model concept can be viewed as a mediator between business strategy and the operational layer of a company. Consequently, some business model researchers consider the business model concept to be a further development of the strategy concept (Umbeck 2009, 50). A good fit is not only critical between the business model and the strategy layer. Business models must also take into consideration corporate realities of the operational layer. Among other things, it is this hard-to-achieve fit which secures competitive advantage for successful companies.

So far, the terms *strategy* and *business model* have been introduced in a rather general, intuitive way. This, of course, neglects the many detailed elements of the latter, and maybe even more so the complex nature of the former – given that a whole academic field has been created around it: the field of strategic management.

Mintzberg et al. (2009, 9ff.) propose to use no less than five complementary definitions to capture the term strategy with all its different facets: Strategies can take the form of *plans* for future direction, or *patterns* that reflect the realised course of action that a company took. Strategies can also take the form of *positions* within the competitive environment. For other scholars and practitioners strategy reveals itself as a *perspective* that represents the distinctive way things are done at a given company. Finally, a strategy can be a *ploy*, a manoeuvre to outwit competitors.

Moreover, the authors suggest in their review of the field of strategic management that ten different academic schools can be distinguished. It exceeds the scope of this work to discuss them all. However, one school in particular has to be mentioned as it bears the closest resemblance to the business model concept used in this work: the so-called *configuration school* (e.g., Miller 1986; 1996). The configuration school supports the notion that elements of a business can be combined to reinforce success within the competitive environment. Also, occasional transformations – or re-configurations – are necessary to sustain that success. Like the business model concept, the configuration school integrates elements from various other schools of strategic management: analytical, prescriptive management versus organisational learning; cultural aspects versus power and politics; the resource-based versus the market-based view of the firm (Mintzberg et al. 2009, 317ff.). Although the different strategic management schools are not discussed in detail, various useful ideas from these schools will thus be introduced as appropriate in later chapters.

Like the concept of strategy, the business model concept has historically not been free of ambiguity and alternative interpretations. Michael Porter even stated that "the business model approach to management becomes an invitation for faulty thinking and self delusion." (Porter 2008a, 117). I argue that exactly the opposite is the case – if (and only if) business models are clearly defined and are not seen in isolation from strategy. A detailed definition and explanation of the business model concept is presented next.

4.2 Definitions and Elements of Business Models

One of the few authors who delivered comprehensive works on business models is Osterwalder (Osterwalder 2004; Osterwalder et al. 2005; Osterwalder & Pigneur 2010). Repeating the (ongoing) discussions around the different uses of the term *business models* in literature would be of little use here; for reviews of the matter compare for example Lambert (2003; 2006), Osterwalder (2004), Osterwalder et al. (2005), and Baden-Fuller & Morgan (2010). Instead, the following definition proposed by Osterwalder (2004, 15) which is based on a synthesis of existing literature on (e-)business models is used:

"A business model is a conceptual tool that contains a set of elements and their relationships and allows expressing a company's logic of earning money. It is a description of the value a company offers to one or several segments of customers and the architecture of the firm and its network of partners for creating, marketing and delivering this value and relationship capital, in order to generate profitable and sustainable revenue streams."

Furthermore, Osterwalder (2004) derived nine business model *building blocks*, or business model elements, grouped into four *pillars* (see table 4.1):

Table 4.1 Business model pillars and building blocks according to Osterwalder (Osterwalder 2004, 43)

Pillar	Building block of business model	Description
Product	Value Proposition	A Value Proposition is an overall view of a company's bundle of products and services that are of value to the customer.
Customer Interface	Target Customer	The Target Customer is a segment of customers a company wants to offer value to.
	Distribution Channel	A Distribution Channel is a means of getting in touch with the customer.
	Relationship	The Relationship describes the kind of link a company establishes between itself and the customer.
Infrastructure Management	Value Configuration	The Value Configuration describes the arrangement of activities and resources that are necessary to create value for the customer.
	Capability	A capability is the ability to execute a repeatable pattern of actions that is necessary in order to create value for the customer.
	Partner Network	A Partnership is a voluntarily initiated cooperative agreement between two or more companies in order to create value for the customer.
Financial Aspects	Cost Structure	The Cost Structure is the representation in money of all the means employed in the business model.
	Revenue Model	The Revenue Model describes the way a company makes money through a variety of revenue flows.

The *Product* pillar relates to what is offered to the customer. The *Customer Interface* pillar describes the customer and how the offering is delivered. *Infrastructure Management* deals with value creation aspects of the business models. The last pillar, *Financial Aspects*, summarises how the company plans to make money with its business model.

Osterwalder's classification has been created against the background of e-business. Nonetheless, it is a useful starting point for analysing green business models (or any other). Other authors have provided alternative (and sometimes conflicting) definitions, elements and classifications, many of which Osterwalder considered and synthesised for his own classification (for an overview see Osterwalder 2004, 24).

Another main source are the recent works on business models by Johnson et al. (2008) and Johnson (2010), particularly due to the perspectives on how established companies can transform their business models. The identified business model components are similar to those of Osterwalder, but grouped and accentuated differently (see figure 4.2).

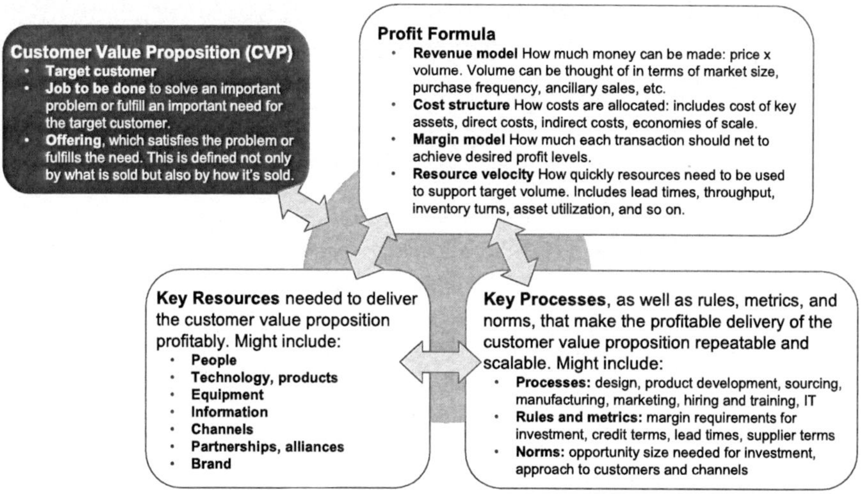

Fig. 4.2 The elements of Johnson et al's business model concept (adapted from Johnson et al. 2008, 54)

Johnson et al. (2008) identify the *Customer Value Proposition (CVP)* as by far the most important part. In contrast to Osterwalder's *Value Proposition* building block, the *CVP* also contains the target customer and its needs, which Osterwalder assigns to the separate *Customer Interface* pillar. Johnson et al. do not use an equivalent to the *Customer Interface* pillar and instead distribute the respective components on different dimensions including *Key resources (Channels)*. Together with the closely related *Key Resources*, Johnson et al. view *Key Processes* as the means to describe how the value is created and delivered to the customer and the firm as defined by the *CVP* and the *Profit Formula*, respectively. With regard to the delivery dimension, the authors seem to follow an understanding of business models that is somewhat overlapping with the operational layer as defined in chapter 4.1. However, they explicitly consider only *key* resources and processes to be part of the business model which in turn is consistent with the view advanced in this work that not the actual implementation, but only the systemically relevant aspects of a business should be considered. This means, for example, that a specific process that influences the overall logic of the business model is considered on a conceptual level, in contrast to generic processes that (most) businesses employ in a similar fashion.

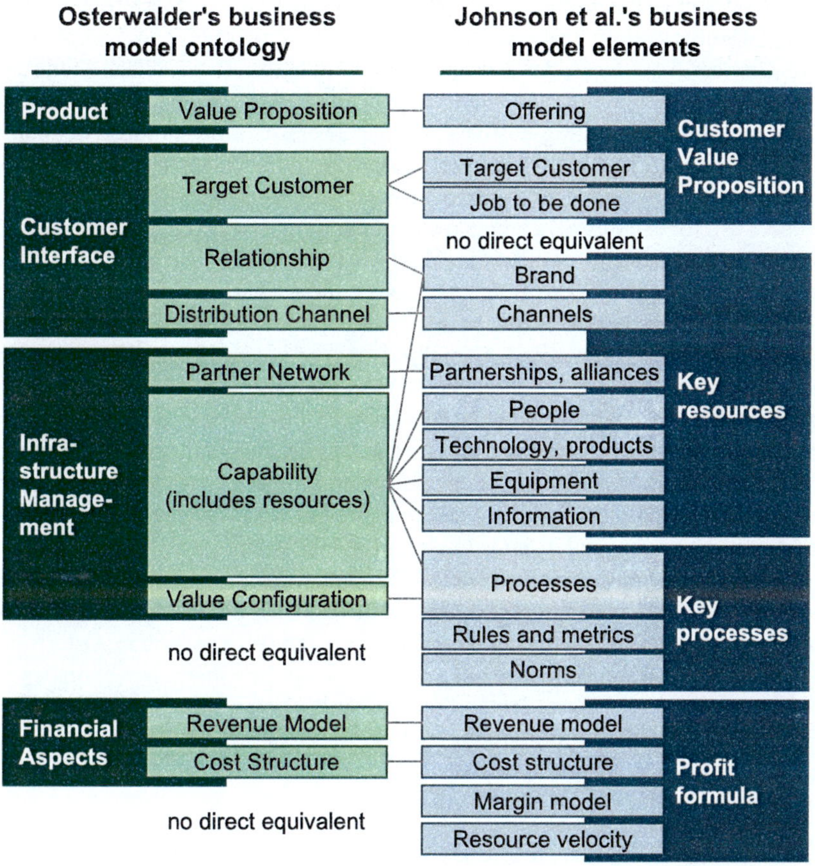

Fig. 4.3 Matching of business model elements of Osterwalder (2004) and Johnson et al. (2008)

Figure 4.3 presents an attempt to visualise a matching of the business model elements of Osterwalder (2004) and Johnson et al. (2008). Although figure 4.3 suggests that most elements used by the two authors are equivalent, it has to be noted that while element *categories* which Osterwalder and Johnson et al. use are largely the same, on a detailed level there are some important differences. The differences in the conceptions of business models can be explained with the different background and purpose of their works. Osterwalder heavily draws on e-business model literature and aims to develop an ontology – a "rigorous conceptual model of business models" (Osterwalder 2004, 2) – that can be used as a basis to develop (software-based) management tools. Osterwalder also points out that the four pillars are in line with the four perspectives of the well-established *Balanced Scorecard* management tool by Kaplan & Norton (2000). However, in the practice-oriented book "Business Model Generation" (Osterwalder & Pigneur 2010), the authors rely on the nine building blocks only, presumably because the four-pillar model has turned out to be too generic to be of practical use.

In contrast to Osterwalder's original work, Johnson et al. detach themselves from the e-business notion and intend to explain if and how incumbent companies should change their existing business models. Therefore, the approach of Johnson et al. provides a good fit with the research in this work. Tobias Umbeck, who also synthesised and extended previous business model conceptualisations with their practical use for established companies in mind, proposes to consider an additional dimension, that contains competitive elements (competitive barriers, inimitability, uniqueness) (Umbeck 2009, 60ff.). Although these characteristics are important and receive due attention in the management framework of chapter 9, they are considered strategic outcomes (of combinations) of business model elements, and not elements themselves. Furthermore, due to the centrality of the human factor for the success of almost any business model, Umbeck places employees, corporate culture, and organisation at the centre of business model elements. While the motivation is comprehensible, this again mixes different levels of consideration. For instance, corporate culture influences many business model elements, but as a phenomenon goes beyond a single business model design.

In conclusion, the combined works of Osterwalder and Johnson provide a well-suited basis to conceptualise business models as a unit for analysis for management science. While Johnson et al. cover a slightly larger scope, Osterwalder discusses each of the business model elements including their characteristics and interrelationships in greater depth. Consequently, Osterwalder's dissertation will serve as the theoretical starting point and be used to structure the detailed discussion of business model elements in the following.

4.3 Recapitulation of the Business Model Ontology by Osterwalder

The nine business model building blocks presented in table 4.1 represent the core of Osterwalder's business model ontology[7]. However, in some cases, these elements are decomposed even further into sub-elements. Osterwalder describes each of these components in a highly structured (technical) manner, including its definition, type of element, attributes and relationships to other elements. The chosen level of detail of the following recapitulation is necessary to develop a

[7] For clarification purposes, Osterwalder's pillars, building blocks, (sub-)elements and attribute categories will always be written capitalised and italic in the following; attribute *values* will be written in normal font with parenthesis if necessary for differentiation.

substantiated understanding of business model elements for the purpose of this work. It will also be helpful for practitioners who intend to design business models themselves. However, readers who are only interested in gaining an overview of the resulting composition of business models as used in this book may wish to jump directly to chapter 4.4.

4.3.1 The Product Pillar

The *Product* pillar (Osterwalder 2004, 48ff.) contains the *Value Proposition* element which consists of one or more *Offerings* (see figure 4.4). These provide value to the *Target Customer(s)* and are based on the *Capability(ies)* elements (that belong to the *Customer Interface* and *Infrastructure Management* pillars, respectively). Attributes of an *Offering* include *Reasoning*, *Value Level*, *Price Level* and *Life Cycle* (sometimes also referred to as *Value Life Cycle*).

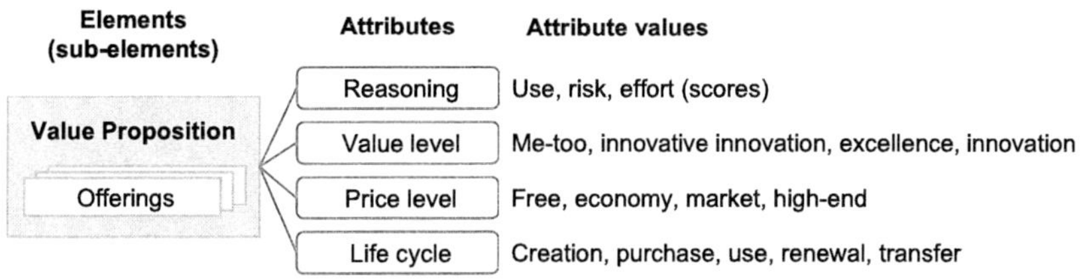

Fig. 4.4 Osterwalder's Product pillar (based on Osterwalder 2004, 48ff.)

Reasoning is the essence of why the *Offering* is expected to be useful to the customer. It consists of several aspects:

- Use: The core of the value; degree of utility depends on the match with the customer's needs.
- Risk: Reducing risks can offer additional value to the customer, including financial risk (e.g., loss of value of owned goods), or risks of poor performance (e.g., compatibility of software, product complexity). For example, a company can offer some sort of insurance and guarantees or provide complementary risk mitigation services.
- Effort: The convenience (or hassle) which the customer experiences in conjunction with the offering. This includes effort and cost along the whole product life cycle.

The second attribute, *Value Level*, refers to a relative, qualitative value scale including the level "me-too" (no differentiation through product features, but potentially offering aggressive prices), "innovative imitation" (selected innovative elements added to a standard *Value Proposition* or *Offering*), "excellence" (perfected value), and "innovation" (entirely new product / service or novel combination of the same; advantage may only be short-lived). The *Value Level* has to be considered in conjunction with the *Price Level* attribute, which is defined with the levels "free", "economy" (lower price than most competitors), "market" (close to average), "high-end" (e.g., luxury segment or innovations). In the case of "free" offerings like Google searches, other revenue streams are exploited instead (e.g., advertisement). Based on Kambil et al. (1996), Osterwalder presents the concept of the *value frontier*, i.e., the currently feasible combinations of value and price (that can be offered given the corresponding cost). Business model transformations can either move a company along the value frontier, or, in the case of a true innovation, shift it further towards the lower right (see figure 4.5):

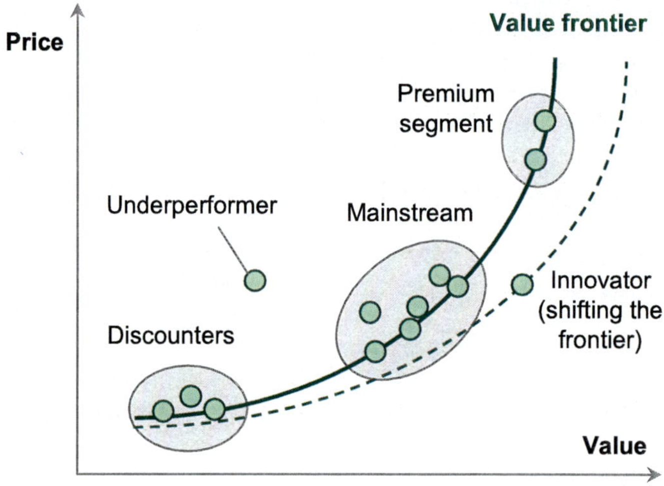

Fig. 4.5 Illustrative example of companies' positioning on the value-price-matrix (adapted from Kambil et al. 1996, 8)

The last attribute of *Offering* that Osterwalder suggests is *Life Cycle*. He distinguishes the value generated at creation, purchase, use, renewal (e.g., product update) and transfer (e.g., re-selling or disposal) of the *Offering*. Although Osterwalder presents the *Life Cycle* attribute to be separate, it could be considered an alternative view to the *Reasoning* attribute: For example, a sophisticated purchasing method reduces effort, an innovative transfer concept can reduce risk (of ownership), and the "use" aspect of *Reasoning* directly corresponds to the use phase.

When comparing all *Value Proposition* attributes to those of competitors, the relative attractiveness should become apparent. However, some factors that Osterwalder does not consider explicitly to be part of a *Value Proposition* can become decisive for the buying decision and customer satisfaction. For instance, the power of brands plays a crucial role for consumer goods. Brand, considered a resource by Johnson et al. and part of the *Customer Interface* pillar by Osterwalder, can influence the value proposition in various ways: For example, brand can lower the perceived risk of product failure with respect to the *Reasoning* attribute. It can also influence the value generated during the purchasing process through reduced product search and evaluation costs, or deliver higher utility during the use phase through prestige. This demonstrates how difficult it is to provide a comprehensive evaluation of the value proposition as the customer would experience it (beyond formal product criteria). However, the lens of the business model concept at least forces analysts to consider different elements of value creation, delivery and capture separately *and* as a whole, thereby reducing the risk of self-serving biases and sugar-coating of its attractiveness.

4.3.2 The Customer Interface Pillar

Customer Interface (Osterwalder 2004, 58ff.) comprises all customer-related aspects including the elements *Target Customer*, *Customer Relationship*, and *Channel* (see figure 4.6). The latter two form a bridge between the *Target Customer* and the *Value Proposition*.

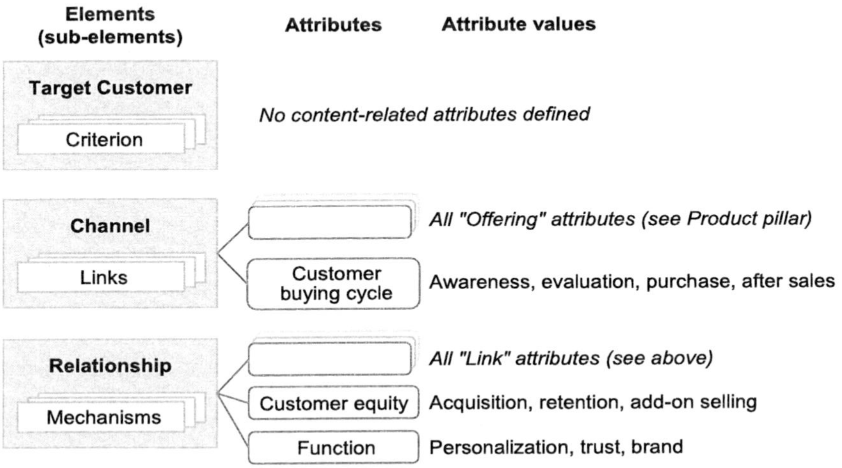

Fig. 4.6 Osterwalder's Customer Interface pillar (based on Osterwalder 2004, 58ff.)

A sound segmentation of customers is critical because it ensures that the *Value Proposition* precisely addresses the customers' needs. To define the *Target Customer*, Osterwalder introduces the *Criterion* element, but gives little advice how to apply it. He simply states that they "could be of geographical or socio-demographic nature" (Osterwalder 2004, 61). Osterwalder & Pigneur (2010, 20) later proposed to segment customers based on different needs, channels, types of relationships, levels of profitability or their willingness to pay for certain aspects.

The notion of customers as the only target group is too narrow for some business models. Particularly for green or social business models, the value proposition often needs to appeal to other groups as well if they have essential influence on their success. For example, regulators can make green business models viable through subsidies or taxes on less clean alternatives.

Through the *Channel* element, the *Value Proposition* is delivered to the *Target Customer*. A firm can use several *Channels* to make contact with the customer. They can be direct or indirect, physical or virtual. Each *Channel* consists of a number of *Link* elements that represent individual marketing tasks. The high granularity of marketing tasks is necessary to understand and avoid channel conflicts due to overlapping activities that can confuse or even disgruntle customers (Moriarty & Moran 1990). Osterwalder (2004, 68f.) proposes to use the *Hybrid Grid* tool presented by Moriarty & Moran (1990) to map channels and marketing tasks to remedy multi-channel challenges. Using Osterwalder's terms, it consists of a matrix with the *Channels* on the vertical axis, the phases of the *Customer Buying Cycle* ("awareness", "evaluation", "purchase", and "after sales") on the horizontal axis and the *Links* as cells of the matrix.

Osterwalder acknowledges that channel *Links* can contribute to a *Value Proposition*. Consequently, the *Link* element can positively influence "use" (e.g., through online help pages), "risk" (e.g., confidence through in-store buying advice), or "effort" (e.g., convenient internet sales) of an *Offering*. Moreover, the value contribution of a *Link* can also be assessed based on the *Life Cycle* attribute introduced in the previous section.

The last building block of the *Customer Interface* is the *Relationship* element. It describes the relationship between the company and its customer segments. Osterwalder distinguishes between acquisition, retention and add-on selling to specify the *Customer Equity* attribute of a *Relationship*. Osterwalder (2004, 72f.) highlights the various trade-offs that companies face in this respect: Customer acquisition tends to be much more expensive than retaining existing customers. However, not all customers are equally profitable for the company, which suggests pursuing a differentiated retention approach. Blattberg et al. (2001, 71ff.), who provided the classification of *Customer Equity* that Osterwalder follows, argue that companies should not take a too narrow view and consider customer satisfaction only. In fact, retention is already influenced during the acquisition phase when customer expectations are created that may or may not be met later on (an argument against too aggressive customer acquisition through discounts, etc.). Most of the other mentioned retention factors are already covered within the *Value Proposition* building block: ease of purchase, value relative to price, product uniqueness and suitability, and customer service. Additional factors include loyalty mechanisms (like airline frequent flyer programmes) and lock-out provisions that increase switching cost (e.g., some products can require re-learning how to use the product when switching).

A further way to increase the profitability is add-on selling as it tends to be much more profitable than selling to new customers. Add-on selling is not limited to selling related products (which is then called cross-selling), but can also use the existing customer relationship to sell completely unrelated products (Blattberg et al. 2001, 95ff.).

Osterwalder introduces the *Mechanism* element to decompose the *Relationship* building block. A Mechanism can also be a *Link* and contribute to the *Value Proposition*. It can have the following *Functions*: "personalization", "trust", and "brand".

Personalisation can be realised in a traditional, "human" way like the in the old corner shop. Alternatively, information technology can economically realise a similar level of personalisation. Possibilities include one-to-one marketing, recommending systems (for personalised product recommendations), or the more product-related mass customisation. Mass customisation means that customers customise products (e.g. through a product configuration website) which are then produced in an automated way (see Da Silveira et al. 2001; Zipkin 2001).

A minimum level of trust is a prerequisite for any business transaction. Trust traditionally depends on "identity, assumed quality or the perception of risk and it deepens over the time of a relationship" (Osterwalder 2004, 75). The e-business age originated new trust mechanisms like community-based ratings and online reputations systems. Third party certification, verification and labelling services have gained importance, for example to generate trust in online payment services. In addition to the application that Osterwalder mentions, these trust mechanisms also play an important role in the sustainability domain. There are myriads of green labels of all kinds, and consumers around the world usually assume that a product is green if it carries a green label. At the same time, however, only few understand the exact meaning of different labels and – especially in Europe – many labels are considered to be misleading (Manget et al. 2009, 18f.).

Osterwalder considers the third *Function*, brand, to be of great importance for the competitiveness of firms. Well-defined brands also help to ensure alignment between business model design and communication to the customer. Osterwalder remarks that every interaction with customers or the environment affects the brand of a company and defines brand mechanisms as "actions that are devoted to identity and brand building" (Osterwalder 2004, 76). However, his understanding of branding may be too narrow and ignore its reciprocal nature: Recent research observed a new form of branding called *user generated branding*. It describes the phenomenon that individuals can publish brand-related content at great reach and speed through the internet – beyond the control corporate brand management. Progressive companies now try to exploit this circumstance by consciously fostering such activities to build trust and a positive brand, for example through corporate blogs (Burmann & Arnhold 2008). This phenomenon is also highly relevant in the sustainability domain, where green reputation often plays a pivotal role for success. Therefore, brand is first and foremost a key resource for many companies, a view that is also advanced by Johnson et al (2008). More generally, companies can maintain purely transactional relationships with their customers, or, at the other end of the spectrum, even encourage them to actively participate in value creation activities.

In line with the argument brought forward with respect to target groups, *Relationships* may also need to be understood more broadly in the context of certain business models. A "target group equity" may, for example, include government aid from supportive regulators, or the idealistic help of NGOs that promote the business model without any formal agreement. The difference to the partnership element, however, is blurry in such cases – as it is the case when customers participate in building a brand or designing a product.

4.3.3 The Infrastructure Management Pillar

So far, the described building blocks cover the value that is offered to the customer and how it is delivered. The third pillar, *Infrastructure Management* (Osterwalder 2004, 79ff.), deals with how the economic value is actually created by the firm and its business partners. Although not very visible to the outside, this part of the business model is maybe the most complex and difficult to get right. Therefore, it is not surprising that *Infrastructure Management* is the most comprehensive of Osterwalder's four pillars (see figure 4.7).

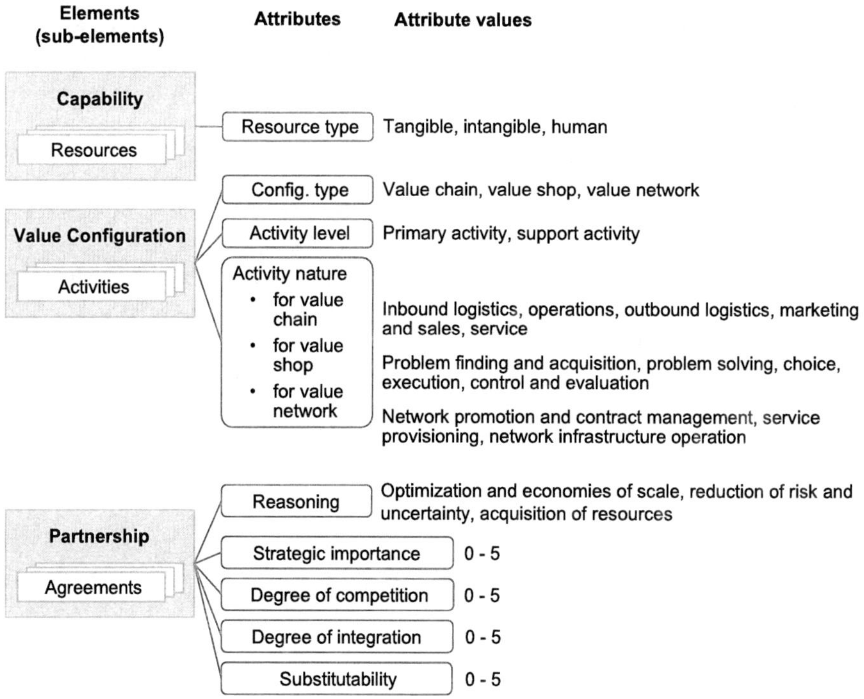

Fig. 4.7 Osterwalder's Infrastructure Management pillar (based on Osterwalder 2004, 79ff.)

Infrastructure Management comprises three building blocks: The first is *Capability* that represents the ability to use resources necessary to create value. The second building block, *Value Configuration* describes the way the company arranges its activities to support or perform the creation of value. The last building block, *Partnership*, covers the agreements which the company has with external parties to create value. As Osterwalder points out, the partner dimension has gained special importance in recent years due to simplified cooperation through electronic means. This fact gains further importance in the context of sustainability which covers very diverse knowledge domains and in many instances requires very specific capabilities that few companies can or want to build internally (Senge et al. 2008).

Capabilities are understood as "repeatable patterns of action in the use of assets to create, produce, and/or offer products and services to the market" (Osterwalder 2004, 79; based on Wallin 2000). Osterwalder (2004, 81) presents Wallin's business capabilities framework that distinguishes two dimensions: internal vs. external and resources vs. customer focused capabilities. The resulting four capability types are (1) *resource-integration* capabilities to deploy firm-addressable resources (external, resource-related), (2) *generative* capabilities like execution and innovation (internal, resource-related), (3) *customer-interaction* capabilities (external, customer-related), and (4) *transformative* capabilities to create offerings that provide sufficiently high value at low enough cost (internal, customer-related).

Capabilities rely on *Resources* that can be divided into tangible, intangible, and human. Johnson (2010, 40) offers a more concrete categorisation into *people, technology and products, equipment, information, channels, partnerships and alliances, funding,* and *brand.* For the business model, only key resources are of interest. The relative importance of each type of *Resource* depends on the industry and company (Osterwalder 2004, 82); for example, consultancies are extremely dependent on their human resources while a commodities producer mostly relies on tangible assets that can be found on balance sheets. Intangible resources are not fully reflected on balance sheets but have seen increasing importance in recent years: According to Brand Finance, a brand valuation consultancy, intangible assets now comprise about two thirds of the global enterprise value (Brand Finance 2007). This also has implications for Green Business Model Transformations. Björn Stigson, President of the World Business Council for Sustainable Development (WBCSD), remarks that "at a normal global company, the majority of the assets underpinning market capital are intangibles. That is a different story than in the past. The business case for sustainability is now increasingly connected to brand and reputation." (Economist Intelligence Unit 2008b, 33)

The second building block of this pillar is *Value Configuration* consists of a set of *Activities* and is divided into three *Configuration Types*: The classic value chain, as defined by Porter (1985), and two extensions by Stabell & Fjeldstad (1998). The value chain consists of *primary activities* that create and deliver value to the customer, namely *inbound logistics*, *operations*, *outbound logistics*, *marketing and sales,* and *service.* In addition, *supporting activities* are necessary that allow the *primary activities* to take place. They are: *procurement*, *technology development*, *human resource management*, and *firm infrastructure* (Porter 1985). Stabell & Fjeldstad (1998) extended Porter's mainly sequential value chain framework with two alternatives: *value shop* and *value network.*

Value shops are supposed to better represent service provider business models (e.g. consultancies and venture capitalists) that create value by solving unique problems for customers. Value networks correspond to intermediary business models (e.g., stock exchanges and telecom operators) that "sell mediation between customers or places" (Fjeldstad & Haanœs 2001, 5).

Osterwalder's distinction into *Capabilities* and *Value Configuration* building blocks is largely equivalent to that used by Johnson et al. (2008) into *key resources* and *key processes.* Both authors emphasise the strong link between the two: processes use or share one or more resources and hence are only useful in combination. The ability to create or adapt such combinations (e.g., responding to changing market conditions) is referred to as *dynamic capabilities* (Teece et al. 1997).

The third building block, *Partnership*, consists of a set of *Agreements*. These *Agreements* have the purpose to allow having *Capabilities*, *Resources* and *Activities* of external partners at the firm's disposal. Consequently, partnerships enable companies to tap into many additional ways of value creation (Doz & Hamel 1998; Gulati et al. 2000). Partnerships can hence create social capital, and represent a valuable resource in itself.

The topic has been discussed in management literature under many different terms including *strategic networks*, *b-webs*, *value networks*, *alliances*, and so on. Osterwalder (2004, 90f.) elaborates on four theoretical perspectives that can inform thinking about partnering:

1. *Transaction cost economics* (TCE) (see Coase 1937; Williamson 1981) suggests comparing whole transactions performed internally versus through the market, and thereby considering production *and* transaction costs.
2. The already mentioned *resource-based view* (RBV) of the firm emphasises how firms can acquire and use external resources (e.g., customer data, brand names, patents or technologies) to gain competitive advantage.
3. *Organisational learning* (see Argyris & Schön 1978; Senge 1990; Nonaka & Takeuchi 1995) implies that, by working together, companies can exchange knowledge (e.g., production methods, market expertise) to their mutual advantage.
4. *Co-opetition* (see Brandenburger & Nalebuff 1996) asserts the possibility that cooperation between competitors can create win-win situations, e.g. by working together to grow or capture a new market.

All perspectives are useful when analysing new green business model designs, as the case examples in chapter 7 will demonstrate. These perspectives should be viewed as complementary. Because they emphasise different aspects of value creation (e.g., cost versus capabilities), the answers to questions like which value chain activities to perform internally, or what kind of relationships to enter with external organisations, may not always imply the same courses of action (Eisenhardt & Schoonhoven 1996, 137).

As mentioned earlier, partnerships play a pivotal role for the success of many sustainability initiatives. Particularly greening supply chains has become a key priority (and challenge) for many companies. For instance, retailers – above all Wal-Mart – have already created significant value this way (Makower 2009, 138ff.).

Partnerships are characterised by the *Agreement* element which is encompasses five attributes: *Reasoning*, *Strategic Importance*, *Degree of Competition*, *Degree of Integration* and *Substitutability*. Osterwalder (2004, 93) distinguishes between three broad kinds of *Reasoning*: "optimization and economies of scale" (e.g., through outsourcing or shared infrastructure), "reduction of risk and uncertainty" (e.g., by using external market expertise or co-opetition), and "acquisition of resources" (to externally leverage the business model). The remaining attributes are supposed to be operationalised with an ordinal scale (0 to 5), but – with the exception of *Degree of Integration* – Osterwalder offers no advice on how to set the score.

4.3.4 The Financial Aspects Pillar

The fourth and final pillar, *Financial Aspects* (Osterwalder 2004, 95ff.), is about the economics of the business model. It contains the *Revenue Model* and *Cost Structure*, which together determine the profitability of the business model as a whole (see figure 4.8).

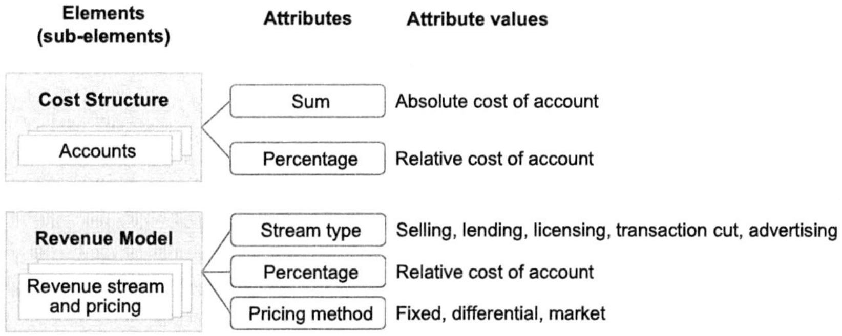

Fig. 4.8 Osterwalder's Financial Aspects pillar (based on Osterwalder 2004, 95ff.)

The *Revenue Model* consists of one or more *Revenue Stream and Pricing* elements. Osterwalder (2004, 97f.) defines five different *Stream Types*:

- Selling: Involves a partial or complete transfer of ownership (e.g., a physical CD is sold, but not the property rights for the music).
- Lending: The use of goods that are made available to the customer on a temporary basis (e.g., money lending or car rental). Osterwalder & Pigneur (2010, 31) additionally distinguish two more related revenue streams: usage fees (e.g., hotel stays) and subscription fees (e.g., gym memberships)
- Licensing: The permission to use intangible property; in contrast to lending, parallel licensing of the same intangible is possible. Licensing enables a business to generate revenue without direct involvement in production or commercialisation activities (e.g., through patents, copyrights, franchising)
- Transaction cut: Enabling or performing a transaction (e.g. investment banking deals, e-business platforms)
- Advertising: Allowing the placement of commercial messages on owned media (e.g., TV or newspaper ads)

One important revenue stream-related concept in the sustainability context that Osterwalder does not mention is *servicising*. The concept aims to substitute selling physical products and material use with the provision of services instead. Not only does it help the environment by facilitating a more efficient use of resources, it has also the potential to increase revenue and profits (Rothenberg 2007).

Business models often involve a mix of several types of revenue streams. Information technology has facilitated to diversify revenue streams and advance underlying *Pricing Methods*. Furthermore, greater transparency of prices through the internet has increased the customer's power as a price maker, which in turn induces some companies to move away from simple towards more complex (and harder to compare) pricing mechanisms. Osterwalder (2004, 98ff.) distinguishes three generic methods and 12 mechanisms for pricing: (1) fixed pricing (pay-per use, subscription, list price/menu price); (2) differential pricing (product feature dependent, customer characteristic dependent, volume dependent, and value-based); and (3) market pricing (bargaining, yield management, auction, reverse auction, and dynamic market).

It is important to realise that pricing can have a direct impact on the environmental sustainability of a business model. For instance, subscriptions encourage consumption while pay-per-use pricing incentivises customers to use as little product – and therefore resources – as possible.

The last building block, *Cost Structure*, consists of a set of *Accounts*, each of which represents a specific type of expenditure. Osterwalder simply defines *Sum* and *Percentage* of the total *Cost Structure* as its attributes. He remarks that *Accounts* can be defined very detailed or as aggregates of expenditures. As an example, he mentions cost of goods sold and operating expenses that could be split further into R&D, marketing and sales, and general and administrative expenses. It is surprising that Osterwalder leaves the discussion of the greatly important cost element at this rather shallow level. After all, one of the important problems that many (large) companies face is not aggregated cost and revenues considered in isolation, but that they do not know which part (e.g., product lines, customer segments) of their business make money and how much. However, declining profitability of certain business segments can remain undetected given satisfactory overall profitability. Not only can this become a problem in the long run if the segment gains in relative importance. But it is also critical to have this kind of information when a business model is supposed to be transformed, as this usually leads to shifts in the revenue, cost, and customer structure.

Johnson et al. (2008) suggest the introduction of a *margin model* in addition to cost and revenue models. This way, choices to accept losses in some parts of the business in favour of long-term success or the overall logic of the business model can be made more explicit, e.g., supported by sophisticated sensitivity analyses. Another element that Johnson et al. (2008) propose regarding the money-earning logic of a business is *resource velocity*. This encompasses "how fast we need to turn over inventory, fixed assets, and other assets – and, overall, how well we need to utilize resources – to support our expected volume and achieve our anticipated profits" (Johnson et al. 2008, 53). This item may have certain relevance for some business models, but is not considered central and thus not incorporated explicitly in the following.

4.4 The Business Model Composition Applied in this Work

As outlined above, Osterwalder's business model ontology provides a suitable starting point as a unit of analysis for management science. However, two main factors limit its applicability in the context of this work: Firstly, the ontology has been created with a strong e-business focus, thereby neglecting some aspects important for this work that thus had to be supplemented. Secondly, definitions and arrangement of business model elements are done in a way that is appropriate for the translation into a formal modelling language (in this case the eXtensible Markup Language XML). However, some business practitioners that Osterwalder interviewed raised concerns about its complexity. One CEO of an industrial company suggested to adapt the complexity to "specific managerial levels of decision making" (Osterwalder 2004, 135). In the practitioner-oriented book "Business Model Generation" (Osterwalder & Pigneur 2010), labelling of elements has therefore been made more intuitive.

The final conceptualisation of the business model as a unit of analysis is thus principally a blend of the works of Osterwalder (2004) and Osterwalder & Pigneur (2010), Johnson et al. (2008) and Johnson (2010), and own considerations (based on interviews and multiple iterations in the development of the management framework in chapter 9).

Despite the focus on green business models, the resulting business model conceptualisation is applicable to green and non-green business models alike. No specific elements representing sustainability aspects have been added. The business model concept can be used to analyse and manage the business logic of a firm, as well as – a main function in this publication – for the exploration of future opportunities (Osterwalder 2004, 19ff.). There is no reason why these uses should focus only on financial and operational viability, and not also include environmental performance. Sustainability originates from the very same elements as growth or profitability. If a quantified analysis is desired, the cost and revenue models could be extended by environmental cost (impact) and benefit calculations (see Schaltegger et al. 2006 for a detailed discussion). Because this practice is optional, and not obligatory to successfully manage Green Business Model Transformations, I refrain from incorporating this aspect explicitly.

Figure 4.9 (see below) provides an overview of the business model composition that forms the basis of all references to the business model concept in the remainder of this book.

At the heart of the understanding of business models is the division into two value perspectives: ***capturing*** and ***creating value***. Each perspective comprises two components. ***Value propositions*** and ***target groups*** constitute the value capture perspective, ***key resources*** and ***key processes*** the value creation perspective. The fifth component, ***financial logic***, integrates the two. It is critical to understand that all components are interconnected, as indicated by the puzzle shapes of the two value perspectives in figure 4.9. Each component is further divided into several business model elements that are briefly outlined in the following sections. Depending on the requirements, analyses can thus be conducted on the level of the two value perspectives, the five business model components, or a greater number of business model elements.

In addition, a business model is embedded in the ***external business environment*** that influences – and is influenced by – the business model. In addition to competition, this includes general economic, political and societal factors as well as activities by stakeholders of the firm that are not directly participating in creating and capturing value for the firm (for a broad discussion of the nonmarket environment, see Baron 2009). This dimension must not be ignored when analysing or designing new business models. An unfavourable external business environment can spoil an otherwise sound business model, e.g. due to defensive moves from competitors, or political opposition (also see Osterwalder & Pigneur 2010, 200ff.).

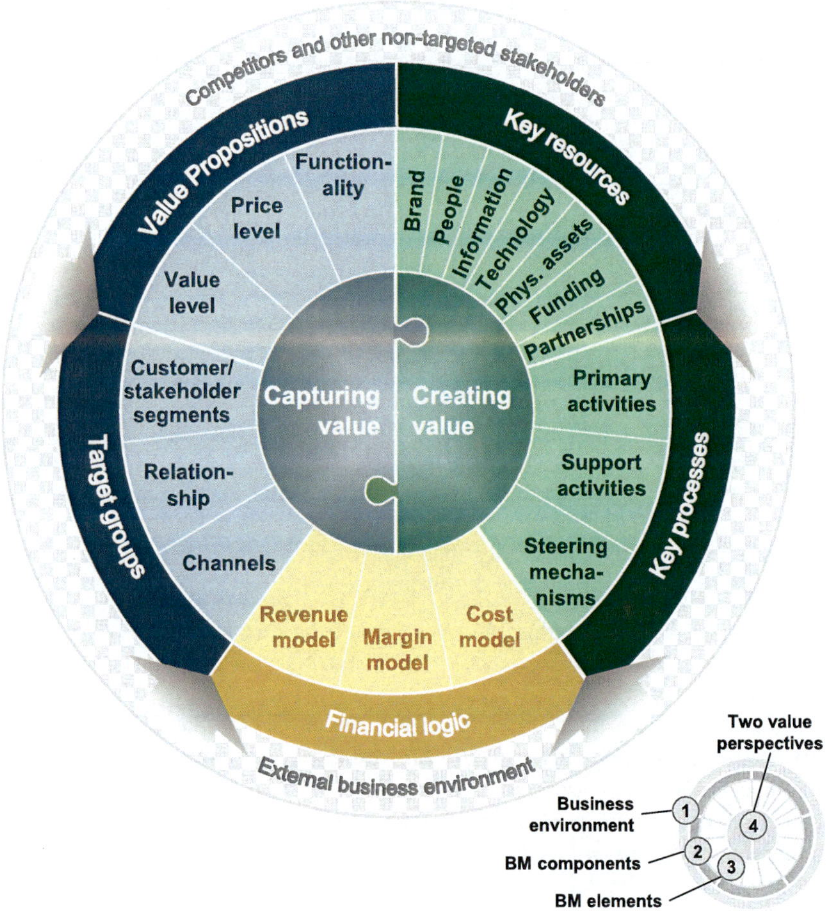

Fig. 4.9 Overview of business model elements

4.4.1 Value Propositions

The value propositions component is largely congruent with Osterwalder's *Product pillar* (see 4.3.1), but uses a more associative name (cp. Johnson et al. 2008; Osterwalder & Pigneur 2010). The ***functionality*** element refers to the value which a customer derives from the offering, as operationalised by Osterwalder's *Reasoning* and *Life Cycle* attributes. The elements of ***price level*** and ***value level*** correspond to the attributes of the same name by Osterwalder. As they are to be seen in conjunction with rivalling offerings, they bring in the competitive perspective. However, quality standards or absolute price points are of importance too. For example, even without a rivalling product, the price point impacts the adoption rate.

Green business models can distinguish themselves in various ways in terms of value proposition: For instance, a customer may derive reputational value from a green product (functionality), be willing to pay extra for a "high end" product (price level), or consider it more innovative than competitors' offerings (value level). Considerations are thereby not limited to simple substitutions of product characteristics: Rethinking the entire logic of achieving a certain function (e.g., by selling a service instead of a product) should be considered. Even manipulating prevalent customer needs as such is conceivable (e.g., by making the shortcomings of current offerings explicit) (for an example of a respective multi-level management framework (COSY) see Schneidewind 1995, 23ff.). Value propositions are thus always to be seen in conjunction with the next pillar, *target groups*.

4.4.2 Target Groups

Johnson et al. and Osterwalder differ on whether the target groups component should be subordinated to the value proposition. Osterwalder's variant that places them on the same level is preferable because it makes the duality between the two explicit. After all, attractive value propositions alone are not sufficient for market success without a sound understanding of target groups, and vice versa. An example in chapter 7.2 will illustrate how such neglect had almost catastrophic consequences in the course of a Green Business Model Transformation in the food sector.

The target group component is thus very similar to Osterwalder's original pillar, the *Customer Interface* (see 4.3.2). However, its scope is extended to also incorporate critical stakeholders that have direct impact on the two value perspectives of the business model (e.g., regulators for subsidised business models, NGOs that are needed to support business models in developing markets).

The first element, ***customer/stakeholder segments***, corresponds to Osterwalder's *Target Customer* element. It represents the firm's view on how the market should be segmented, and which groups the value propositions are intended to appeal to. Green business models can thereby create value by offering superior value to groups that are already addressed, or by facilitating access to new customer segments.

The ***relationship*** element is to be understood as explained above. It may, especially for consumer goods industries, gain in importance through Green Business Model Transformations.

Although the ***channels*** element may not play a role as crucial for green business models as for e-business models, it must still be designed appropriately and can be a source of innovation (e.g., see the case example in chapter 7.3).

The channels element has been assigned to the target groups component due to its direct involvement in delivering value to the customer. However, sophisticated channels can also be a key resource, one of the two value creation components. The same is true for target group relationships. These items are not listed twice, however, in order to avoid confusion.

4.4.3 Key Resources

Resources are the "stocks of available factors that are owned or controlled by the firm" (Amit & Schoemaker 1993, 35). Adapted from Johnson (2010, 40), seven main types of resources are distinguished:

Brand is worth dozens of billions of dollars for some companies, in certain cases also driven by green branding (Interbrand 2009). But it is maybe also the most vulnerable of all, with environmental blunders being a major source of value destruction (Vranica 2010).

People, i.e. employees with their "training, experience, judgment, intelligence, relationships, and insights" (Barney 1991a, 101), are a crucial resource for any company. While human resources can often be hired at short notice, some critical functions may be challenging to staff or need long-term planning (see Strack et al. 2008). In addition, an inspired, well-motivated workforce does not only greatly improve efficiency, it can also compensate for business model design flaws through commitment and creativity. Corporate culture is also a factor to consider, as certain business models may be incompatible with the existing culture of a company.

Information can relate to any business model element to be of value. Examples include information on customer preferences, or the firm's environmental footprint (often a prerequisite for managing a green business model). Some highly successful business models (e.g., Google) are built around this resource. Information is often dependent on sophisticated IT to be useful, but is frequently complemented by the tacit knowledge stored in employees' heads.

Technology can relate to both products and processes. It includes tradeable knowhow like patents or licenses, and the systems that a firm uses to run its business. Technology obviously plays a critical role in all clean technology business models. Examples in this book include green chemistry, electric cars, and even low-tech solutions like solar-powered lighting (see chapters 7.3, 7.6, and 7.8).

Physical assets include property, plant, and equipment (Amit & Schoemaker 1993, 35). Physical assets can be a powerful source of competitive advantage, but in some cases seriously impair business model transformations (Johnson 2010, 162ff.).

Funding may not play a central role for many business model transformations, as opposed to start-ups. It can still be important, however, if a new business model promises long-term profits at the expense of short-term cash flows, or if the firm is in turnaround situation.

Partnerships allow firms to access external resources. The topic has been discussed at quite some length (see 4.3.3), and Osterwalder considered Partnerships a building block on the level of, and separate from, resources (*Capability*). Partnerships are a special case in that they can provide access to all other resource types listed above. On the other hand, partnerships – e.g., think of a deep relationship to a key partner or a sophisticated supply chain – can be argued to represent a valuable resource in itself. Furthermore, partnerships "can be vehicles for new organizational learning, helping firms to recognize dysfunctional routines, and preventing strategic blindspots" (Teece & Pisano 1994, 545). Despite their pivotal role for the success of certain business models, the same can be argued for other resources, e.g. people (Umbeck 2009, 63). Hence, I follow the approach of Johnson et al. and view partnerships as an element of the key resources component.

The list of seven types of resource above may not be exhaustive, and for certain business models a different categorisation may be more appropriate. However, it has proven practical for general purposes as applied in the management framework of chapter 9.

In contrast to Johnson et al., Osterwalder combines key resources (*Capabilities*) with key processes (*Value Configuration*) in the *Infrastructure Management* pillar. Amit & Schoemaker (1993, 35) contrast resources and *capabilities* which "refer to a firm's capacity to deploy *Resources*, usually in combination, using organizational processes, to effect a desired end." This definition contradicts Osterwalder's separation into Capabilities and Value Configuration, but more importantly, it combines the distinct concepts of resources and processes. The concept of capabilities may be serviceable for the purposes of the RBV. Yet, a business model can fail to create value either because key processes do not work, because key resources are lacking, or because the interplay between the two is poor. Capabilities could blur these important distinctions. Therefore (analogously to the argumentation for the value capture perspective), the duality of resources and processes is made explicit, and the concept of capabilities is not used.

4.4.4 Key Processes

A process is "a specific ordering of work activities across time and place, with a beginning, an end, and clearly identified inputs and outputs: a structure for action." (Davenport 1993, 5). The term *key processes* is used instead of *key activities* (Osterwalder & Pigneur 2010, 17), because the concept is established in management practice and emphasises the conceptual nature of the component, as opposed to the atomic term *activity*.

Porter's distinction of ***primary*** and ***support activities*** is used (see 4.3.3), as it is well-established in management literature (Mintzberg et al. 2009, 108f.).

Depending on the industry or specific characteristics of the business model, a more granular or even different breakdown into elements may be useful for analysis purposes (e.g. to explicitly include environmental management). For instance, Osterwalder (2004, 83f.) has included the concepts of *value shop* and *value network*, and Holweg & Pil (2006) propose to use a *value grid* instead of a chain to gain a multidimensional perspective. Porter himself suggests using *activity-system maps* to study whether the activities of a company show strategic fit and how they can reinforce themselves (Porter 1996, 71ff.). Zott & Amit (2010) distinguish the design elements *content* (what activities), *structure* (links and sequence) and *governance* (who performs activities) for activity systems. The latter also includes that activities may be performed outside of the boundaries of the focal firm. To keep it simple, however, I use the classic value chain in figure 4.9. The non-linear, holistic nature of key processes will receive due attention by means of the systems thinking approach which is introduced in the next section (4.5).

An additional third element, ***steering mechanisms***, is located on the rules level (see Morris et al. 2005, 731f.) and represents important formal or informal means of influencing business practices and decision-making. This element follows the thoughts of Johnson (2010, 45ff.) who speaks of *business rules*, *behavioural norms* and *success metrics*. Because business model transformations change the fundamental logic of doing business, a search for new or adapted *routines* (Winter & Nelson 1982) is often essential for the business model to be executed effectively. Although this is not an entirely conscious, controllable process, inconsistencies with the new business model can presumably be identified in most instances by systematic efforts. Sometimes, higher-level rules that govern modifications of lower level rules (see Cyert & March 1963) may need to be changed, too. Necessary changes may differ greatly in magnitude, ranging from simple policy changes (e.g., hiring guidelines or sales incentives) to far-reaching organisational changes (e.g. founding a new unit to foster experimentation).

4.4.5 Financial Logic

The financial logic contains three elements: the ***revenue***, ***cost***, and ***margin model***. The cost and the revenue model are understood analogously to Osterwalder's explanations (see 4.3.4). As suggested by Johnson (2010, 36f.), the margin model (target unit margin) is treated as a separate item. It is important as it implies certain conditions (e.g., expected economies of scale and target volumes) for the business model to break even. For example, a new business model serving poor people in developing countries may generate satisfactory overall profits through larger volumes, despite lower gross-unit margins. Increased resource velocity (see Johnson 2010, 37ff.) can have similar effects.

As mentioned above, business models are characterised by systemic properties that can be addressed with the systems thinking approach. Systems thinking can be understood as a general problem solving approach that follows systems theory principles and will be introduced in the following.

4.5 Understanding System Behaviour

General systems theory (von Bertalanffy 1950) provides a shared platform for scientists from many academic fields as diverse as physics, biology, sociology, economics, and management science (Boulding 1956). As a result, practical tools and techniques based on systems theory – systems thinking being one of them – can be applied to a broad range of problem domains, including not only business models, but also sustainability and organisational change dynamics. It can thus help practitioners to design more potent green business models and to implement these business models more effectively.

4.5.1 Background and Basics of Systems Thinking

System dynamics research attempts to help us better understand complex systems by studying the interactions within and between systems. It originates the understanding that conventional, analytical decomposition of problems can lead to solutions that neglect the interaction between the individual parts of the problem (Forrester 2003). Even systems with a small number of components or options for decisions (low *combinatorial* complexity) can be highly complex due to interactions of system components over time (*dynamic* complexity) (Sterman 2001, 11).

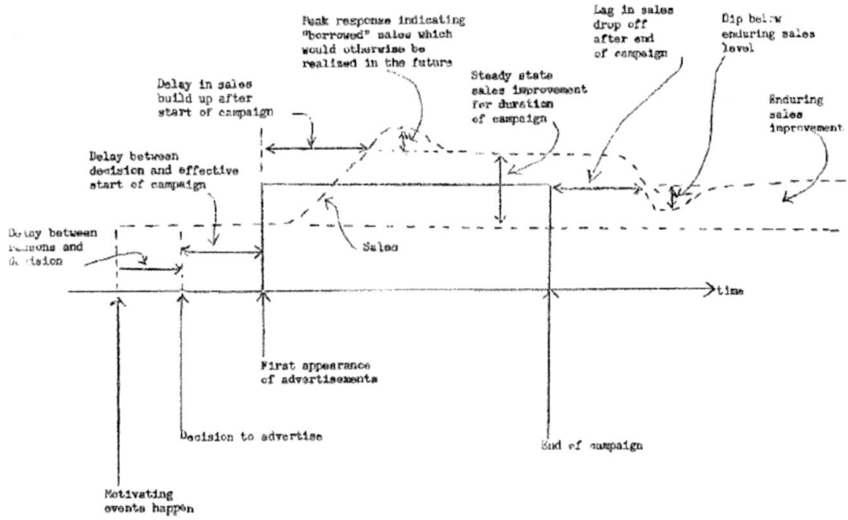

Fig. 4.10 Timing and extent of effects of an advertising campaign (Forrester 2003, 340)

Recognised as "the genesis" of system dynamics, Jay W. Forrester in a 1956 article discusses the shortcomings of existing models for economic systems and industrial organisations (Forrester 2003). One major shortcoming he identified was the negligence of *closed loop systems* of flows of money, labour, goods and information that amplify or dampen disturbances and influence the time of occurrence of events that affect them. These systems are characterised by resistance to change (also see chapter 6.3.4), accumulation (e.g., fluctuating inventories, liquid assets), delays (e.g., between the occurrence of events, resultant decisions and their taking effect), quantising (e.g., periodic reporting), and policy and decision-making criteria (Forrester 2003, 332). The alternative, that is applying an *open loop system* (or event-oriented) *view*, can lead to short-sighted decisions, like boosting short-term profits through a cost cutting programme that – as an unanticipated side effect – hurts important capabilities and long-term profits (Sterman 2001). Forrester (2003, 340) illustrates often overlooked closed-loop system effects with the example of an advertising campaign (see figure 4.10). A company that is unaware of systemic delays and feedbacks may easily misinterpret changes in demand and consequently reach poor decisions with respect to production and inventory levels. There are other, more complicated cases than the advertising campaign depicted above. However, individuals and organisations alike regularly fail to grasp the dynamic nature of even the simplest systems (Sterman 2001).

Systems thinking aims to counter the above-mentioned problems. According to one of its most prominent proponents, Peter Senge and his co-authors, it simply means "stepping back and seeing patterns that are, when seen clearly, intuitive and easy to grasp" (Senge et al. 2008, 23). They suggest that people conceive reality based on four levels of the *Systems Thinking Iceberg* (Senge et al. 2008, 172ff.): The top layer comprises observable *events* that we can react to. The second layer consists of *patterns and trends* which we can anticipate. These in turn are formed by *systemic structures* (the third layer) that we can design. Unlike events, these two layers already tend to be hidden below the surface. The fourth layer concerns our *mental models*, i.e. core beliefs and basic assumptions, that may need to be transformed in order to see and adapt the system to work in a more favourable way.

One simple yet alarming real-life example out of the sustainability domain that Senge et al. (2008, 170ff.) use to demonstrate the adverse effects of ignoring system dynamics is the depletion of fish stock due to overfishing: Following high demand, fishing companies increase their fleets and employ more effective fishing techniques. However, while the output grows and grows, at some point, the regenerative capacity of the fish stock is reached. The problem is that this tipping point is not accompanied by observable events (such as reduced catch). To the contrary, the fish catch grows well beyond that point until, at its maximum, the decline of the fish stock is the greatest.

The example of overfishing reveals a class of serious problems that can prevail even after the long-term consequences of issues like that of overfishing are understood: *the tragedy of the commons* (Senge 2006, 397f.). If one fish company reduces its catch, the others can – and likely will – catch more (see Senge et al. 2008, 170ff.).

With his seminal book "The Fifth Discipline" (Senge 1990), Peter Senge has made systems thinking popular to a much wider audience than ever before. *The fifth discipline* (the term he uses to refer to systems thinking) integrates the other four disciplines – *personal mastery*, *mental models*, *building shared vision*, and *team learning* – which he deems necessary to achieve the organisation's ideal state of the *learning organisation*.

Senge (2006, 57ff.) lists 11 "laws" of the *fifth discipline* which managers should keep in mind when dealing with systemic problems. They state that ...

- ... current problems often stem from past interventions
- ... systems tend to "push back"; one needs to work *with* the system's forces
- ... low-leverage interventions improve system behaviour before it deteriorates
- ... easy-to-see solutions often do not solve the underlying problem
- ... a (non-systemic) "solution" can even make things worse
- ... lasting solutions may take time
- ... the link between cause and effect may be non-obvious and comprise a longer delay
- ... small (but often non-obvious) interventions may have big effects
- ... trade-offs may disappear if interrelated goals are considered over time
- ... some problems are indivisible; seeing only parts of the problem does not suffice
- ... in systemic problems, there is no one to blame (everyone is part of the system)

Although the issues around system behaviour are known in academic circles for half a century by now, systems thinking and new mental models have not been adopted broadly: Politicians have been found to show little consideration of system dynamics, managers still use heuristics that are based on simplistic cause-and-effect relationships (Sterman 2001), and even well-educated elite business school students perform very poorly in respective tests (Sweeney & Sterman 2000). Nevertheless, a yet limited but increasing number of managers practising systems thinking may change this in the not so distant future (Senge 2006, 256f.). The business model concept naturally invites even managers not familiar with systems thinking to take a holistic perspective.

4.5.2 Practical Application of Systems Thinking

In order to recognise cases where systems thinking should be applied, one must first identify dynamically complex systems. They are characterised by the following (Senge 2006, 71): (1) short-term effects of actions differ significantly from long-term effects; (2) consequences differ locally from those in other parts of the system; (3) "obvious interventions produce non-obvious consequences". Another hint for dynamic systems being at work can be if interventions show no significant effect at all, despite large efforts (Senge 2006, 86). In the following, the most important tools and techniques are presented that can be used in management practice.

4.5.2.1 Causal Loop Diagrams

In order to understand and manipulate system dynamics it is helpful to use graphic representations of the problem. A very simple yet effective way is to use so-called *causal loop diagrams (CLDs)*. They comprise type and direction of feedback loops, including potential delays that are important as they can mask the system's underlying patterns.

There are only two types of feedback loops (Sterman 2001, 17): (1) positive – or self-reinforcing, and (2) negative – or self-correcting, balancing loops. Examples for self-reinforcing feedback loops are nuclear arms race, price wars, or the diffusion of a new product through word-of-mouth. Balancing feedback loops include limits to regenerative capacity of resources (e.g., in the fish stocks example), and market saturation (growth stalls once saturation approaches 100 percent at the latest). Figure 4.11 shows a graphic representation in the form of a CLD for the introduction of a new product with a self-reinforcing feedback loop (marked with an "R") that represents contagion through word of mouth, and a balancing feedback loop (marked with a "B") that represents market saturation. Plus and minus signs next to the arrow indicate the relation between cause and effect. That is, a *plus sign* means that the two variables move in the same, a *minus sign* that they move in the opposite direction:

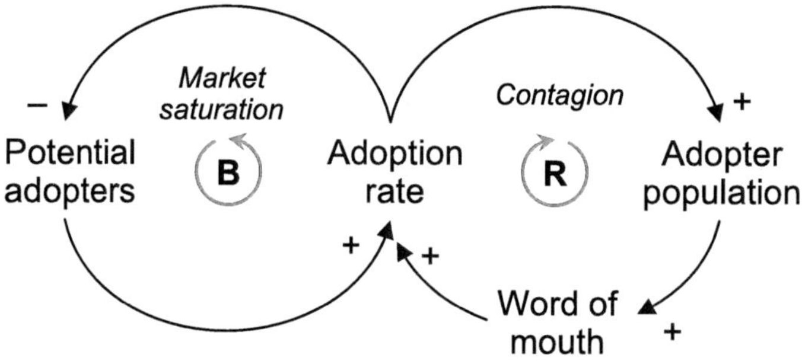

Fig. 4.11 Causal loop diagram for the adoption of a new product (Sterman 2001, 18)

On the one hand, the adoption rate is self-reinforcing since new adopters cause others to adopt it as well, who again spread the word (all "+"). This can result in exponential growth of adopters. However, at some point, the balancing feedback loop kicks in: Because of the growing number of adopters, the number of remaining potential adopters is decreasing ("–"). Eventually, the adoption curve will flatten out and converge to a certain saturation point (Sterman 2001, 18ff.).

4.5.2.2 Using Simulations to Understand System Behaviour

CLDs are very valuable in making the systemic structure explicit. In practice, there will be more relevant feedback loops than the two shown in the example of figure 4.11 (e.g., price reductions enabled through economies of scale, cannibalisation with other products, and new entrants). In some cases there will be *many* more relevant feedback loops. Then, it is virtually impossible to determine their cumulative effects without simulation. This can be done by using dedicated simulation software – some of which already offers very easy to use interfaces for non-scientists (Sterman 2001, 19ff.).

However, it should be emphasised that it is often already a big step to make decision makers aware of systemic structures and the potential patterns these might produce. It makes them aware that their decisions may be too short-sighted or based on unrealistic (static) assumptions. On the other hand, because even small dynamic systems can behave in ways that qualitative systems thinking cannot predict, this approach is oftentimes not sufficient to understand what a good decision would look like. Then, simulation through building sophisticated formal models is a necessary next step (Richardson 1996, 7f.; Sterman 2000, 37ff.; 2002, 524f.). However, even large mathematical models are often highly sensitive to estimated parameters and initial conditions – which can severely limit their accuracy in predicting the system's behaviour (along with simplifications of the model itself, of course). Nevertheless, simulations can expose dominant feedback loops and turn attention to some dynamics that may be subtle or counter-intuitive at first glance (Sterman 2001; Sterman 2008).

When a company wants to employ simulation models, an augmented causal diagram including underlying equations needs to be constructed. In the example above, the CLD can be enhanced by showing the stocks (potential and actual adopters) and flows (the conversion of potential into actual adopters) of the adoption model. Below, the word-of-mouth process is modelled in a more detailed way (see figure 4.12). It now depends on the frequency of social contacts and the share of adopters among these as well as the conversion rate after such contacts.

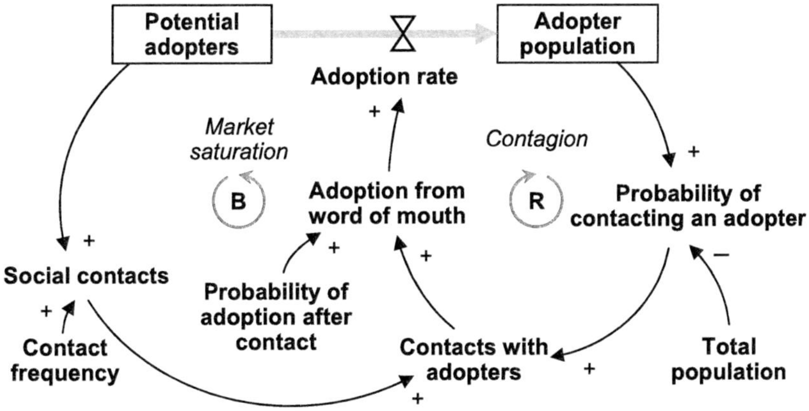

Fig. 4.12 Causal diagram showing stocks and flows of product adoption (Sterman 2001, 18)

Parameters like the *probability of adoption after contact* need to be estimated, e.g. through market research, expert interviews, or data from analogous cases (Sterman 2001, 21).

The respective equations that represent the causal diagram and translate it into the formal model are as follows (Sterman 2001, 19, formatting changed):

- *Adoption rate = Adoption from word of Mouth [+ Adoption from other sources]*
- *Adoption from word of mouth = Contacts with adopters * Probability of adoption after contact*
- *Contacts with adopters = Social contacts * Probability of contacting an adopter*
- *Probability of contacting an adopter = Adopters / Total population*
- *Social contacts = Potential adopters * Contact frequency*

A causal diagram like the above is equally suitable to model the adoption of a new business model by customers. It is thereby critical to concentrate on the dominating loops only that represent the most important success factors. One way to mitigate the risk of overwhelming complexity is to first try to understand the type of system the company is dealing with.

4.5.3 Systems Archetypes

Many problems that are based on dynamic complex systems can be allocated to certain classes of frequently occurring *systems archetypes* like the already mentioned tragedy of the commons. Once a problem is correctly recognised as one of the systems archetypes, it is considerably defanged since systems archetypes can be addressed with the same kinds of solutions (although real-world problems often contain additional variables; system archetypes should therefore be considered basic building blocks only).

Consequently it is useful to briefly summarise the most important systems archetypes and suitable solution strategies. The following explanations are largely based on Senge (2006, 389ff.) and specifically put into context of issues related to business models, environmental sustainability, and organisational change.

Balancing process with delay

This most simple of the systems archetypes only consists of one *balancing loop with a delay* (Senge 2006, 389f.) (delays are represented graphically by two parallel lines that break the respective arrow). This causes individuals, groups or organisations to overshoot corrective actions during the time when the expected impact has not materialised yet. Examples include production and inventory cycles with inadequate responses to market reactions.

Managers facing this archetype can either wait patiently until their corrective actions take effect or improve the system's responsiveness (Senge 2006, 390). New green business models may thus require a redesign of interactions between certain elements. For example, a retailer that changes food offerings from conventional to organic produce may find it difficult to ensure availability due to – formerly unproblematic – delays within the supply chain of fewer and smaller suppliers.

Shifting the burden

Shifting the burden (Senge 2006, 391ff.) describes a case when a symptomatic, short-term "solution" is chosen to deal with a problem instead of applying a fundamental solution. This behaviour is problematic in several ways: First, the solution usually only cures the symptoms, and hence the problem comes back. Second, repeatedly relying on a symptomatic solution often impedes the ability to apply a fundamental solution (side effect).

In a business model context, a company that observes deteriorating results may initiate a cost cutting programme instead of reconsidering its business logic or strategy. For a while, the cost cutting efforts may restore profits, but after some time results will deteriorate again if the business model itself was inadequate (and not just inefficient). Worse even, the reliance on the symptomatic solution – imprudent cost cutting – often leads to a loss of valuable people and competencies, increasing the difficulty of finding a fundamental solution in the form of a new business model. As a result the organisation can become reliant on external experts or consultants to turnaround its business. This special variant of the systems archetype is called *shifting the burden to the intervenor* (Senge 2006, 393f.).

Ironically, the approach proposed in this work de facto also follows a shifting the burden structure: Almost by definition, a company that undergoes its first Green Business Model Transformation does not possess all the competencies to manage this task internally, and hence relies on external help. However, as long as these capabilities are built successively during and after the transformation there is no significant risk involved of falling trap to shifting the burden issues.

In the sustainability domain, shifting the burden structures can manifest themselves in the form of end-of-pipe solutions that ignore fundamental redesign options. For instance, instead of changing the process to prevent the creation of toxic waste, it is treated afterwards. Finding a solution that prevents pollution then becomes even more unlikely as the company has already invested in expensive treatment equipment.

To deal with this systems archetype effectively, one must concentrate on the fundamental solution and ideally weaken the symptomatic one. Applying a symptomatic solution may be necessary due to short-term pressures. However, the long-term consequences of choosing the (usually easier) symptomatic solution have to be articulated clearly in order to gain support for developing a fundamental solution (Senge 2006, 109f.).

In certain cases, sustainability is not part of the problem, but represents the fundamental solution to a shifting the burden structure: Consider the case of the European premium frozen foods producer Frosta AG (see 7.2 for a more detailed discussion): Ever increasing margin pressure led the company to use an increasing amount of inferior ingredients in order to save production costs. However, eventually the quality of the food was approaching that of the discount competitors. As a result, not only did the company offer less healthy and tasty food, but it also started to deprive itself of its market differentiation. The underlying problem was that the relatively small difference in product quality to the products of the discount competitors did not allow for defying the margin pressure from retailers. To solve this problem in a lasting way, the company had to improve its products and decided use only healthy, high-quality ingredients. This way, the difference to the low-cost alternatives could be demonstrated more clearly to the consumer, the company was able to justify a significant price premium with its retailers, and significantly improved overall profitability.

Eroding goals

In another related structure called *eroding goals* (Senge 2006, 394f.), the short-term solution represents sacrificing long-term goals to reduce the gap to the aspired condition. The fundamental solution, namely taking actions to improve the condition and thereby reducing the gap, is associated with a delay that pressures management to let the goals erode. The eroding goals systems archetype contains the archetype balancing loop with a delay. However, in this case the problematic reaction is inactivity that threatens long-term performance rather than overshooting.

Eroding goals can also have adverse effects on sustainability as it can lead to sliding environmental targets. One way to mitigate eroding goal structures is to set up smart business model monitoring, combined with early actions on observed performance gaps.

Fixes that fail

Fixes that fail (Senge 2006, 399) are characterised by short-term success combined with unintended (longer-term) consequences. This systems archetype sounds very similar to shifting the burden but differs in that fixes that fail directly worsen the problem while shifting the burden emphasises dependence on solutions that help to relieve the symptoms but ignore the underlying problem. For example, when a company gets bad press because it is attacked by an NGO for low environmental performance, the company may consider the problem a public relations issue and rely on denial or greenwashing. However, this behaviour directly worsens the problem as the NGO then feels urged to intensify its efforts to point out existing grievances and look for additional ones. Obviously, fixes that fail should be avoided, although in certain situations it may be sensible to use them consciously to gain time to develop a fix that works (Senge 2006, 399).

Escalation

Escalation (Senge 2006, 395f.) describes an unhealthy form of competition in which each party defines its results relative to the results of competitors, leading to a rat-race. Although the same structure can incite rivalling athletes in sports to great performances through more effort and training, escalation structures can have devastating effects in business, e.g. in the form of price or advertising wars or unhealthy fighting for resources within companies (for example, through inflated budgets). The escalation system archetype is best represented as two connected balancing loops – one for each contester (in contrast to one reinforcing loop as described earlier). This way, it is easy to grasp why the escalation structure can be so treacherous: each party is caught in its own balancing loop, eagerly trying to improve one's own relative position. Each party will thus usually feel to act defensively and blame the other side for the escalation.

Luckily, the vicious circle can often be ended unilaterally through (potentially symbolic) non-aggressive acts. Victims of this structure should search for win-wins to stop the escalation (Senge 2006, 395). Moreover, companies that want to introduce new green business models should also be aware of the possibility that competitors that feel threatened may start an escalation that destroys the viability of the business model (for all). If this possibility poses a significant threat, contingency actions should be considered, e.g. through some form of cooperation with the most important competitors (also see chapter 8.3).

Accidental adversaries

Even with partners, however, escalation-like dynamics can unfold: partners may become *accidental adversaries* (Senge 1994, 145ff.). This system archetype starts with an intended virtuous circle of mutually beneficial activities. Then, one party applies a fix to improve its own performance, unknowingly obstructing the partner's success. This sets in motion a spiral of escalation that – if not suspended early enough – can destroy all synergies and benefits of the partnership. The better the mutual understanding and the more powerful the shared vision of the partnership, the less likely are escalation structures to become severe or start in the first place. To end an escalation, partners should jointly seek to understand the adverse effects of their seemingly isolated fixes and renew their partnership on this basis.

As explained earlier, partnerships and alliances are of specific importance to many green business models because some required capacities will (should) not be available in-house. Some partnerships will involve unfamiliar types of partners like NGOs, in which case partnerships are especially vulnerable to accidental adversary processes (Dahan et al. 2010, 336). Firms should also be careful not to exploit their stronger power base vis-à-vis partners inconsiderately. Recent far-reaching sustainable supply chain initiatives by mighty IBM and Wal-Mart could potentially hurt suppliers and set in motion unwanted dynamics. IBM therefore allows some flexibility in the advanced environmental management that it now requires from its suppliers. Both Wal-Mart and IBM also intend to help their suppliers in meeting their powerful customers' demands in order to ease the potentially thorny adaptation process of suppliers (IBM 2010a; Wal-Mart 2009b).

Limits to growth

Limits to growth (Senge 2006, 390f.) is a central systems archetype for the introduction of new business models. The adoption of a new product as illustrated in figure 4.11 is an example for limits to growth. The mistake often made when growth slows is trying to strengthen the reinforcing loop; instead the limiting condition should be removed or weakened. In the example, the new product could be introduced to new geographic markets or customer segments.

For example, German organic beverage producer Bionade, which saw wild growth in recent years, ran into such a limit to growth: it had difficulties sourcing organic lychees, a key ingredient. Bionade started to add some conventional fruits to cope with the shortage and, as a result, got criticised in the media and tainted its green brand (Arnann 2009).

Two often cited limits to growth in the renewable energy domain are grid capacities required for large-scale solar or wind generation, as well as missing electricity storage capacities (Hoffert et al. 2002).

Limits to growth patterns can also be observed in many change initiatives: After great initial success, strong resistance suddenly appears within the organisation. Due to the success of the initiative, some people start to feel threatened and try to reverse the change (Senge 2006, 98f.). Senge et al. (1999) identified several specific types of balancing loops that endanger sustained success of change initiatives. They are: objections of control-oriented managers, costs of change that become visible before the benefits, strong polarisation between converts and resisters, and fragmented management structures in which innovators stay disconnected. Again, Senge (2006, 99ff.) recommends removing the source of limitation or resistance, and not pushing harder. Sometimes several limiting factors may be at work in parallel; in this case it is necessary to understand how they relate to each other and in which order and with which priority they should be addressed. Ideally, the limits to growth are removed even before the balancing loop takes effect.

Growth and underinvestment

Growth and underinvestment (Senge 2006, 399f.) describes a case in which an activity gets allocated too few resources – paradoxically despite its initial success. It contains the limits to growth archetype and is often associated with eroding goals: Following high demand, the company initially shows strong growth, but does not invest (early) enough to sustain its performance level (e.g., delivery times, product and service quality). Gradually, the deteriorating performance negatively affects demand. Ironically, the diminishing demand is then used as an argument *not* to invest; instead, the original growth targets are adjusted downwards. In retrospective, it seems like initial demand forecasts were unrealistically high – de facto, however, the company was itself responsible for the decline (also see Senge 2006, 114ff.).

This systems archetype is a real danger for new green business models. Even if it has been tested with a pilot, growth and underinvestment dynamics can still prevent its breakthrough on a larger scale. Management needs to keep an eye on relevant performance levels and proactively plan for capacity extensions. This involves upfront cost and some risk-taking. External demand indicators can help to objectivise capacity investment planning for this purpose.

Success to the successful

Success to the successful (Senge 2006, 396f.) means that one of two (or more) activities, individuals, or groups perform better than the other, and therefore get a large share of the resources – which further increases the performance gap, and so on. The initial performance difference may be random or structural, e.g. if product A is an immediate market success and B, although equally promising in the long run, needs more time to develop. An example where this system archetype did *not* unfold its harmful effects is the Toyota Prius (also see chapter 7.8): The Toyota

Prius was introduced in 1997 in Japan and three years later in Europe, North America and other regions (Toyota Motor Corporation 2008). Although it showed poor financial returns at first (Maynard 2007; Toyota Motor Corporation 2008), Toyota believed in the hybrid concept and resisted temptations to abandon the Prius and jump on the sports utility vehicle bandwagon – like American car producers did. A decade later the conclusion of the story is clear: Toyota dominates the growing hybrid market and shows huge ensuing gains in brand value: According to brand management firm Interbrand, Toyota has gained US$ 15.2 billion or 81 percent of corporate brand value from 2000 to 2008 (Interbrand 2001; 2008). Interbrand stresses the strong contribution of the Prius to this gain. For example, the 2008 report states: "Toyota continues to benefit from the 'green halo effect' of the Prius, which is still widely talked about in the media." (Interbrand 2008, 26). Meanwhile, struggling competitors like General Motors are criticised for their short-sighted product development strategies (Vlasic & Bunkley 2008).

Especially because of the long-term benefits of green initiatives it may be sensible for companies to decouple them from other initiatives to prevent unhealthy competition for resources. However, as the saying goes, one should not throw good money after bad; judging whether the struggling activity really is promising or not will always stay one of the most difficult management tasks. Nevertheless, short- and long-term interests of the firm must be balanced as is seen by the decline of General Motors.

The tragedy of the commons

The tragedy of the commons systems archetype (Senge 2006, 397f.) describes how each party tries to maximise its returns through increased activity. However, the gain *per* activity diminishes – with a delay – due to the resource limit. This systems archetype has already been introduced in the context of overfishing. In fact, all types of common resource depletion through extraction and ecosystem destruction through pollution (including climate change) are subject to this structure. The structure also applies when organisational units use a shared resource (like a pool of experts) in an uncontrolled way.

To overcome the tragedy of the commons, Senge explains that the "commons" need to be managed, e.g., through education, self-regulation and peer-pressure, or by introducing a respective formal mechanism with suitable sanctions.

Within industry sectors that heavily rely on resource usage or cause significant pollution, there are many opportunities to take a leadership role and innovate green business models that help to control tragedy of the commons structures.

4.5.4 Dynamics for Growth with New Business Models

Of course, not all system dynamics are harmful and treacherous. Sterman (2000, 364ff.) provides an excellent overview of the dynamics ("engines") that determine corporate growth. Companies seeking to transform their business models can thus exploit non-linear behaviour of markets while avoiding the dampening effects and vicious circles mentioned above. Table 4.2 summarises ten growth engines adapted from Sterman (2000, 364ff.):

Table 4.2 Engines for corporate growth and related levers, market effects, and strategies

Growth engine	Levers (means)	Related competitive strategies (examples)
1. Product awareness	▪ Advertising ▪ Direct sales effort ▪ Word of mouth ▪ Media attention	▪ Hype-based strategies, viral diffusion ▪ Defend saturated market share
2. Unit cost	▪ Shared fixed cost (e.g., development) ▪ Production unit cost (economies of scale & scope, learning curve, process improvement)	▪ Classic cost leadership strategy to dominate through price ▪ Low initial price to encourage broad adoption and grow market volume
3. Product differentiation	▪ Improved value proposition features	▪ Classic differentiation strategy to reduce price-based competition
4. New products	▪ Uniqueness ▪ New uses, new needs	▪ Innovation leadership ▪ Contain imitators (keep their margins low)
5. Network effects	▪ Community and compatibility effects ▪ Complementary goods (vertical or horizontal expansion)	▪ First or fast second mover ▪ "Get big fast" strategies (e.g. with free products and openness)
6. Market power	▪ Market leadership, monopolies ▪ Buying power ▪ Lobbying ▪ Public engagement	▪ Monopolistic strategies ▪ Changing the "rules of the game" ▪ Standard setter
7. M&A	▪ Mergers, acquisitions, joint ventures, etc. for vertical or horizontal integration	▪ Eliminating (future) rivals ▪ Internalising value chain profits ▪ Portfolio extension, buying market share ▪ Buying key resources
8. Employer attractiveness	▪ Employer branding ▪ Corporate culture ▪ Wages and other benefits ▪ Job security ▪ Career opportunities	▪ Employee development strategies ▪ Good corporate citizen strategy
9. Cost of capital	▪ Investor relations	▪ Eliminate environmental risks
10. Aspirations	▪ Stretch targets	▪ Moonshot visions

Each of the ten growth engines listed above is powered by one or more self-reinforcing feedback loops. Hence, once these loops work in favour of the firm, it enjoys sustained growth and profitability. However, balancing loops sooner or later limit growth dynamics. Moreover, the dominance between the positive feedback loops can shift (and start to work in favour of competitors instead). Worse even, self-reinforcing loops can become vicious circles once the dynamics have reversed, leading to substantial losses or even bankruptcy (Sterman 2000, 382ff.).

The first growth engine listed in table 4.2, *product awareness* (Sterman 2000, 365ff.), encompasses four principal levers that that can be used to make the product a best seller: Advertisement can be used to jump-start a market, and the generated revenue can be reused for additional advertising. This is particularly effective for new products (ideally, a hype is initiated) and in mature, consumer-oriented markets where every point in market share counts (e.g., soft drinks). In other markets (especially business-to-business), direct sales force efforts are a bigger leverage. Increased customer awareness can both increase market share and the size of the total market. Once a product becomes "hot", word of mouth and media coverage often greatly contribute to the virtuous circle of success. However, in contrast to advertising and direct sales effort, these are much harder to control. Nevertheless, they can create extraordinary success stories: For example, Bionade, the German maker of an organic beverage of the same name managed to create exponential growth from a couple of million bottles per year to 200 million in just four years – roughly tripling sales every year. Bionade created this growth in a highly competitive market without classic print or TV advertising by using a viral marketing strategy that fosters word of mouth and media coverage (Treumann 2007; Dengel 2008).

The second growth engine, *unit cost* (Sterman 2000, 367ff.), is based on several effects. If more units are sold, fixed upfront cost for development can be split on more units. If this (expected) unit cost reduction is transferred to the customer through lower prices, more units are sold and unit cost fall even further. Hence, an early, aggressive pricing policy can result in significantly larger revenue over the course of the product lifecycle. However, the company needs the financial strength to cope with negative cash flows at the beginning and carries the risk to remain unprofitable if the expected volumes do not materialise. Sterman points out that apart from spread fixed cost, unit cost is driven down by economies of scale and scope, learning curves, and process improvements. If production capacity, distribution channels, expertise and other resources can be shared for certain product lines or units, economies of scope arise. Moreover, accumulating experience in production usually leads to better procedures with less errors and higher throughput. Low unit cost provides the basis of Porter's classic *cost leadership* strategy (Porter 1980).

The third growth engine corresponds to Porter's second classic strategy, *product differentiation* (Sterman 2000, 371ff.). It follows a different rationale: The generated revenue is not used to reduce the product's price, but to invest in developing new product features that make it more attractive for customers, including new ones. Differentiation tends to reduce direct comparability of products and thus eases pressure on prices. Alternatively, companies can pursue Porter's third and last generic strategy, *focus* (Porter 1980, 38f.). By giving up industry-wide growth aspirations, a company may tailor its products to the needs of a specific customer segment or market and therefore serve these customers better or more efficiently.

New products (Sterman 2000, 373f.) can also offer significant price premiums. Some firms manage to enjoy high growth and profitability at the same time due to a successful innovation leader strategy (e.g., Apple). New products can also be used as a means to generate replacement purchases. For example, ever increasing hardware requirements of software helps to shorten buying cycles. Another viable strategy is to use a margin model that uses profits from new products to keep prices for older products low. This way, the profitability of imitators is kept at a low level. Hence, their financial power and ability to develop high-end products remains limited.

The fifth growth engine is *network effects* (Sterman 2000, 370f.). Network effects are often highly non-linear and create winner-takes-all markets. They become relevant once a certain threshold of installed base is reached and dramatically increases the utility of a product. Examples include fax machines, VCRs, or social networks. Successful strategies include becoming big fast, e.g. by offering the product for free (Facebook), or by incentivising other participants to create value (e.g., Apple's App Store for the iPhone). Once the market leader reaches a certain size, it is very difficult to break its dominance. However, aggressive followers can be successful if they act early enough (e.g., Apple's Macintosh vs. the technically inferior Wintel platform).

Firms that have enjoyed growth through some of the above-mentioned growth engines will eventually command significant *market power* (the sixth engine) (Sterman 2000, 374f.). This market power further strengthens the position of these companies: Suppliers are forced to grant them preferential treatment and better contract terms, they can introduce industry standards that are in line with the firms' best interests, and customers are often locked in (e.g., due to high switching cost or network effects). Some corporations also become opinion leaders or use lobbying to exert significant influence on public debates and political processes (Sterman 2000, 380). However, too arrant exploitations of market power are subject to anti-trust proceedings and may cause backlashes from consumers and civil activists.

Market power and other growth engines may result from organic growth or be bought through *M&A* (Sterman 2000, 375f.). Types and motivations of M&A can differ greatly: Some companies buy market share and swallow smaller competitors to improve their market position (e.g., Oracle Corp.). Other deals are supposed to create synergies and economies of scale (e.g., the (failed) merger of Daimler-Benz and Chrysler), or to reduce cost and internalise value chain profits through vertical integration (M&A of suppliers or customers). Firms may also buy market access to certain regions or customer segments (e.g., the acquisition of The Body Shop by L'Oreal) or buy know-how and competencies (e.g., Daimler's acquisition of stakes in electric car pioneer Tesla). While M&A can lead to the virtuous circles described above, many deals turn out to be unsuccessful. One common challenge is to prevent the best people from leaving the firm.

A high workforce quality leads to better products and market success. The generated excess revenue can be used to pay higher wages and invest in employee development, further increasing the *employer attractiveness.* A successful firm also offers good career opportunities which increases loyalty even more (Sterman 2000, 376ff.). Being a good corporate citizen and operating green business models also fosters identification of employees with the firm.

A successful firm also has another advantage over its less successful competitors: It has *lower cost of capital* (Sterman 2000, 378f.). As the capital markets gain trust in the financial performance of the firm, risk premiums (i.e., interest rates) decrease and the company can raise more capital. One possibility, as mentioned earlier, is to avoid any environmental risks (Sharfman & Fernando 2008).

The last of the growth engines, *aspirations*, is largely a psychological phenomenon (Sterman 2000, 380ff.): Organisations have produced extraordinary results when their leaders showed great ambition and set stretch objectives. The scope of such objectives can range from organisations as a whole to the individual level. Like all of the growth engines, aspirations are a double-edged sword: If the tension becomes too great, frustration, burn-out or even unethical behaviour can be the result.

Traditional economic theory implies that balancing loops dominate the economy, thereby preventing companies from enjoying substantial above-average returns over long periods of time. However, empirical evidence indicates that self-correcting market forces function only partially, and instead frequently follow the positive feedback structures presented above (Sterman 2000, 382).

Furthermore, winner-takes-all conditions are path dependent, i.e. small, seemingly insignificant events at an early stage can determine who that winner will be (Sterman 2000, 349ff.). Hence, it is critical to intervene early enough while market dynamics are still responsive to respective policies. For example, it becomes increasingly difficult to compete on price with a larger competitor once a certain threshold of units produced is reached and economies of scale become significant. An early price reduction or advertising campaign can turn the tide in favour of a follower while respective efforts would be in vain in a more mature market.

Sterman (2000, 382ff.) remarks that successful firms tend to use several growth engines in combination. It is vital to understand which growth engines are relevant in the market at which point in time. Moreover, not all growth engines can be used together. For example, product differentiation and unit cost are based on conflicting dynamics. This does not mean, however, that a firm cannot use them sequentially: In an early phase it may be sensible to seek for low unit cost to penetrate the market and contain early imitators. Later, when certain once unique capabilities have become readily available to imitators, product differentiation can be used to evade price-based competition and acquire more demanding customer segments.

All growth engines will stall sooner or later. In the case of beverage producer Bionade, the business model got into several difficulties at once: Not surprisingly, the success of Bionade attracted imitators that tried to steal market share. The scalability of the business model also became a problem (availability of organic raw materials) (Arnann 2009).

In summary, the implications of system dynamics for successful business model transformations are threefold: (1) Firms need to understand and exploit dominant business dynamics before launch, (2) respond timely to changing dominance of feedback loops, and (3) remove limits to growth, ideally before they are reached.

4.6 Business Model Change and Evaluation

The importance for business model change and innovation can hardly be overestimated: According to Johnson et al. (2008, 52), "fully 11 of the 27 companies born in the last quarter century that grew their way into the Fortune 500 in the past 10 years did so through business model innovation." In chapter 6, the theoretical basis of organisational change and innovation is discussed in detail. Hence, only a short overview is presented as the conclusion of this chapter.

The first task is to define what kind of organisational changes constitutes a *business model transformation*. Linder & Cantrell (2000, 10ff.) distinguish four different approaches to business model change, which they refer to as *change models* (see figure 4.13):

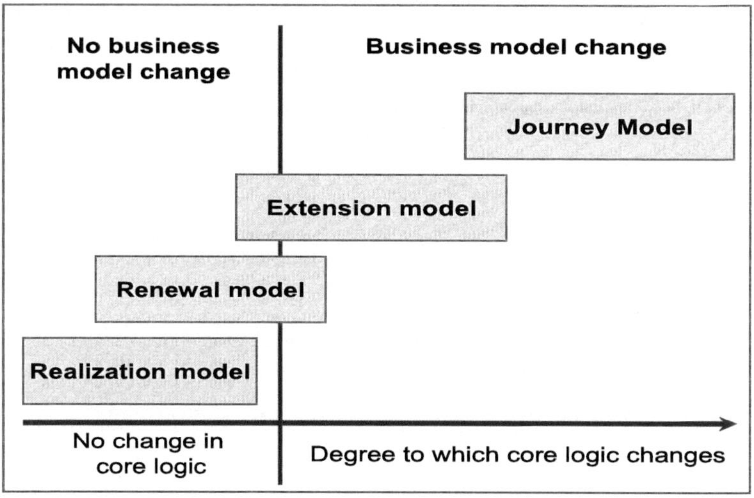

Fig. 4.13 Change models (adapted from Linder & Cantrell 2000, 13)

Realization models do not change the core logic and hence represent no business model change. They are an attempt to maximise profits from the existing business model. This can include brand maintenance (e.g., advertising), product line extensions, geographic expansion, penetration (share of wallet), additional sales or service channels, or roll-up (consolidation of firm fragmentation).

The second change model, the *renewal model*, usually encompasses isolated innovations or adaptations that do not represent a true business model change either. New service offerings or brands, entry in untouched (and different to existing) markets, new retailing formats, or disruptive new product or service platforms all belong to this category.

The third change model, the *extension model*, usually brings along significant change to the core logic of the firm. It adds new – rather than replaces – business operations. It can be realised through value chain integration, i.e., backward (e.g., through private label brands), forward, or horizontal. Alternatively, a company externalises an internal capability.

Linder's and Cantrell's last change model is called the *journey model* and is the most radical. It comprises commoditization (which changes the nature of the value proposition focus from product to price), avoiding commoditization (e.g., through services or solution offerings), globalization (coordinated, global reach and capabilities), moving upmarket in products (e.g., through speed and agility), and moving upmarket in services (e.g., through brand or expertise). The journey model thus represents a new strategy that implies a lasting change of the firm's overall direction.

In summary, Linder & Cantrell (2000, 10ff.) offer a long list of means to change elements of a business model, yet two of the four identified types of change models are ambiguous with respect to whether or not the business model as a whole is changed fundamentally.

The following definition aims close that gap:

> *A* ***business model transformation*** *is a fundamental change to the logic of doing business with the purpose of improving or sustaining financial performance.*
>
> *Value is thereby created and / or captured in a fundamentally different way than before. That is, at least one of the business model components – value propositions, target groups, key resources, key processes, or the financial logic – is changed considerably, and the interplay of business model elements is consequently altered, too.*

Business model transformation is no end in itself – business models need to be evaluated in order to identify the need to adapt or transform them in the first place. Johnson et al. (2008, 57) name five strategic circumstances that can necessitate a business model transformation:

1. Opportunities from so far unserved customer segments, e.g. in emerging markets.
2. Opportunities to introduce new technology into an existing market or vice versa.
3. Opportunities from unmet customer needs neglected by established players.
4. "The need to fend off low-end disrupters."
5. "The need to respond to a shifting basis of competition."

Osterwalder & Pigneur (2010, 216ff.) provide a catalogue of questions that can be used to evaluate the strengths, weaknesses, opportunities, and threats (SWOT analysis) of the current business model with respect to each business model building block.

Several authors have addressed the issue of what the process of business model transformation should be like. Earlier approaches are often specific to business models of the digital economy. The ones discussed by Osterwalder (2004, 37f.) include Tapscott et al. (2000) and Gordijn (2002) who propose approaches following a deconstruction-reconstruction logic. Petrovic et al. (2001) emphasise the connection between transforming business models and underlying mental models. Papakiriakopoulos et al. (2001, 453ff.) propose changing e-business models in a four-stage process that considers stakeholders, value flows, competitive forces, and feedback chains (i.e., adaption to market needs).

More generally applicable is the three-step approach of Johnson et al. (2008, 54ff.): The first step is to think of the opportunity to satisfy a customer's needs. Only in the second step the company should start to think about a blueprint, i.e. business model design, for how to fulfil that need profitably. The third step is to compare the blueprint to the existing business model and identify what needs to be changed. Johnson (2010, 107ff.) emphasises that business model design and implementation thereby have a strong element of experimentation as opposed to rigid execution.

Osterwalder & Pigneur (2010, 244ff.) present a five-phase process for designing new business models. The first phase, called *mobilize*, is intended to set the stage. The second phase, *understand*, is used to create the knowledge necessary to successfully carry out the *design* phase. The business model design is then executed during the *implement* phase. Finally, the business model's further evolution is fostered during the *manage* phase. The process described by Osterwalder and Pigneur differs in some important aspects from the management framework presented in chapter 9, inter alia, because it does not focus on green business models. However, it follows a similar underlying logic and with its many illustrations and additional examples their book may serve as an informative complement for practitioners.

Eventually, the old business model and the new design have to be compared in economic terms. In this context, Osterwalder (2004, 38f.) contrasts the approaches of Hamel, Gordijn, and Afuah & Tucci. Hamel (2002, 99ff.) determines a business model's "wealth potential" by assessing four factors: Efficiency, uniqueness (which allows high prices), consistency of elements, and usage of means to boost profits like competitor lock-out and strategic economies. Gordijn (2002, 120ff.) proposes to create a profitability sheet for all involved actors and the valuation of objects they exchange in order to evaluate e-commerce ideas. He further proposes using "evolutionary scenarios" related to possible future events and potentially wrong assumptions to represent expected changes of profitability over time. Afuah & Tucci (2001, 79ff.) assess business models in terms of profitability measures (earnings and cash flows), profitability predictor measures (profit margins, market share and revenue share growth rate), and provide benchmark questions to evaluate individual business model components.

Johnson et al. (2008) suggest testing the market "fit" of the aspired value proposition with the current business model, i.e. evaluate if it still works with respect to the *Profit Formula,* if it can be realised using most or all existing resources and processes, and if the current core metrics, rules and norms are still valid. Johnson (2010, 162) lists the following business model metrics for consideration:

Table 4.3 Examples of business model metrics (Johnson 2010, 162)

Financial metrics	Operational metrics	M&S / R&D / HR metrics
▪ Gross margins ▪ Opportunity size ▪ Unit pricing ▪ Unit margin ▪ Time to break even ▪ Net present value calculations ▪ Credit items	▪ End-product quality ▪ Supplier quality ▪ Owned versus outsourced manufacturing ▪ Customer service level ▪ Channel options ▪ Lead times ▪ Throughput	▪ Pricing ▪ Performance demands ▪ Product-development life cycles ▪ Basis for individuals' rewards and incentives ▪ Brand parameters

The considerations presented above for the most part relate to identifying economically viable design changes in business models. However, realising business model changes in practice is more than just an analytical challenge (Chesbrough 2010). Comprehensive business model transformations require a much deeper analysis of organisational change, starting with a thorough understanding of organisations itself. Before the theoretical background of this field of research will be laid out in chapter 6, I complete the theoretical examination of business models with the development of a taxonomy for green business models.

Chapter 5
Towards a Taxonomy of Green Business Models

As management science still lacks a universally accepted definition of business models it comes at no surprise that theoretical research on *green* business models is virtually non-existent at this point in time.

First attempts to conceptualise "sustainable business models" or "business models for sustainability" have been made (Stubbs & Cocklin 2008; Lüdeke-Freund 2009). However, a robust body of research on sustainable business models – let alone with an environmental focus – is not yet available. As this work attempts to provide managers with a practical methodology to manage Green Business Model Transformations, the question what actually distinguishes green from non-green business models has to be answered first. In addition to providing a theoretical, abstract distinction, means to recognise green business models in the real world need to be developed as well. This can also help to objectify discussions about greenwashing which is likely to continue to play a role in the coming years. A generally applicable approach to identifying green business models is thereby a worthwhile research question in its own right. However, the aim of this chapter is to start a discourse on the topic, rather than to conclude it. As will be demonstrated in the following, a taxonomy for green business models raises a number of theoretical and practical challenges whose resolution requires contributions that vastly exceed the scope of this work, including but not limited to improved methods for measuring and comparing environmental impacts as well as industry-specific examinations.

5.1 Existing Business Model Classifications

Existing business model classifications provide a starting point for developing a taxonomy of green business models. Osterwalder (2004, 26ff.) has gathered a number of business model classifications that have been created during the dot-com boom (Timmers 1998; Applegate 2001; Linder & Cantrell 2000; Tapscott et al. 2000; Weill & Vitale 2001; Rappa 2001). Although all of them, with the exception of Linder and Cantrell, created their classifications specifically against the background of e-business, the dimensions used for categorisation differ greatly (see table 5.1). The same is true for the other examples of the same table that have been collected. Some use only one, rather simple dimension (e.g., Johnson 2010, 131), while most authors use multiple, sometimes complex dimensions. The various kinds of taxonomies also differ in their level of analysis, i.e. firm vs. network (Lai et al. 2006).

A. Sommer: Managing Green Business Model Transformations, SPLCEM, pp. 93–108.
springerlink.com

Table 5.1 Categorisation dimensions of selected business model classifications

Author(s)	Categorisation dimension(s)	Resultant categories
Timmers (1998)	▪ Degree of innovation ▪ Degree of functional integration (from single function to multiple functions / integrated)	11 generic e-business models
Applegate (2001)	▪ Product offering (focused distributor, portal, producer, infrastructure provider)	4 e-business model categories plus subtypes
Linder & Cantrell (2000)	▪ Core, profit making activity (providing, channel, or intermediary) ▪ Relative position on the price/value continuum (high-value, premium-priced innovations to low-priced, standardized offerings)	9 operating business model types plus subtypes
Tapscott et al. (2000)	▪ Control (hierarchical vs. self-organising) ▪ Degree of value integration	5 types of b-webs
Rappa (2001)	▪ Position in value chain and revenue source	9 generic e-business models
Weill & Vitale (2001)	▪ Main functionality / type of value	8 atomic (modular) e-business models
Moore (2005)	▪ Complex systems vs. volume operations	2 business model types
Rajala & Westerlund (2005)	▪ Degree of involvement in customer relationships ▪ Level of homogeneity of offering	4 software industry business model types
Schweizer (2005)	▪ Value chain constellation (existing vs. innovating) ▪ Market power of innovators vs. owners of complementary assets ▪ Total revenue potential	4 business model types
Chesbrough (2006)	▪ Innovation process (none, ad hoc, planned, externally supportive, connected to business model, identifies new business model) ▪ Intellectual property management (n/a, reactive, defensive, enabling asset, financial asset, strategic asset)	6 business model types
Lai et al. (2006)	▪ Type of assets involved: financial, physical, intangible, or human ▪ Type of rights being sold: creator, distributor, landlord, broker	16 business model types (14 existing in practice)
Johnson (2010)	▪ Financial logic	19 business model types
Wirtz et al. (2010)	▪ Classical business activities (on internet markets)	4 prototypical internet business models

Furthermore, all five components of the business model framework have been used for categorisation: value propositions (e.g., Linder & Cantrell 2000), target customers (e.g., Rajala & Westerlund 2005), key resources (e.g., Lai et al. 2006), key processes (Schweizer 2005), and financial logic (e.g., Johnson 2010). Some authors use dimensions that are outside the business model framework. For example, the maturity of a business model is used as a criterion (e.g., Chesbrough 2006).

Differences in categorisation dimension can be observed both for classifications specific to the e-business domain and for general classifications. The large heterogeneity originates in part from the fact that most of the listed classifications are typologies rather than taxonomies. According to Lambert (2006, 7f.) typologies are classifications that serve specific purposes, and are usually derived deductively based on a few qualitative criteria. Taxonomies, in contrast, consider a broad range of characteristics and use a quantitative basis. As a result, taxonomies classify business models according to their similarity in all aspects, and classes may change based on empirical findings.

As the study of existing green business model candidates (some of which are documented in chapter 7) has revealed, they can differ greatly from each other and from conventional business models in numerous aspects, while they are alike in others. The levers for value creation though environmental sustainability (see 3.2.3.1) are widely distributed across virtually all business model elements. Conversely, not all components need to be inherently green to qualify the business model as a whole. As a result, no simple logic of classification reveals itself from examining real-world business models.

The only definite commonality across all green business models that can be identified at this point of the exploration is their superior environmental performance – whatever form this might take.

5.2 Basic Considerations for Developing a Taxonomy of Green Business Models

There are a number of fundamental issues that need to be dealt with in order to create a meaningful taxonomy of green business models. These include:

- Paradigm: The weak versus strong sustainability paradigms have to be considered.
- Comparability: The question which business models are to be compared, i.e. to what degree their customer need satisfaction is equivalent (substitutable).
- Scope: The considered scope – or system boundary – of the business model determines which environmental impacts are included in the assessment.
- Level of abstraction: The actual versus theoretical impact of a business model, i.e. the influence of its realisation may lead to very different conclusions in certain cases.
- Detail of assessment: There is a trade-off between the level of detail of the categorisation and the effort in completing the evaluation.

First, the taxonomy to be developed needs to be evaluated regarding its consistency with weak and strong sustainability. Balancing of economic versus natural capital in the sense of the weak sustainability paradigm (see Neumayer 2003) is rejected. Otherwise it could be argued, for instance, that business models with high impact like building luxury sports cars are green just because prices per unit are even higher than that of mainstream cars. Moreover, a price increase would make a business model more green if environmental impact is measured relative to monetary output (see also Verfaillie & Bidwell 2000, 17). As a result, the financial output (e.g., net sales) of business models is disregarded in this context. Conversely, this also means that policy decisions (e.g., about subsidies or tax breaks) cannot be based on the proposed green business model taxonomy alone: Even if a business model is categorised as green, there may be more economically efficient alternatives that politics should support instead. Yet, economic efficiency is still in force anyway in the sense that inefficient green business models will usually be eliminated by competition or remain niche.

On the other extreme, few if any volume business models to date satisfy the requirements of the strong sustainability paradigm, or could do so realistically. Hence, the taxonomy is not based on either of the extremes. Instead the taxonomy follows an intermittent view, possibly best described as "sustaincentric" (see Gladwin et al. 1995).

Second, the sports car example reveals another difficulty: Should the environmental impacts of the business model of a sports car producer be compared with that of other sports car producers; or with that of car producers in general, or even with that of mobility goods or service providers including, for instance, airlines, railway companies, or car sharing companies?

The related question of what constitutes an ecological (green) product is helpful in finding an answer. Schaltegger & Sturm (1990, 283) argue that a green product must fulfil three criteria:

1. highest product-based ecological efficiency
2. highest functional efficiency
3. volume consumption of product does not cause permanent damage to natural systems

Product-based efficiency relates to the environmental impact (full lifecycle) of a product with a specific functionality (need fulfilment). In practical terms, one car can be more or less efficient than another car with comparable functionality. Functional efficiency only relates to the "job-to-be-done", e.g. transporting people over a given distance and time. Here, different modes of transport would be compared (e.g., driving a car, riding a train, or flying) while assuming equivalent utilisation and technological maturity (i.e. an average car is compared with an average train, both half full). One unfortunate implication is that the same product can be green and non-green – depending on the use case. Schaltegger & Sturm (1990, 284) remark that analogous considerations are valid for production processes, too.

Business models usually consist of bundles of offerings and numerous primary processes. Hence, to make a statement about the business model as a whole, a certain degree of abstraction and aggregation is necessary. However, even individual offerings may satisfy a range of needs (e.g., short vs. long distance travels by car). To compare business models they must therefore cover a similar range of customer needs. As such, the analysis may reveal that both a producer of low-emission cars and a car sharing company using ordinary cars can both have a green business model, if their environmental impact is significantly lower compared to the average car producer and seller business model. At the same time, a bicycle manufacturer's business model is not suitable to satisfy the needs of most car buyers, and thus is not eligible for comparison in the sense above. This also implies that the relative "greenness" of business models with differing needs satisfaction profiles cannot be sensibly compared. In order to operationalise what has been referred to as "similar coverage of needs", it may be serviceable to define a set of functional characteristics that satisfy *basic functional needs* (e.g. number of seats in a car) vs. *non-basic functional needs* (e.g., motorisation). This distinction can generally be expected to be controversial no matter how it is set. Grey areas will be significant for many industry sectors, they depend on cultural attitudes, and differ between customer groups. One could argue that a Toyota Prius and a Rolls Royce fulfil equivalent basic functional needs; most owners of a Rolls Royce, however, will probably disagree. Yet, despite such issues, without such a distinction of sets of needs, basically every single product would have to be considered non-substitutable.

Third, the chosen system boundary of a business model can heavily influence the conclusion regarding its relative environmental impact. A business model may create a comparatively low impact within the company doors, but show disastrous characteristics when the whole value chain is considered. For example, the use of certain chemicals may make a production process cleaner, but produce strong additional environmental impact when the customer uses or disposes the product. Conversely, most manufacturers of environmental technology do not have low impact production methods per se – the environmental benefits of the business models are found at their customers' sites. Hence, it is sensible to argue that any meaningful approach needs to include suppliers' and customers' environmental impacts, as long as they can be attributed to the business model under examination. This may also include the suppliers' suppliers and customers' customers, and so on. This way, green claims which are based on shifting "dirty" parts of the value chain to customers, suppliers or other partners can be refuted, too.

More difficult is the question whether systemic, macro-level environmental effects should be considered. In some cases, well-meaning business models can turn into an environmental (and public relations) nightmare on a macro level. For example, biofuel producers that were once praised for providing a fuel that releases no additional CO_2 into the atmosphere got under scrutiny. Prospects of large profits led to the clearing of rainforests in order to grow biofuel crops as well as replacing food with fuel crops. As *Time* wrote in its cover story on April 7, 2008: "Hyped as an eco-friendly fuel, ethanol increases global warming, destroys forests and inflates food prices. So why are we subsidizing it?" (Grunwald 2008b).

It needs to be noted, however, that macro-level effects are very difficult to assess, let alone predict. In fact, even the detailed assessment of a business model's value chain is afflicted with significant effort and methodological difficulties as practical experiences with life cycle assessment show (Schaltegger 1997). Such assessments are associated with considerable uncertainty due to the complexity of interacting system elements, and because relevant effects often lie far in the future (Sterman 2000). What can be done, however, is to determine whether a business model has a strong tendency to lead to adverse environmental effects on a macro-level, or if these patterns are balanced or negligible, or if the business model even has ecologically beneficial macro-level characteristics. The latter can be the result of desirable behavioural changes in addition to direct environmental advantages. Hence, macro-level effects of business models will be discussed in the following, but are not part of the formal distinction of green versus non-green business models.

The fourth fundamental issue concerns the question to what degree the actual *realisation* of a business model should be considered. By definition, the business model and its implementation are two distinct matters. However, the question remains at what level of abstraction a business model should be evaluated. For example, the evaluation of a generic car sharing business model may conclude that car sharing models are green based on some assumptions about characteristics that ordinary car sharing business models have in common. However, if the car pool of a particular car sharing business model consisted of fuel-guzzling SUVs, it may be worse than the traditional car ownership business model based on some average fleet fuel efficiency.

A specific non-green version of a business model should not exclude its most common occurrence from being considered green – in general. A car sharing model with SUVs is very different to the ordinary business model in many important aspects (e.g., value proposition, customer segment, physical assets) and therefore needs to be evaluated separately. This rule may seem too malleable to some observers. However, the complex nature of the task at hand requires some common sense from analysts in order to keep the approach manageable in practice.

This leads to the last of the five fundamental issues: the trade-off between accuracy and effort. Generally, the more detailed the rules and distinctions underlying the taxonomy, the more accurate it can be. However, at some point applying these rules either becomes unreasonably time-consuming or difficult (thus resulting in limited accuracy, too). On the other hand, if the chosen approach is too simplistic, the result of an assessment is of little theoretical or practical value.

There are two basic types of errors: false positives (non-green business models are declared green) and false negatives (green business models are declared non-green). Both errors are problematic: The former invites companies to greenwashing and may lead to hypocrisy. The latter would discount worthwhile efforts towards greening the economy. Since false positives could seriously harm the credibility of the proposed approach, it is designed to minimise both types of errors, yet making sure there are no glaring false positives, and that the effort remains within acceptable limits.

5.3 Four Generic Business Model Types Differentiated by Aggregated Environmental Impact

The taxonomy of green business models proposed in this work is based on the distinction of four generic types, based on their overall environmental impact compared to output. Only business models of type 2 and 3 (not type 0 and 1) are considered to be *green* in a dichotomic sense. Figure 5.1 depicts the trend of the *overall environmental impact I* over time (see below).

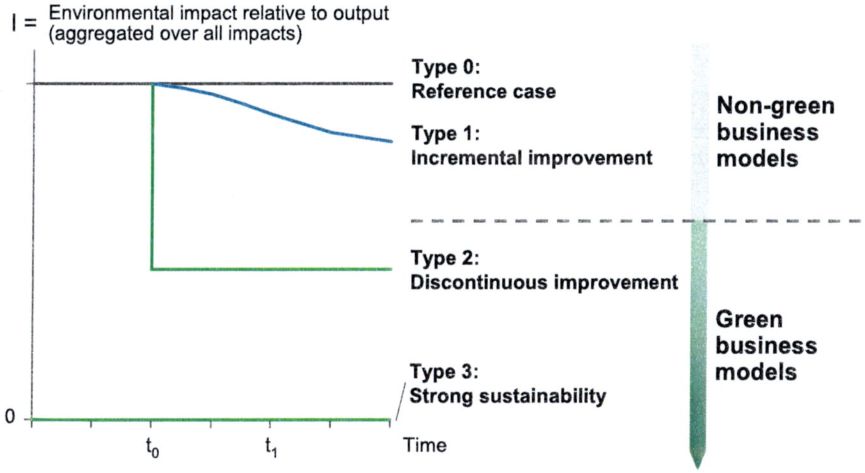

Fig. 5.1 Case-based distinction of business models related to their relative environmental impact

I is defined relative to a business model's output over a certain period of time (e.g., one year). *I* consists of one or more *individual environmental impacts i*, weighted by their *relative importance w*. For each *environmental issue k* (e.g. water use, NO_x emissions, hazardous waste, etc.), $i_{k,t}$ therefore refers to the impact regarding that issue during *time period t*, as measured in *unit u* and divided by the *output o*. As explained above, *o* should be defined in terms of volume (e.g., number of product units sold) or function delivered (e.g., passenger-kilometers), not monetarily (see also Verfaillie & Bidwell 2000 for various issues with value indicators). For example, a utility that generates power from a mix of fossil and nuclear energy may have an environmental impact $i_{CO2,2010}$ = 600 t CO_2 per GWh. However, the utility's aggregated impact I_t will also include impacts from other environmental issues like air pollution and nuclear waste. The various impacts are added by making them comparable using weights w_k (in $1/u_k$). Hence, the vertical axis of figure 5.1 is defined as

$$I_t = \sum_k w_k * i_{k,t}$$

The four business model types are explained in the following. After that, challenges of their application in practice are discussed.

5.3.1 Type 0: Reference Case

Type 0, the *reference case*, can be interpreted as the current standard practice within a given industry sector. For example, figures for average CO_2 emissions per generated MWh of utilities are readily available for most parts of the world (e.g., IEA 2009b). If there are large regional differences, advanced economies should serve as a reference. Some industries, however, do not have a generally accepted standard practice – at least not for all relevant environmental impacts. In such cases, it may be necessary to compare a business model to that of a key industry player; or, in the case of a transformation, to its own past performance at a given date. While there are obvious issues of comparison between different companies when using reference cases defined the latter way, it will usually be sufficient to determine whether a particular business model is to be considered green or not.

The reference case in figure 5.1 shows constant environmental impact over time (in relative terms, i.e. compared to the business model's output).

This is a simplification as normal efficiency improvements alone should usually slightly reduce the relative environmental impact over time. Hence, type 0 is supposed to represent business-as-usual developments, that is, when a company does not particularly concentrate on reducing its environmental impact. As the standard practice gets adapted from time to time, type 0 business models will thus take a staircase form with a gentle downward slope over longer periods of time. For the sake of simplicity, however, the reference case is assumed to show no improvements in the considered timeframe, and weights are considered constant.

5.3.2 Type 1: Incremental Improvement

By now, many if not most large companies have initiated some sort of environmental programme. The majority of these programmes as featured in corporate sustainability reports will be of an *incremental improvement* type, also referred to as *type 1* in the following. Environmental targets announced by companies sometimes sound impressive. Yet, they usually do not represent real breakthroughs, but rather reflect many small measures that add up over long periods of time. Incremental improvements of a given company can be larger or smaller than an industry's average. Over time, strong incremental improvement can lead to significant advancement compared to the reference case (that follows the industry average). However, if these additional improvements pay off financially, competitors can be expected to imitate, thus eventually raising the bar for all firms that want to stand out. The business models that car manufacturers have used from 1900 through the rest of the century provide a good example for type 1 business models: Business models have not changed much during that time, and despite all the technological advances during the 20th century, fuel economy has only improved gradually.

Therefore, each individual improvement is considered as if it was infinitesimally small (in mathematical terms). Obviously there is no such thing in reality: Every replacement of an incandescent light bulb by a compact fluorescent lamp reduces energy consumption (and thus usually also CO_2 emissions) by a non-infinitesimal quantity. However, compared to the CO_2 emissions of a utility, even the replacement of all light bulbs in buildings of the company would still be negligible and go unnoticed on a chart like that in figure 5.1. In contrast, if the utility were to equip its coal-fired power plants with Carbon Capture and Storage (CCS), the resulting reduction in CO_2 would produce a sudden drop of I_t, the graphical characteristic of a type 2 business model.

5.3.3 Type 2: Discontinuous Improvement

Type 2 represents a fundamentally different approach compared to type 1 in that it features *discontinuous improvement.* While improvements of type 1 are the result of many small steps (doing the same but better), type 2 features big leaps (doing something different). Respective business models are considered green as the improvement can be clearly distinguished conceptually from business-as-usual improvements.

In t_0 the discontinuous improvement takes effect; it represents the "big leap" in environmental performance. From t_0 onwards I_t is shown to be constant (for the purpose of conceptual clarity). However, in reality type 2 business models will generally enter a phase of incremental improvement after the discontinuous change has taken effect. In that respect, the idea of type 2 business models has some similarity with theories of punctuated equilibrium (see chapter 6.2.5).

While the distinction between type 1 and 2 seems clear in abstract terms, in practice it can be difficult to make. There may be relatively long periods during which type 1 business models have a lower actual environmental impact than immature type 2 models – it remains contested during that time whether the latter is in fact a dead end.

In addition, the expected improvement may not materialise right away as new business models will sometimes take considerable time to implement at scale – even when setting aside the maturity aspects mentioned above. The clean step of the curve representing type 2 in figure 5.1 is thus an idealisation, ignoring temporary implementation-related effects.

Furthermore, two alternative business models may have advantages in some, but disadvantages with respect to other environmental issues. For example, compact fluorescent lamps use significantly less energy than comparable incandescent light bulbs. However, the former contain lead. In these cases, determining the weights becomes critical (see 5.4.2).

An issue that has not yet been raised concerns the scope of the environmental footprint needed to qualify business models as green. It can be internal to the company (referred to as *type 2a*). Yet, many green business models have their largest lever in reducing environmental impacts outside of the boundaries of the firm (*type 2b*) – often those of their customers. For example, a utility that replaces all fossil generation capacity with renewable energy sources applies a type 2a business model. The company that produces the utility's wind turbines applies a type 2b business model.

In some cases, it is even the explicit purpose of the business model to reduce the environmental impact of another business model (*type 2c*). That is, these business models have no purpose other than protecting the environment (wind turbines have the primary purpose of generating electricity, and pleasantly do so without pollution). Environmental technologies like pollution filters are examples for this type. In contrast to other business models, the alternative to applying them can be *not* applying them (as opposed to substitute business models). Oftentimes, type 2c business models bring about trade-offs. For example, pollutions filters at coal-fired power plants not only reduce harmful air pollution, they also reduce the efficiency of the plant and thus result in higher CO_2 emissions per GWh. Whether or not certain type 2c candidates are actually green may thus be contested. Their environmental benefit, by definition, will be downstream as this is the core of their value proposition. If there is a market for a type 2c candidate, strong externalities or misaligned economic incentives must thus be present for the business model not to be green indeed.

Type 1 and type 2 have both been defined in relative terms. For each type 2 business model there will thus be a non-green version of type 1. Type 3 is different; respective business models are green in *absolute* terms.

5.3.4 *Type 3: Strong Sustainability*

Type 3 denotes business models that are truly sustainable, i.e. they operate without depleting the capacity of the environment to support its operation indefinitely (strong sustainability). There are very few if any such business models as of today. It would certainly need to use energy from renewable sources, apply closed-loop manufacturing, and ensure that its products do not harm the environment during usage by customers. A weaker form of such a business model can be accomplished if remaining impacts are neutralised by reducing impacts on an equivalent scale elsewhere. Hence, the sum of all impacts needs to be (close to) zero, with the relaxation that positive individual impacts $i_{k,t}$ can be compensated by negative impacts through measures like reforestation. To serve the demands of strong sustainability, however, critical natural capital (e.g., the ozone layer or biodiversity) may not be substituted (see Neumayer 2003).

5.4 Distinguishing the Four Business Model Types in Practice

As indicated above, there are several practical challenges that need to be overcome in practice in order to create robust assessments of business models regarding their categorisation: First, the considered scope of environmental impacts needs to be determined. Second, environmental issues need to be made comparable; transparently determining respective weights for this purpose can prove difficult. Finally, the critical distinction between type 1 and 2 business models needs to be formalised in a reproducible way.

5.4.1 Challenge 1: Define Scope of Environmental Impacts

The scope of assessed environmental impacts needs to be considered carefully, both in terms of business model boundaries and to what extent external environmental impacts of suppliers and customers are factored in.

Consider a steel producer that uses standard production methods. This example seems to be an obvious one for a non-green business model (type 0). However, if the steel producer produced a specialty steel that is exclusively used to build high-performance components for offshore wind parks, it suddenly becomes a type 2b business model. Similarly, if a utility produces only five percent wind energy, but sells this electricity as green energy to environmentally-minded customers, this can be considered a type 2 business model as well – the utility then operates at least two business models, one of which is green. The criterion whether environmental impacts of external value chains are included is *direct attributability*. That is, the steel must be purpose-produced for wind mills to claim the avoided CO_2 in electricity generation, and not be general purpose steel. In this sense, the specialty steel producer inherits its green status from the low-emission generation business model (and so does the wind energy equipment manufacturer). Direct attributability ends at this point, though: neither the supplier of the iron ore, nor the one that delivers the coke offers a product specifically created for building windmills.

Conversely, the CO_2 emissions and air pollution from driving a truck is directly attributable to the truck manufacturer; further impacts resulting from the purpose of using them (e.g., the transported goods) are not. Respective impacts of single-purpose transport vehicles like oil tankers, in turn, are directly attributable. For instance, the average amount and damage of oil spills (per transport volume and distance) could be compared between tankers and pipelines, or between single- and double-hulled tankers. The examples above demonstrate that deep reflections are necessary to arrive at a sound definition of scope.

A company that belongs to an industry that generates relatively low internal environmental footprints but makes products that cause enormous, directly attributable environmental impacts (e.g., car manufacturers) will hardly be a candidate for a type 2a business model. Conversely, companies with a large internal footprint and comparatively low external impacts (e.g., utilities) will have difficulties presenting convincing evidence that they operate a type 2b business model. The obvious reason is of course that the impact reduction in the external value chains will likely be insignificant compared to the overall impact of the business model. There are, of course, some well-founded exceptions as the example of the specialty steel producer above illustrates. Appendix 3 provides an overview of which industry sectors will usually be candidates for the various subtypes of type 2 business models. As a rule of thumb, external value chain segments can be left out of scope if respective environmental impacts are negligible or clearly outside the sphere of influence of the company (with the exception of type 3). For example, a utility cannot be held accountable for the environmental impacts that their customers produce by using the sold electricity.

5.4.2 Challenge 2: Determine Weights for Environmental Impacts

There is a growing body of research that deals with the question how to make different environmental impacts tangible, e.g. by representing them in financial terms (Schaltegger & Burritt 2000; Burritt et al. 2002; Schaltegger et al. 2006). Other approaches aim to measure environmental impact in terms of human welfare or years of life lost (YOLLs) (Droste-Franke 2004). Determining respective weights is a real challenge. They may be subjective considerations informed by current societal knowledge and beliefs. This leads to the unpleasant situation that a business model can be green for one group and worse than the reference case for another. However, this characteristic is not counterfactual to reality: For example, some consider nuclear power a green technology while others think it bears unacceptable risks and produces waste that is more damaging than the CO_2 from fossil power plants it substitutes. In this particular case the nuclear crisis at the Fukushima plant in Japan has apparently changed previous assessments of nuclear power by (some) governments and the public at large quite significantly (e.g., Grunwald 2008a; Kaufman 2011; The Economist 2011; Cooper & Sussman 2011).

One alternative to create weights that are less prone to ideological views could be the introduction of (virtual) prices – for example based on how much it would cost to clean up the environmental impact, or how much the used or depleted ecosystem services would be worth if they had been produced artificially. However, these prices in itself would be a matter of great uncertainty and hotly debated in most instances, too. It may well be that picking different methodologies for different industries turns out to be the most practical approach.

5.4.3 Challenge 3: Distinguish Discontinuous vs. Incremental Improvement

One central problem is the question what the term "discontinuous" means with respect to improvement compared to the reference case. For example, if a utility uses wind power to generate only a part of its electricity, the crucial question is what the threshold should be in order to distinguish between type 1 or type 2. The same is true for "pure" business models that show an improvement of x% compared to the reference – how large does x need to be? How the reference case is determined also has a big impact: One could compare to the installed base, a global average of new installations, use geographical market definitions, etc. Furthermore, a business model can offer a significant improvement for one environmental issue (compared to the most strict reference case), but this improvement gets diluted due to other issues that remain unchanged.

Again, a scientifically robust answer to the questions mentioned above may be provided by an industry-based assessment of issues and their historic incremental improvements.

5.4.4 Proposed Solution: 4-Step Assessment Procedure

No final solution to all the mentioned practical challenges can be offered here. Nevertheless, the following 4-step assessment procedure is proposed to frame the solution:

1. Identify business model boundaries:
    - A company may operate several business models in parallel, some of which are green and some of which are not.
    - Business model boundaries are determined by distinct combinations of value propositions, target groups, and the revenue model (i.e., the value capture perspective); complimentary offerings alone are no separate business model in its own right.
2. Identify components that may potentially qualify the business model as being green:
    - A necessary condition for a business model to be considered green is that at least one of its components qualifies it as type 2 (or 3).
    - The five considered business model components are: value propositions, target groups, key resources, key processes, and financial logic.
3. Confirm the significance of the green characteristics of the business model:
    - All significant, *directly attributable* environmental impacts are considered, including those of up- or downstream value chains (suppliers and customers), or partners.
    - The resulting environmental effect must relate to one or more significant environmental issue(s) of the business model's value chain system.
    - Steady state environmental performance is evaluated; temporary, implementation-related deficiencies are ignored for the assessment.
    - The environmental improvement effect on the issue(s) must be considerable compared to the reference case.
    - The reference case is the average, or most commonly used standard, that would be newly applied in the business model's reference market (i.e., new installations, not installed base; use of the business model's defined market scope).
4. Confirm that the green characteristics are fundamental to the business model:
    - In order to qualify for types 2 or 3, the respective business model components must pre-dominantly *apply* the characteristics in practice (green processes, products, etc.) that distinguish them from a reference case (i.e. beyond a reasonable threshold of x%)
    - The green characteristics must have profound implications for other elements of the business model, i.e. their removal would entail further subsequent adaptations and change the nature of the business model in general.

Commentary: A business model that satisfies rules 1 through 3 but not 4 are considered future candidates for green business models – provided that green characteristics become engrained in the business model and are widely applied.

In conclusion, the proposed taxonomy has a number of conceptual limitations: First, it is not suitable to determine absolute greenness, or to compare non-related business models. Moreover, it neglects efficiency considerations, i.e. how cost-effective environmental improvements are compared to the reference case. The assessment is also limited to quantifiable environmental impacts. For example, food business models that differ in their characteristics regarding animal treatment will be difficult to assess meaningfully based the proposed framework.

Critics may deny that the approach even represents a taxonomy in the narrow sense as taxonomies are supposed to specify homogenous groups. For instance, taxonomies of animals distinguish between groups like fish and mammals based on a number of shared commonalities (e.g., von Linné 1806). The proposed taxonomy of green business models, however, groups otherwise diverse business models based only on their degree of greenness. This, one could argue, is equivalent to classifying animals according to their suitability as foods: some are poisonous, some are edible, and some are delicious.

In summary, a *green business model* can be defined as

> *a business model that represents a significant improvement (discontinuous leap) in overall environmental performance relating to its entire value chain system vis-à-vis that of conventional business models (i.e., the reference case). This improvement is directly attributable to the business model through the alternative design and configuration of business model elements.*

5.5 Green Business Model Prototypes

The means that qualify business models as green are not completely arbitrary. Green business models share common mechanisms to improve environmental performance (see 3.3.3) and capture value (see 3.2.3.1). Often, several of these mechanisms are combined. They include:

- Environmental performance mechanisms: resource efficiency; renewable inputs, low pollution, smart need satisfaction, and sufficiency.
- Value capture mechanisms: cost, quality or reliability, innovativeness, design and style, health, ethics, and political support.

When combining mechanisms of the two value perspectives, prototypical business models can be derived. However, not all realisations will actually be of type 2 (or even type 3) as defined above. Table 5.2 provides a non-exhaustive list of (sometimes overlapping) prototypical business models with green potential. These also cover the business models presented in chapter 7 and others mentioned throughout the book.

Table 5.2 Examples of prototypical business models with green potential

Type	Description	Examples
Low pollution models	▪ Pollution is either prevented or reduced through end-of-pipe solutions ▪ The former can also save cost, while the latter is often reliant on government incentives or other selling propositions	Carbon capture and storage, green chemistry (use of less toxic materials), electric cars
Low waste models	▪ Production processes are made more resource efficient by reducing waste, reusing, or recycling	Remanufacturing of photocopiers, electronic waste recycling
Dematerialisation models	▪ A physical product or process is replaced by a digital offering, thereby saving resources	Video conferencing, paperless office solutions
Smart models	▪ Smart, usually IT-enabled processes allow better coordinated or tailored need fulfilment	Smart metering, virtual power plants, smart traffic systems
Servicising models (product-to-service)	▪ Instead of buying a product, customers buy a function ▪ Environmental benefits from fulfilling customer needs more efficiently, e.g. by sharing physical units	Car sharing, leasing of construction tools
Performance contracting	▪ Guaranteed energy or resource savings ▪ No up-front capital needed, providers benefit from savings	Energy performance contracting for buildings
Renewable models	▪ Use of renewable energy sources (solar, wind, etc.) and materials	Green chemistry (renewable feedstock), renewable energy
Eco consulting models	▪ Firms offer their expertise in environmental protection to other firms	Green logistics services, eco-efficiency services
Base-of-the-pyramid models	▪ Needs of customers at the base of the pyramid are addressed by using adjusted product designs and delivery methods ▪ Unsustainable need fulfilment practices can be replaced	Off-grid, solar-powered lanterns
Do-good models	▪ Customers appreciate ethical conduct with respect to humans, animals and natural resources ▪ Price premiums or functional performance weaknesses	Fair trade, sustainable fishing, socially responsible investing
Health models	▪ Products contain no or less elements that are damaging to human health	Organic food and clothes, green household cleaners
Green meta models	▪ Corporate programmes that foster the creation of green business models, products, or business practices ▪ Include environmental targets, incentives, and initiatives	GE Ecomagination

So far, sustainability in business, business models as a unit of analysis for management science and their intersection – green business models – have been covered. The theoretical part concludes with chapter 6: organisations, change, and innovation.

Chapter 6
Organisations, Change, and Innovation

The body of literature on organisations, change, and innovation is vast and widely ramified. Books and journal articles have been written on almost all conceivable aspects of the topic, and underlying theories are heterogeneous and sometimes even conflicting (Scherer 2002, 2). Hence, the following introduction cannot be exhaustive, nor can it capture every nuance of the various theories I will touch upon. Rather, I will provide a bird's eye view on some key concepts to set the stage and select a few works that I consider useful for the development and application of a management approach to Green Business Model Transformation. By all means, the theories, concepts and frameworks discussed in this chapter are not the only ones potentially suitable for this task and should therefore be considered a suggestion rather than a prescription.

6.1 Organisation Theory

Because of the above-mentioned heterogeneity and the fact that various other disciplines feed into the body of organisation theory literature, a common classification of the various schools – let alone the description of a *unified theory of organisations* seems unrealistic (Scherer 2002). Scott (2002) proposes to view organisations through a five-element model that includes its social structure, participants, goals, technology, and environment. Moreover, he structures organisational studies along three main views of organisations: as *rational*, *natural*, and *open systems*. Scott's classification by type of system view is useful to complement the systems perspective that is integral to the business model concept and will thus also be used to structure the following overview on some of the most influential academic schools.

6.1.1 Organisations as Rational Systems

The *rational systems* view was the first scientific basis to managing organisations in a structured way. It followed a mechanistic understanding of organisations and their constituents that propagated the improvement of their functioning by applying certain "scientific" principles.

A. Sommer: Managing Green Business Model Transformations, SPLCEM, pp. 109–168.
springerlink.com

The following definition of organisations is compatible with the rational systems view:

> *"Organizations are collectivities oriented to the pursuit of relatively specific goals and exhibiting relatively highly formalized social structures."* (Scott 1992, 23)

Under the rational systems view Scott (1992, 34ff.) lists the classic theories of Taylor, Fayol, Weber, and Simon that are briefly outlined in the following.

Taylor is regarded as the founding father of scientific management. In his monograph "The Principles of Scientific Management" he propagates to replace the prevalent – in Taylor's view amateurish – management practices that use rules of thumb with scientific methods. Taylor regards organisations as big machines that can be run much more efficiently through management that is a "true science, resting upon clearly defined laws, rules, and principles" (Taylor 1911, 7).

Fayol's theory of business administration aims to provide principles that can serve for the purpose of rationalisation of the firm. Like Taylor, Fayol propagates a very formal approach to managing organisations. Fayol developed various formal typologies on business activities and entities, including five primary functions of management (planning, organizing, commanding, coordinating, and controlling), and 14 principles of administration (ranging from division of work to esprit de corps) (Fayol 1929). In contrast to Taylor, he suggests a top-down approach through coordination and specialisation (Scott 1992, 36).

Weber's contemporaneous but independent theory of bureaucracy represents only one part in his comprehensive, posthumously published work "Wirtschaft und Gesellschaft" ("Economy and Society") (Weber 2002, first published 1921/22). It includes a description of *ideal-type* (i.e., pure) bureaucracies that he considered to be not necessarily likable, but more rational and efficient than other (older) forms of organisations which rely on charismatic or traditional rather than legal authority (Scott 1992, 38ff.).

The ideas described above have greatly increased efficiency of economic activity and remain influential to this day. At the same time, however, the focus on decomposition and specialisation hinders firms to address sustainability issues holistically.

Herbert Simon is the founder of the theory of administrative behaviour (Simon 1947) and can be considered the pioneer of the modern understanding of organisational decision-making. Simon was an outspoken critic of the earlier attempts of administrative theorists to define general management principles. He rejected the unrealistic assumption of the *economic man* and instead proposed to consider the more human *administrative man* (Scott 1992, 45) who is subject to the concept of *bounded rationality*. This means that although individuals try to act rationally, their decisions may not be consistent with objective rationality because of incomplete information, uncertainty of the assessment of future events, and the limited choice of options (Berger & Bernhard-Mehlich 2002, 140f.). Bounded rationality can explain many of the serious challenges arising from Green Business Model Transformations.

6.1.2 Organisations as Natural Systems

The second broad view that Scott identifies is to consider organisations as *natural systems*. In contrast to the rational view, the natural view considers social systems as evolved entities with an informal structure that has not been designed. Moreover, the natural systems view differs in its consideration of the nature and complexity of organisational goals, i.e. that stated and actual goals are not necessarily the same, and that – in addition to their *output* goals – "each [organisation] must expend energies maintaining itself" by pursuing *support* goals (Scott 1992, 52). Furthermore, Scott stresses that natural systems analysts emphasise the primacy of the goal of survival over all other goals that an organisation may have.

The following definition of organisations is compatible with the natural systems view:

> *"Organizations are collectivities whose participants share a common interest in the survival of the system and who engage in collective activities, informally structured, to secure this end."* (Scott 1992, 25)

Scott reviews several natural systems schools, including the following three: Mayo and the human relations school, Barnard's conception of cooperative systems, and Selznick's institutional approach (Scott 1992, 56ff.).

Mayo (1945) presented one of the most influential interpretations of the famous Hawthorne experiments[8] which implied that individuals and their motives were much more complex than the rational systems view suggested. The emerging human relations school thus puts the human being at the centre of reflections and differs from the rational systems view in some of its basic assumptions, namely that (Scott 1992, 59f.; based on McGregor 1960):

- By nature, most people do not dislike work
- Other means exist to reach organisational goals than control and coercion
- The most effective rewards address people's ego and desire for self-actualisation; people do not generally lack ambition, avoid responsibility, or seek primarily for security

The human relations school with its focus on the individual had significant influence on the topic of change that is discussed in more detail in chapter 6.2 and the following. However, despite its name, the human relations school has been criticised as "cow sociality" that is geared towards making workers more

[8] The Hawthorne experiments aimed to find the optimal degree of illumination to maximise the productivity of workers. However, during the experiments, more, less and even no change to lighting each had a positive effect on productivity – supposedly because workers felt motivated simply due to the attention given to them (Mayo 1945).

productive in a subtle, incapacitating way – just like happy cows were assumed to produce more milk. Maybe more severely, many of the approach's promised positive effects on organisations could so far not be proven conclusively by empirical evidence (Scott 1992, 61).

Barnard's theory of organisations (Barnard 1968, first published 1938) considers them to be systems of cooperation that integrate a purpose with the individual motives of its members. Barnard's ideas have much in common with the rational view, but are set apart by his notion of "the nonmaterial, informal, interpersonal, and, indeed, moral basis of cooperation" (Scott 1992, 63). According to Barnard, organisations consist of three elements: communication, willingness to serve, and common purpose. Survival of organisations – which, according to Barnard, is rare to last long, though – depends on its effectiveness (i.e., the accomplishment of its purpose) and efficiency (i.e., the satisfaction of individual motives exceeds the sacrifices) (Barnard 1968, 82ff.). Since Green Business Model Transformations also serve greater ends than just financial success, this insight can be used to create a level of commitment that would be unimaginable for most other business transformations like mergers or offshoring programmes.

Selznick views organisations as "cooperative systems, adaptive social structures, made up of interacting individuals, sub-groups, and informal plus formal relationships" (Selznick 1948, 32). Selznick believes that some critical decisions change the character of organisations and make them unique; they become *institutionalized* (Selznick 1957). Individuals will resist being reduced to de-personalised *roles* and instead continue to act as *wholes*. Hence, due to individuals' goals, delegation will not lead (solely) to the desired results through the application of formal control mechanisms alone. Selznick suggests the use of *cooptation* as a means of informal leadership to address disturbances between consent and control, and to react to destabilising pressures of powerful individuals and interest groups (Selznick 1948). Smart use of a broad repertoire of behaviours to deal with resistance to change is often essential for the success of large transformations.

6.1.3 Organisations as Open Systems

Boulding (1956) introduced a hierarchy of systems based on their complexity. Organisations reside at the eighth of nine levels of complexity[9]: 1 through 3 are

[9] The nine levels of system complexity are (summarised based on Boulding 1956, 202ff.): 1. Frameworks (e.g., the patterns of electrons around a nucleus), 2. Clockworks (systems with predetermined, necessary motions like the solar system), 3. Thermostat (control mechanism based on the transmission and interpretation of information), 4. Cell (open, self-maintaining systems, i.e. simplest parts of life), 5. Plant (differentiated and mutually dependent parts, determined by a genotype, or blueprint), 6. Animal (increased mobility, teleological behaviour, and self-awareness), 7. Human (self-consciousness, perception of time and history, ability to interpret symbols), 8. Social organisations (includes roles, value systems, etc.), 9. Transcendental (unknowables) .

physical systems, 4 through 6 biological ones, and 7 and 8 are human social systems. However, Boulding remarked that the theory of his time could just deal with level two or three. With the adoption of the *open systems* view, organisation theory has reached level 4 (Scott 1992, 78f.). This means that organisations are seen as self-maintaining. Through the injection of energy (i.e., the throughput of resources) from its environment, organisations can evolve further and avoid their otherwise inevitable degradation (analogous to the second law of thermodynamics in physics). As such, the common perception that an organisation must protect itself from the influence of its environment is considered misleading – it is this interrelation that enables survival in the first place (Scott 1992, 83ff.). Consequently, other than the closed systems perspectives described before, the open systems view concentrates on the relationship of organisations with their environment. Yet, the open systems view can be combined with both the rational and the natural systems perspective.

The following definition of organisations is compatible with the open systems view:

> *"Organisations are systems of interdependent activities linking shifting coalitions of participants; the systems are embedded in – dependent on continuing exchanges with and constituted by – the environments in which they operate."* (Scott 1992, 25)

The open systems view highlights two further characteristics of organisations:

First, the fact that they are (often) *loosely coupled:* To a large extent, system elements can act independently of other elements – especially if their hierarchical proximity is low. One the one hand, this fact makes individual organisational units capable of acting and enables the organisation as a whole to be adaptive (Scott 1992, 82). On the other hand, loose coupling can also result in the often criticised practice of "departmental thinking", which is particularly prevalent in large corporations (and a problem for corporate transformations). Loose coupling is even present on the level of an individual's goals and actions (March & Olsen 1976), which can have profound impact on change management and resistance to change as it reduces predictability of behaviour.

Second, they are embedded in and contain (an abstract form of) *hierarchies* (see Scott 1992, 85f.). That is, when undertaking organisational change one needs to consider change at multiple levels: Change does not only happen intra-organisational, it can also take place on the level of the industry sector, the broad economic and political context, on the nation state level, or even on a global competition level (Pettigrew et al. 2001, 699). Within organisations, levels range from the organisation as a whole to divisions, departments, etc. down to individuals and their roles (Scott 1992, 85). In conclusion, organisations as open systems can be described as a "loosely linked coalition of shifting interest groups" (Scott 1992, 82; based on Cyert & March 1963; Pfeffer & Salancik 1978).

A number of schools from the open systems view are of interest for this work. Early ones include *systems design*, *contingency theory*, and Weick's *social-psychological model of organising* (see Scott 1992, 86ff.).

Systems design is based on general systems theory as already covered in chapter 4.5. It does not only seek to describe and understand organisations, as Scott remarks, but also aims to change and improve them. As explained before, the approach follows the understanding that systems are not reducible to its components. The analytical complexity of looking holistically at organisations is often offset by limiting the analysis to the relevant parts only. Consequently, systems design looks at inputs and outputs of elements but often treats the elements themselves as black boxes. Systems design also has well-known limitations: Analysts must be careful in applying analogies from less complex systems correctly and always be aware that it is not possible to predict complex systems' behaviour exhaustively (Scott 1992, 86ff.). This implies that the management framework in chapter 9 should allow for flexibility to cope with surprises.

Contingency theory can be viewed as a branch of systems design that is described by the following three assumptions (Galbraith 1973, 2; Scott 1992, 89):

- "There is no one best way to organize."
- "Any way of organizing is not equally effective."
- "The best way to organize depends on the nature of the environment to which the organization relates."

Lawrence & Lorsch (1967) argue that organisations face different demands from their respective environments, depending on the uncertainty and tempo of change. Because organisational subunits operate in different sub-environments, they need to be adapted independently (differentiated). For the success of the organisation as a whole, subunits must be well integrated as well – which, obviously, is an antagonistic goal. Naturally, large corporate transformations trigger disputes over where the right balance lies, both for individual subunits and the company as a whole.

Various aspects of contingency theory have been criticised. One argument of specific interest in the context of this work is that contingency theory assumes that the external environment is beyond the company's sphere of influence (see Kieser 2002, 185ff.). However, business model transformations almost always change not only the organisation itself, but also influence its customers, partners, or competitors – potentially even the larger business environment of the company. In fact, seeking for planned change of the external environment can offer much more rewarding results than just adapting the organisation to the external environment.

In his model of the *social psychology of organizing*, Weick (1979, first published 1969) argues that, instead of examining *organisations* and their structures, one should look at the process of *organising*. He believes that human beings organise in response to the uncertainty they are confronted with. They do so by the means of *interlocked behaviours* that are developed and maintained between actors. Weick distinguishes the stages of *enactment*, *selection*, and *retention*. Enactment emphasises both the subjective perception as well as the individual exercise of influence on the environment. Selection is based either on *rules*, or, in the case of less standardised circumstances, on *communication-behaviour cycles* in which information is exchanged and interpretations are formed. The developed responses can then be repeated in comparable situations and become retained (Scott 1992, 91). Weick's view of organisations implies that individuals' behaviours are largely unpredictable and that its interpretation ("sense-making") frequently occurs only retrospectively (Weick 1979; 1995).

To conclude the overview on relevant organisation research, another three relevant open system theories that were introduced after 1970 are summarised in the following: *agency theory*, *population ecology*, and *resource dependence* (see Scott 1992, 105ff.).

Agency theory (e.g., Alchian & Demsetz 1972; Jensen & Meckling 1976; Eisenhardt 1989) deals with the question how it can be ensured that decision makers act in a way that is in line with an organisation's goals. For this purpose, agency theory considers the structure of principal-agent relationships, in which the principal (e.g., an executive manager) needs the agent (e.g., a subordinate manager or worker) to carry out activities to achieve a desired outcome. Both parties are assumed to act in their own personal interest – which may not be aligned. Incentive systems are used to make the agent act in a way desired by the principal. Due to the problem of information asymmetry, these incentive systems should ideally be designed to ensure that the agent is incentivised to act congruently to the principal's interests without the need for costly control systems (Scott 1992, 105). For example, a company wishing to build a green product portfolio may introduce a guideline that for each division at least 50 percent of newly developed products need to fulfil certain environmental criteria. However, since these criteria may be regarded as a burden that puts additional strains on development budgets, the divisions may come up with unimaginative, hastily developed modifications of conventional products that offer inferior value propositions. Alternatively, the corporate centre could decide to provide a special budget for the development of innovative green products. Because the divisions now compete for additional resources, they genuinely *want* to develop new green products. Then, the results of such efforts can be expected to be much more inspired.

Organisation theorists have provided many useful insights that can explain the behaviour of individuals and groups. Some of these perspectives – like scientific management and the human relations school – represent very different underlying philosophies and thus imply different courses of action. Theorists have also argued about why some firms survive and others do not. The view adopted on this question has profound consequences for the courses of action that can sensibly be recommended in any management framework. It is therefore necessary to look into the most important of these rivalling perspectives, starting with *population ecology* (Hannan & Freeman 1977; Aldrich 2007, first published 1979).

Population ecology has its roots in the work on natural selection by Darwin. It aims to explain why certain types of organisations survive and multiply while others do not. The underlying mechanism is assumed to be selection of the fittest: A population of organisations produces variety, the fittest mutants are selected and reproduced or duplicated (retention). Consequently, changes of populations of organisations are not assumed to be primarily the result of adaptation of individual entities, but of their replacement. Moreover, like in biological evolution, change is not equivalent to progress (Scott 1992, 113f.). Although the common unit of selection used in population ecology theory is the organisation as a whole, other levels like routines and competences are also used (Aldrich 2007, xxi). On this level, competencies of successful companies have a higher chance of selection and spread across a population of organisations through mechanisms like best practice sharing or consulting projects (Kieser & Woywode 2002, 259).

In contrast to population ecology, the *resource dependence* approach (e.g., Pfeffer & Salancik 1978) assumes that organisations can use mainly adaptation in order to survive. They do so by engaging in exchanges with their environment to acquire resources. Despite the dependence on resources, organisations are considered active entities that employ strategies to manage not only the organisation, but also the environment (Scott 1992, 114f.).

Even more degrees of freedom for managers are propagated by *strategic contingency* theorists (e.g., Child 1972) that stress the existence of strategic choice. Decision-makers face environmental constraints, but their choices are not determined by the environment (Scott 1992, 112) – they may even manipulate it (Child 1972).

Modern institutionalists (e.g., Meyer & Rowan 1977; DiMaggio & Powell 1983) would argue that these constraints mainly stem from pressures by social and political forces rather the dependence on efficient usage of resources (Zucker 1987). Hence, organisations are assumed to imitate practices with known legitimacy, whether they are economically efficient or not (Kanter et al. 1992, 29ff.).

6.1.4 Complexity Theory and Organisations

As noted above, organisation science has reached level 4 of Boulding's hierarchy of complexity by applying open system models, still falling short of the actual level of complexity of the object under study. Thus, researchers have attempted to use insights from complexity theory to better understand organisations, e.g. by viewing them as *complex adaptive systems* (e.g., Anderson 1999). There is no consensus yet in what way complexity theory can and should be applied to organisations (Stacey et al. 2002). Nevertheless, Anderson (1999, 217f.) has put forward six insights that are useful for organisation science, and which he considers to be well-established scientifically:

(1) Many dynamical systems do not reach a state of equilibrium – neither a static nor cyclical one (see Dooley & Van de Ven 1999). (2) Seemingly *random* processes may be *chaotic* instead. *Coloured noise* (constrained randomness) and

white noise (randomness) differ from chaotic systems in that the latter in fact imply the presence of a high degree of order and deterministic pattern – which is obscured by their non-linearity, though. As such, they are bound to limited patterns of "fractal" behaviour by revolving around a *strange attractor*; yet, specific occurrences still appear irregular and remain unpredictable (Stacey et al. 2002, 87ff.). (3) Small differences in initial conditions can have a large impact over time. This phenomenon of chaos theory is often referred to as the *butterfly effect*, where the wing flap of a butterfly can later cause a storm on the other side of the planet. (4) A holistic (instead of reductionist) view is often necessary to understand organisations. (5) Agents that follow a few simple rules can produce complex patterns of behaviour. (6) Complex systems that start from a random state tend to move towards order, i.e. they are *self-organising* (Kauffman 1993).

Scholars of complex adaptive systems have suggested that organisations should exist at the *edge of chaos*, i.e. in between a state of stable equilibrium and chaos. Stable organisations are only capable of incremental changes. That is, on their *fitness landscape*[10], they are capable of hill-climbing but cannot leave a local peak (stable attractor). In contrast, organisations in a chaotic state can move to other (potentially higher) peaks but are unable to stay at them. However, at the edge of chaos, organisations undergo frequent incremental and infrequent radical change – a condition that is considered ideal for survival (Kauffman 1993; Anderson 1999, 223f.). The shape of the fitness landscape is determined by the connections (i.e. conflicting constraints) within and between organisations in such a way that more connectivity produces more *rugged* landscapes (which have more but lower peaks) (Kauffman 1993; Levinthal 1997; Levinthal & Warglien 1999; McKelvey 1999; Stacey et al. 2002, 110ff.). Agents (or groups of agents), unable to foresee the systemic consequences of their actions, coevolve by trying to improve their own, dynamic fitness landscapes, thus creating emerging patterns of self-organisation (Anderson 1999, 223).

I argue that many companies have enjoyed a selection advantage in the past as they depleted resources and treated stakeholders in an unsustainable way. I further argue that the underlying fitness function is in the process of changing, but has not yet led the mainstream population of companies evolve towards more sustainable attractors.

The brief synopsis of organisation theory presented above inevitably omits many important contributions to the topic. Nevertheless, the different perspectives presented already point to important aspects to be considered when developing a framework for management practice. Moreover, the view of organisations as systems provides a bridge between the business model concept and the transformation aspect: the critical subject of organisational change.

---

[10] The definition of "fitness" is non-trivial. It may include various performance measures (potentially associated with a trade-off), but for the sake of simplicity the metaphor it is usually assumed to be a single (or aggregated) measure like profitability (Anderson 1999, 225).

6.2 Theoretical Perspectives on Organisational Change

Some aspects of organisational change have already been touched upon briefly in the course of the introduction of organisation theory. Although representing only a part of the broader topic of organisations, organisational change is of high relevance for management practice and has thus led to the creation of vast amounts of literature on the topic. In order to filter out the most relevant aspects I will first present the various dimensions, characteristics, and types of change relevant to the scope of this work.

6.2.1 What Is Organisational Change?

The phenomenon of change is one of the oldest known themes in the history of philosophy: The Chinese ancient text *Yijing* ("Book of changes"), one of the five classics of Confucianism, views change as the result of "the dynamic balance of opposites", *yin* and *yang* (Wikipedia 2009b). Change is considered the continuous and inevitable process of movement (see Wilhelm & Baynes 1967, lv). In Western culture, the Greek philosopher Heraclitus is well known for his aphorism "Nothing endures but change." (quoted in Kanter et al. 1992, 9)

Unlike this philosophical perspective on change, in today's business world *organisational change* is typically seen as a deliberate, planned transition from some original state to a new one (Kanter et al. 1992, 9). In most cases, organisational change is associated with explicit goals whose achievement is later reviewed once the change process (or phases thereof) is completed (Burke 2002, 2f.). This is much in line with Lewin's (1951) "unfreeze-change-refreeze"-model of change. Although criticised for being overly simplistic, the discrete view on change is still prevalent in current management practice (Kanter et al. 1992, 9f.). An overview of more elaborate theories on organisational change is presented in chapter 6.2.3. Before, some common typologies of organisational change are described.

6.2.2 Typologies of Organisational Change

For the preliminary scoping of this work in chapter 2.3, I have used a combination of the nature and scale of change, dubbed "radicalness" (see figure 2.3). Weick & Quinn (1999) assert that the nature of change, i.e. whether it is continuous or episodic, is a widely used distinction in change research. Various other labels have been used to describe this dichotomy, including evolutionary vs. revolutionary; incremental vs. transformational; first vs. second order change. Other dimensions that have been used to categorise change are the style of change leadership and people involvement, the unit of change, the content of change, or the firm attitude towards change.

For example, Dunphy & Stace (1988) propose a process model of change based on the nature of change and leadership style dimensions to define four types of change: the combination of incremental vs. transformative and collaborative vs. coercive change. They also provide a more detailed version of that typology, now consisting of four leadership styles (collaborative, consultative, directive and coercive) and a range of four degrees of scale from fine-tuning to corporate transformation (Dunphy & Stace 2000, 381ff.).

Brown & Eisenhardt (1998) relate organisational change to the change of the external environment; the strategic challenge is to manage change in this respect. This can be done on three "levels": reactive, anticipating, or leading. Essentially, the three levels of change relate to distinct types of goals. Reactive change aims to counter competitor's actions or to exploit observable changes of the environment (e.g., policy changes or changed customer demands). Anticipating change aims at getting an edge vis-à-vis competitors by acting before the external change occurs or becomes evident. And finally, leading change means to start actively influencing the external environment itself in a favourable way, i.e. shaping the external environment of others (Brown & Eisenhardt 1998, 4f.). Green Business Model Transformations require a firm to adopt a proactive approach, be it out of a deliberate desire to lead or because the firm's survival is threatened.

Kanter et al. (1992, 14f.) have identified three "kinds of movement" (macro-evolutionary, micro-evolutionary, and revolutionary change) and further distinguish three "forms of change": identity, coordination, and control changes. Identity changes redefine the relationships of the organisation with its environment or - in the extreme case – completely change the nature of the organisation's business. Roughly relating to the micro-evolutionary type of motion, coordination changes refer to a "deliberate reshaping or revitalizing" of the organisation's internal working. And finally, control changes are triggered by shifts in ownership or governance, i.e. "makeover through takeover" (Kanter et al. 1992, 15). Green Business Model Transformations will usually entail identity changes, and definitely require coordination changes to make the new business model work.

As a last example, Van de Ven & Poole (1995) use a two dimensional typology based on the unit and mode of change. The unit of change differentiates between change that focuses on a single entity and change that relates to the interactions between multiple entities. The mode of change distinguishes between prescribed and constructive change, i.e. whether the change occurs within a given frame and sequence in a predictable way, or is created by breaking the existing framework. The typology results in four basic types of change generating mechanisms, or process theories, that will be used to present a systematic overview on theories of change in the following.

6.2.3 Four Generic Types of Process Theories for Change

Van de Ven & Poole (1995) have made a convincing attempt to systemise the various organisational change theories based on four generic types of process theories (primitives) for change. The four theory types are considered "motors" that explain why and how the change process occurs; their matching with the various change theories helps to comprehend and compare them (and to scrutinise further research in the field, which is not part of this work, though). The four "motors" are (Van de Ven & Poole 1995, 512ff.):

1. *Life cycle theory:* Life cycle theory assumes that organisations undergo a set of cumulative stages of organic growth, starting with the organisation's birth and ending with its death. This means that the sequence of stages is fixed based on some imminent programme or rules. Moreover, the characteristics developed by organisations – at least in their basic forms – are thought to be predetermined based on the respective life cycle stage (Van de Ven & Poole 1995, 513ff.). There are various models of organisational life cycles. Quinn & Cameron (1983) developed a summary model with four stages based on nine earlier models: the *entrepreneurial stage*, the *collectivity stage*, the *formalisation and control stage*, and the *elaboration of structure stage*. Since this work specifically focuses on established and mature corporations, respective insights gained from life cycle change models are given special attention in a separate section (6.4.3) within innovation – a capability that is often considered a particular challenge for mature organisations.
2. *Teleological theory:* This school of thought follows the understanding that all movement of organisations is guided by its purpose and goals (and adaptations of the same). The purposeful planning character of organisations highlighted by teleological theory has its roots in the rational systems view. Van de Ven & Poole (1995, 514) name Weber and Simon as two of its pioneers. However, the concept of goal enactment (see Weick 1979) through consensus on means and cooperation (see also March & Olsen 1976) reveals that teleological change is also consistent with the natural systems view (see Scott 1992, 102ff.).

   In contrast to life cycle theory, teleological theory propagates much higher degrees of freedom for which types of changes an organisation can (and should) undergo at a given point in time. Only the external environment and the availability of resources constrain an organisation's change efforts. These can thus be assessed based on the organisation's progression towards the goal or envisioned end state; however, the theory does *not* assume that there is only one valid path the organisation must follow (Van de Ven & Poole 1995, 516f.).
3. *Dialectical theory*: Whereas teleological theory emphasises the common purpose of organisations, dialectical theory propagates conflict as its dominating force. Contradictory forces – thesis and antithesis – finally lead to synthesis (which may then replace the thesis). Stability and change are hence results of the balance or unbalance of power. While synthesis can create win-win outcomes, overthrowing the thesis with an antithesis (or the failure of such an attempt) is often a win-lose event. Change is generally considered neutral; it can be for the greater good (but not necessarily) (Van de Ven & Poole 1995, 517).

4. *Evolutionary theory*: In this school of thought, competitive survival is the driving force behind change. It follows a repetitive cycle of probabilistic variation, selection through competition for scarce resources, and retention through forces like inertia and persistence. On the level of competing organisations, the survival of particular ones cannot be predicted, but the general population dynamics can be explained. The theory is also applicable on a sub-organisational level, i.e. to determine how strategy making and organising within organisations works (Van de Ven & Poole 1995, 517ff.).

Figure 6.1 summarises the four process theories on organisational development and change:

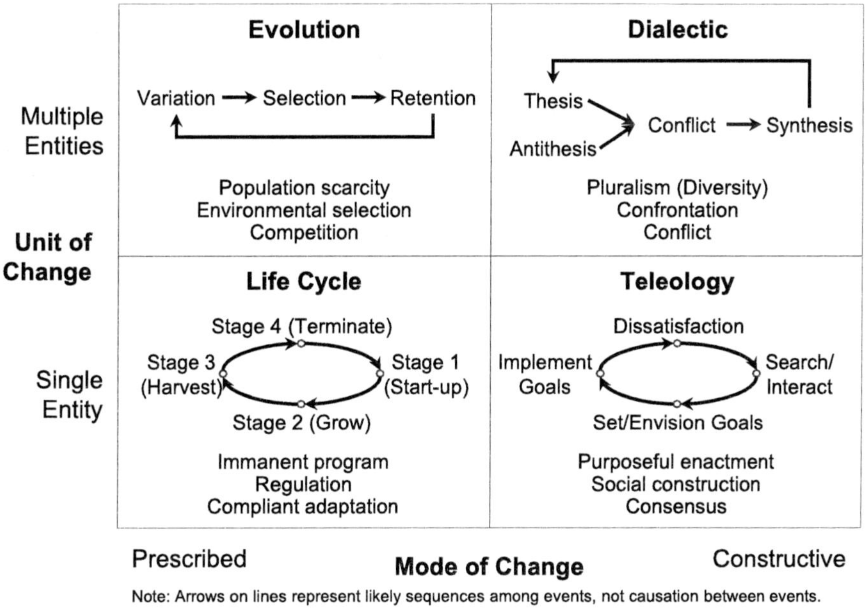

Fig. 6.1 Process theories of organisational development and change (Van de Ven & Poole 1995, 520)

The matrix shows the respective cycles of change events and underlying generating mechanisms of the four process theories. They are classified based on their respective mode and unit of change: Teleological and dialectic theories produce constructive, directly influenceable change, whereas evolutionary and life cycle change follow prescribed patterns. Moreover, evolutionary and dialectic changes depend on multiple entities that interact (compete or antagonise each other). In contrast, life cycle and teleological theories mainly consider a single entity (which is under secondary influence of its environment, though).

Although managers tend to view their change programmes from a teleological perspective, it can be helpful to consider them from the other three theoretical angles as well. Some well-justified change efforts may fail simply because they contradict the life cycle stage of the company: they require a more fundamental organisational renewal in order for its members to *live* the change. Or, instead of coercively trying to force new strategic goals onto the organisation, it might be worthwhile to strive for a mutually beneficial synthesis with the coalition of the "old" organisation. And finally, instead of determining the means of goal fulfilment centrally (and potentially somewhat removed from practice), management may give some room for experimentation accompanied by a clever selection process for new routines and competencies. After all, it will be difficult to conceive the right way of doing things beforehand for such complex efforts like Green Business Model Transformations.

Real-world change will rarely occur as a result of only one of the four change motors; there will likely be overlaps and potentially even interrelations. For example, the implementation step in teleological change may trigger the antithesis in the dialectical change cycle. Change motors may operate at the same time or alternately; they may work on the same level of analysis or be nested within each other (e.g., one motor explains change on a sub-organisational level, the other for the organisation as a whole). Therefore, specific change theories that were created to explain particular change contexts tend to be more complex and incorporate several of the four basic process theories (Van de Ven & Poole 1995, 525ff.).

In the following, I will thus describe three specific change theories that are of interest with respect to Green Business Model Transformations of established companies:

Greiner's model of *organisational growth and crisis states* (Greiner 1998) warns against an approach that ignores the history and specific characteristics of the organisation. This model combines life cycle and dialectic change motors (Van de Ven & Poole 1995, 528). Second, a combination of teleological and evolutionary change, *organisational punctuated equilibrium* (Tushman & Romanelli 1985) is presented as the main metaphor for Green Business Model Transformations on a macro-level. And finally, the *garbage can model* of organisational change (Cohen et al. 1972) demonstrates how change is often *not* orderly and sometimes requires exploiting random processes rather than mechanistic planning. This change model is a special case as it is not considered to revert to any of the four basic change motors (Van de Ven & Poole 1995, 528).

Together with the organisation theories presented in chapter 6.1 (some of which also encompass aspects of change), the three change theories provide a fairly broad coverage of the 16 possible combinations of the four change motors (see table 6.1 for an overview[11]):

---

[11] Van de Ven & Poole (1995, 529) remark that "the authors of these exemplary theories or models may not agree with our classification, because they did not have our framework in mind when they developed their theories."

Table 6.1 Specific change theories and their change motors (based on Van de Ven & Poole 1995, 528)

	Incorporated motors (generic change mechanisms)			
Specific change theories	**Life cycle:** "Immanent program"	**Teleology:** "Purposeful enactment"	**Dialectic:** "Conflict and synthesis"	**Evolution:** "Competitive selection"
Teleology (March & Simon 1958), see also 6.1	-	✓	-	-
Population ecology (Hannan & Freeman 1977), see also 6.1	-	-	-	✓
Org. growth and crisis stages (Greiner 1998)	✓	-	✓	-
Adaptation-selection models (Aldrich 2007), see also 6.1	✓	-	-	✓
Org. punctuated equilibrium (Tushman & Romanelli 1985)	-	✓	-	✓
Social psychology of organizing (Weick 1979), see also 6.1	✓	✓	-	✓
Garbage can model (Cohen et al. 1972)	-	-	-	-

6.2.4 Organisational Growth and Crisis Stages

Greiner (1998, first published 1972) proposed a model of organisational development that is based on alternating periods of growth and crisis. The organisation thereby passes through five life cycle phases of 3 to 15 years that mainly depend on the maturity and age of an organisation as described in table 6.2. A high growth industry thus results in shorter periods of stability between the crises. Moreover, a high-profit industry can delay the company entering a crisis stage, which also tends to be softer compared to those in poorer industries. Crises are generally caused as old solutions become inadequate for the new problems (Greiner 1998, 58). On the one hand, an (upcoming) crisis may be used to redirect the sense of urgency for change towards a Green Business Model Transformation. On the other hand, dysfunctional organisational structures and political fights during crisis states can jeopardise a clean transformation process. In any case, change managers should consider the organisation's specific life cycle characteristics. While infant companies in the first phase are not in scope of this work, phase-2 firms need to make sure that they do not follow too much of a top-down approach, thereby neglecting many important details and losing commitment. Firms in the delegation phase might in fact encounter the greatest challenges as Green Business Model Transformations tend to require extensive coordination (if not planning). Companies in the coordination phase, in contrast, need to ensure that things get done, that the transformation does not get stuck in the bureaucratic machinery. Greiner's description implies that firms in the collaboration phase should be the easiest to transform. However, one may argue, that the most mature organisations also carry the most preconceptions, which are hard to come by because of their subliminal nature.

Table 6.2 Greiner's organisational growth and crisis stages (based on Greiner 1998)

Phase	Description	Crisis
1. Creativity	▪ Informal leadership and communication ▪ Growth leads to increasing inefficiencies in operations ▪ Usual solution: installing professional business leadership that gives the company direction	Leadership
2. Direction	▪ New functional organisational structure including specialisation and formal working procedures ▪ Increasing complexity and diversity eventually makes the directive leadership inappropriate ▪ Usual solution: Stronger delegation	Autonomy
3. Delegation	▪ Newly found motivation of lower managers in profit centres once senior managers let go and lower managers learn to handle the amount of responsibility ▪ Autonomy of the sub-units leads to a crisis of control; attempts to reinstall central management practices usually fail ▪ Usual solution: special coordination techniques	Control
4. Coordination	▪ New formal planning procedures and company-wide programmes of control and review; centralisation of certain technical functions; careful management of capital investment ▪ Headquarter and operating units lose touch and start to distrust each other ▪ Usual solution: replacing formal, bureaucratic systems with strong interpersonal collaboration	Red tape
5. Collaboration	▪ Team-based problem solving, interdisciplinary cooperation, and a matrix-type structure (mainly for senior management)	?

Greiner's growth and crisis states are rather simplistic and should be considered a stimulus for thought rather than exact science. In a commentary 25 years after the original publication of the model, Greiner speculates about what the fifth crisis may turn out to be. Other than in the original article, he identifies the lack of an internal solution to growth, and brings up a potential sixth phase of networked organisations (Greiner 1998, 65).

It seems that many organisations today enter a crisis of integrating growth and sustainability. "Sustainability" could thus provide an alternative interpretation for a sixth phase to be added.

6.2.5 Punctuated Equilibrium Theory

Like population ecology, the *punctuated equilibrium* theory is a paradigm that has its origins in Darwin's theory of evolution. However, Gould & Eldredge (1972; 1977) presented a conception of evolution that disavows the original notion of incremental, cumulative change. Instead, they suggested a model that comprises "relatively long periods of stability (equilibrium), punctuated by compact periods of qualitative, metamorphic change (revolution)" (Gersick 1991, 12). Punctuated equilibrium has served as a guiding metaphor in many fields as diverse as adult development and the history of science. In the field of organisational change it has been introduced by Tushman & Romanelli (1985). Conceptually, there are three main components that form the basis of the theory: *deep structure*, *equilibrium periods*, and *revolutionary periods* (Gersick 1991, 13).

Gersick (1991, 16) compares deep structures with "the design of the playing field and the rules of the game". This descriptions sound very similar to some definitions of business models. However, Tushman & Romanelli (1985, 176ff.) go beyond that scope and consider five facets (which Gersick identifies as deep structures): (1) core beliefs and values, (2) strategic aspects related to products, markets, technology and competitive timing, (3) the distribution of power, (4) the organisational structure, and (5) the control systems. Hence, while the business model is an important part of deep structure, it also includes the organisation's culture, strategy, and its implicit and explicit structure.

During *equilibrium periods*, the deep structure remains untouched; instead, firms become more internally consistent and more skilled in what they do. As deep structures tighten and increase interdependencies of their parts, inertia grows (Tushman & Romanelli 1985; Gersick 1991, 16ff.).

It is during *revolutionary periods* when *deep structures* are fundamentally realigned. In the most radical cases, even core beliefs and values are changed, something that analysts have warned against (e.g., Collins & Porras 1996). The transition from system equilibrium to revolution originates from two possible sources: Internal changes that lead to misalignment internally or with the environment, or "environmental changes that threaten the system's ability to obtain resources" (Gersick 1991, 21). For example, customer demands can change in a way the firm is not prepared for. Internal change could relate to organisational growth which, at some point, destabilises the system. It should be noted that, while the growth argument sounds very much like a life cycle effect, punctuated equilibrium does not propagate directed change towards an end state of organisations. Theorists stress the point that dismantling the old deep structures and building new ones are two separate tasks. Therefore a *need* for change is not sufficient to trigger a revolution to occur. Misalignment and failure merely set the stage and provide energy which then needs to be channelled (Gersick 1991, 21f.), e.g. towards a Green Business Model Transformation.

Punctuated equilibrium is the main metaphor used to explain and justify Green Business Model Transformations: Companies have long optimised their operations incrementally based on a stable set of goals that reflected fitness of organisations in the marketplace. However, this fitness function has begun to change in recent years when sustainability gained prominence and stakeholders like NGOs discovered additional means of influence through the internet. Sustainability has become a pivotal factor in the competitiveness of many firms (Haanœs et al. 2011), in some cases due to new market demands, in some cases due to novel, more effective and efficient ways to create and deliver value (Senge et al. 2008, 121ff.). These new circumstances make organisations ripe for radical teleological change – as there is a clear need for change.

However, the punctuated equilibrium theory says little about the transition from equilibrium to revolution (Van de Ven & Poole 1995, 531). The topic of organisational inertia and ways to overcome it is the crux of Green Business Model Transformations; hence it is enlarged upon in a separate section (6.3.3).

6.2.6 The Garbage Can Model

The *garbage can model* is a special case: according to Van de Ven & Poole (1995) it does not make use of any of the four change motors. It is meant to explain organisational change in highly ambiguous settings, referred to as *organised anarchies* (Cohen et al. 1972). They are characterised by (1) problematic (inconsistent and ill-defined) preferences of decision makers, (2) unclear technology that is associated with a poor understanding of decision processes, and (3) fluid participation in which active decision makers change and devote varying amounts of energy (Cohen et al. 1972). The conditions above are frequently present in organisations like universities, but also in for-profit corporations. In fact, an organisation undergoing Green Business Model Transformations can be expected to fit the three characteristics of organised anarchies pretty well: At the outset, decision makers will rarely follow a consistent set of goals, and these may change as well once the implications are better understood. The decision process is also unclear due to the high complexity and uncertainty of the subject matter. Finally, people eligible as decision makers will show changing interest in participation based on how strong they feel affected and depending on competing issues vying for their attention.

Cohen et al. (1972, 2ff.) build their model based on four, relatively independent streams of *problems, solutions, participants*, and *choice opportunities* ("garbage cans"). *Problems* are concerns of external or internal people looking for attention. *Solutions* look for problems and are sometimes created before a suitable problem exists. *Participants* dump problems and solutions into available *garbage cans*. However, for a decision to be made, participants need to devote a sufficient amount of energy – depending on the attached amount of problems. Hence, a decision is made when – in a more or less random occasion – all streams meet in a certain way. Consequently, there are three styles of decision: (1) *by resolution*, meaning the problems attached to the choice are resolved, (2) *by oversight*, meaning that the decision can be made before problems are attached to it, (3) *by flight*, meaning that problems leave the choice for a more attractive one and the decision can finally be made without solving problems (Cohen et al. 1972, 8). The simulation of garbage can models imply rather poor results of decision-making of such organisations: Although most decisions are made within the simulated time period, relatively few problems are resolved (Cohen et al. 1972, 9ff.). Moreover, the model suggests that the effectiveness of decision-making represented by problem activity (i.e. the degree of conflict), problem latency, and decision time cannot be improved in all dimensions simultaneously (Berger & Bernhard-Mehlich 2002, 151f.).

Although the explanatory power of the garbage can model is clearly limited to specific circumstances or organisations (Cohen et al. 1972; Olsen 2001), its similarity to Green Business Model Transformations yields some valuable insights on the level of intra-organisational change:

First, it should be considered that many solutions necessary to operating a more sustainable business model may already have been found – there just were not suitable choice opportunities or problems available at the time. And indeed, this thinking fits very well with the broad argument I follow to explain why the vast majority of corporations have not yet adopted more sustainable business practices: Not because it is unfeasible or because the business environment would not reward it, but because companies didn't need to – their current business model has never been regarded as a *problem*, to which a solution had to be found. Consequently, the proposed change methodology in this work needs to recommend spending sufficient time in searching for *existing* solutions; this is likely to be much faster than inventing all new solutions from scratch.

Second, an appropriate provision of choice opportunities and level of participation is crucial: If participants are distributed to few, high-ranking garbage cans (e.g., large steering committees), the decision process is slow or may even come to a halt. Are there too many garbage cans (e.g., project groups), the energy of participants may be too low for reaching a decision, resulting in a severe loss in momentum. Moreover, limiting the access of participants to the decision-making process is not a solution either because detailed problems stay latent this way and threaten implementation (Berger & Bernhard-Mehlich 2002, 152). I have seen all these effects come into play during major transformations of otherwise well-functioning large corporations. Finding the right balance of participation and speed, delegation and leadership is a challenging matter indeed.

6.2.7 The Theoretical View on Change Applied in this Work

For the purpose of this work, understanding change in organisations is not an end, but a means to accomplishing Green Business Model Transformations. Hence, I must not decide which specific theoretical perspective to follow, or even to develop my own, new perspective. Instead, I can apply insights from all four basic process theories as needed. And indeed, as long as it is a conscious choice, the application of various relevant theoretical perspectives on specific aspects of Green Business Model Transformations can be expected to be particularly insightful, despite – or even because – I do not restrict myself to a single, self-contained theory.

More precisely, I will consider punctuated equilibrium as the main metaphor to justify why Green Business Model Transformations can improve the fitness of established firms. I will use results from life cycle theory to gain a deeper understanding of what specifics should be considered for radical transformations in mature organisations. The transformation process will largely be structured according to the teleological approach. And finally, I will consider dialectic theory with respect to change management and external stakeholder management in the wake of the high-profile nature of the change efforts.

6.2.7.1 Limits to Planning and Control of Organisational Change

Radical representatives of population ecology theory believe that organisational change of individual entities is minor or random. All major change is assumed to occur on a population level through new entities and replacement. If this were the case, then green business models – if effective at all – would be spread through innovative, small companies that eventually replace the unsustainable giants of the past.

However, I believe that some large companies contradict this one-dimensional view. IBM is one such example – a company with long tradition that survived serious crises and repeatedly transformed itself in many ways (IBM 2011a). For example, IBM transformed itself purposely from a pure play hardware producer to big player in the professional services business. It changed its structure from centralised to decentralised to a global matrix. From 2002 to 2007, IBM has increased its Indian workforce from less than 5,000 to 73,000 employees (IBM 2008b), which obviously has massive implications on corporate culture, leadership, and product delivery. One might argue that some of the revolutionary changes are to be attributed to life cycle phases (e.g., structural changes), random evolutionary variation (e.g., certain product and technology innovations), or just imitations of "institutionalized" management practices (e.g., building up workforce in India). However, some of these radical changes were highly controversial at the time and unprecedented. Hence, a constructive, deliberate element of change can hardly be denied in the case of IBM.

A trickier objection may be that constructive transformation is mostly random, meaning that organisational change is "only loosely coupled with the desires of organizational leaders and with the demands and threats of environments" (Hannan & Freeman 1984, 150). This perspective does not deny the possibility of major endogenous change, but questions whether it can be directed and controlled. Instead, purpose could be alleged to have been added retrospectively once actions turned out successful (see Weick 1979). The notion of complex adaptive systems (see section 6.1.4) leads to similar implications: The manager's role is reduced to changing the fitness landscapes for local agents as well as the organisational architecture that frames adaptation within the organisation (Anderson 1999, 228f.). Some scholars would even deny that these interventions are safe due to the unpredictability of long-term and system-level effects at the edge of chaos (see Stacey 1995; Stacey et al. 2002). Attempts of purposeful change would then be stricken with large failure rates (at least in the sense of the originally envisioned results). And indeed, I agree in so far as it is naïve to believe that a linear envision-plan-implement approach could suffice.

6.2.7.2 Towards a Considerate Approach to Organisational Change

A perspective on organisational change developed by Mintzberg & Westley (1992) may provide a path to reconcile the contradicting views on change as implied by complexity science (see Stacey et al. 2002) and the attempt to manage Green Business Model Transformations in a planned, controlled manner. Mintzberg & Westley (1992, 43f.) propose to condense the various possible approaches to organisational change to the following three basic approaches to change:

- *Procedural planning*: This change type is deliberate and deductive, i.e. conceptualised change programmes are translated into concrete, actionable steps. It corresponds to the terms "strategic planning" or "organisational development" frequently found in the relevant literature.
- *Visionary leadership*: A single, often charismatic leader (e.g., the CEO) conceives or adopts a new vision for the organisation. If the vision appeals to the wider organisation it can become a *shared vision* (Senge 2006) and set in motion revolutionary change. While the vision may be a deliberate choice or just the synthesis of signals from within the organisation, the details of its subsequent realisation will often emerge informally.
- *Inductive learning*: This process is informal, decentralised and can yield completely unexpected results. It corresponds to the "sand pile metaphor" used by complexity scholars (e.g., Anderson 1999): Much like a pile of sand on which grains of sand are dropped in a random fashion, most small change events will have no or little effect on the organisation as a whole. But sometimes, once a sufficient amount of such minor changes have occurred, a single grain of sand may create an avalanche and impact the behaviour of the entire organisation (for a broader discussion of organisational learning and its representatives see Fiol & Lyles (1985) and Crossan et al. (1999)).

Mintzberg & Westley (1992) view the three change approaches as complementary rather than alternative. They propose that the approaches are often combined in a sequential order: Inductive learning gives rise to a new vision which is then translated into concrete actions through procedural planning. Variations include the *informal* and *implicit change process* in which the procedural planning (informal) or the visionary leadership (implicit) are bypassed. Moreover, the learning can be done by external parties (e.g., consultants); however, the authors warn against bypassing the internal vision process in this case – they claim it leads to "mindless" change (Mintzberg & Westley 1992, 45f.).

The fact that inductive learning frequently brings about surprises does not mean – as some authors suggest – that the learning has to occur completely random. In contrast to biological evolution, so I argue, humans have the ability to distinguish more promising fields of learning from others through conclusion by analogy or imitating competitors and peers. Critics may object that this prevents true novelty from emerging (Stacey et al. 2002). However, I argue that novelty is merely restricted to certain search fields (that are more likely to produce "fit" solutions) – as long as learning from the past or peers is not understood as pure copying. In fact, many of the most notable innovations (think of Apple's iPhone) were in some ways simply reconfigurations of existing components (also see chapter 6.4).

Considering Kauffman's fitness landscape model (see chapter 6.1.4), the perspective on change brought forward above could be translated as follows: Inductive learning relates to "hill climbing". For incremental change, no visionary leadership is required; learning can succeed without a new overarching direction. Procedural planning may assist in implementing the results of inductive learning in an efficient manner (i.e., climbing the hill faster, or on a more direct route). Perhaps, inductive learning has accumulated in a way that reveals some information about other peaks of the fitness landscape. Early on, the effects of jumping to another peak will be largely unknown. This includes the height of the peak, as well as the non-linear changes of the fitness landscape(s) resulting from new, discontinued and shifted connections within the organisation and to other network entities. Systems thinking can help to understand some of these dynamics (but certainly not all). Visionary leadership would then correspond to finding the direction and distance of the jump. Procedural planning would be responsible for the "technical" execution of the jump and efficient hill-climbing once landed (as the jump will most certainly not happen to hit the top of the new peak).

Authors who consider the outcomes of planned change to be entirely unpredictable, and who insist that the power law underlying the edge of chaos produces extinction events completely beyond the control of human intervention, will heavily object at this point. However, I argue that some developments are to a certain degree foreseeable (e.g., that global warming will have an increasing effect on the global economy) and can hence be exploited by "informed guessing", combined with adaptation once the 'factual' future unfolds itself. Moreover, to use another metaphor, snow avalanches can neither be predicted in terms of when, where or how they occur; however, targeted detonations can be used to trigger artificial avalanches instead, thus reducing the risk, and – in Kauffman's terms – avoiding most of the large extinction events.[12]

In conclusion, I believe that prescribed change motors are real and limit which aspects of an organisation its corporate leaders can and should influence. Moreover, managers will never be able to accurately predict the longer-term future or foresee the full range of implications of their planned change initiatives. However, I sincerely believe that astute corporate leaders and their staff have the capacity to lead radical change efforts in an auspicious direction and overcome the complete randomness of biological evolutionary systems by the capacity that distinguishes humans from other biological organisms – the cognitive process of reasoning.

6.3 Change Management

"Change is not made without inconvenience, even from worse to better."
Richard Hooker (1554–1600), Anglican theologian (Wikiquote 2011a)

The definition of "success" in organisational change is a challenge in itself. It may be some function of "the quantity, quality, and pace of change" that has been achieved, often involving a trade-off (Pettigrew et al. 2001, 701). Up to now, descriptive, theoretical aspects of change were in focus. Yet, in order to minimise any trade-off and achieve the goals of their Green Business Model Transformation, firms need to develop a practical understanding of the change process and be able to take appropriate courses of action. The following sections thus outline how change in organisations can be managed.

In contrast to the descriptions of structured and smooth corporate change processes that can be found in many books and articles, the real thing tends to be messy and gruelling (Kanter et al. 1992, 369ff.). Change initiatives frequently take longer than initially planned, severely strain employee morale, and consume a

[12] Stacey et al. (2002) have brought forward the argument that using different causal frameworks for different levels or participants of change is problematic (and inconsistent). For example, they argue that if researchers associate managers with teleological change (i.e., as observers outside the system) whereas organisations are treated as uniform organisms that are subject to evolutionary change, they neglect the fact that organisations consist of humans as well, and hence cannot be treated as if they were not. Stacey and colleagues might be right from a philosophical perspective (they use Kant to back up their arguments) and also from a "technical" perspective (e.g., on a micro level, respective metaphors for change might break down). A detailed exploration of their argument would go far beyond the scope of this work; however, they miss the point here: It is often sufficient to know *roughly* what reality (or the future) looks like to make 'good enough' decisions. Like Newton's laws are sufficient to determine the approximate motion of objects at speeds and scales that exist in everyday life, I reason that organisational change is best managed by applying "dirty" approximations rather than discarding planned change completely (note that the physics analogy frivolously refers to a system of lower complexity).

large amount of management attention (Kotter & Schlesinger 2008). This raises the question whether the problem is rooted in poor change management or simply the uncontrollable nature of change. The theoretical perspectives on change presented previously suggest that the truth probably lies somewhere in between: On the one hand, many dos and don'ts seem to be neglected repeatedly by the responsible managers (Kanter et al. 1992, 386). On the other hand, change theories based on evolutionary and life cycle metaphors suggest that change cannot be constructed at will. In addition, systems theory implies that organisations are very complex entities; hence even well-planned change is certain to bring about surprises. Consequently, while it is important to consider the limitations of change management, it is an imperative to make the most of its potential.

In the following, I will therefore enlarge upon some of the most important strategic factors that significantly influence the success or failure of change initiatives:

- The adequacy of the nature of the change initiative in reference to its goals
- The pacing of the change initiative
- The understanding of the roles of change
- The handling of inertia and resistance to change.

6.3.1 Choosing the Appropriate Type and Pace of Change

As explained earlier, one of the most common typologies of change contrasts evolutionary with revolutionary change (Burke 2002, 11). In accordance with this distinction, March (1991) presents two alternative strategies for change: (1) *exploitation* of old certainties, and (2) *exploration* of new possibilities.

Exploitation is based on the "refinement and extension of existing competences, technologies, and paradigms" and promises "positive, proximate, and predictable" returns (March 1991, 85). In contrast, *exploration* is associated with "uncertain, distant, and often negative" returns (March 1991, 85). Because both strategies compete for the same resources, managers need to choose – and they understandably often opt for exploitation (March 1991, 71ff.). March further claims that exploitation can be self-destructive in the long run. For example, he argues that it can degrade organisational learning and also diminish competitiveness in critical dimensions where it is important to be among the best of many competitors: Due to reduced performance variability, an exploitation-improved mean performance may not suffice.

Green Business Model Transformations correspond to March's exploration strategy. Hence, there is no need to enlarge upon strategies of evolutionary change; merely, it is important to keep in mind that the alternative to business model transformations (i.e., optimising the current one) will often be considered to be safer and more desirable. Therefore, conflict may not only erupt on the question whether there is a need for change or not; opinions can also diverge on the beliefs about which type of change is more appropriate (Crossan & Berdrow 2003).

Every change effort is different. However, scholars have formulated some broad insights that seem instructive for managing revolutionary change in general (Weick & Quinn 1999, 374; based on research on effective large group interventions by Bunker & Alban 1992):
Systems theory serves as the theoretical reference for such change efforts. Compared to traditional approaches the organisation engages more actively with its environment and information is shared widely. Organisational learning is not limited to the level of individuals or groups, but also includes the organisation as a whole. Furthermore, a common understanding of the change together with a more participative approach can speed up the change effort compared to slow, hierarchy-oriented models. This means that accountability is spread throughout the organisation and does not exclusively rest with senior management. At the same time, consultants play a less prominent role in the change process.

In addition to the question whether change is to occur in many small or few large steps, managers must decide how much of the change is to be realised in which timeframe, i.e. the pace of change. The pace of change has two sometimes conflicting perspectives: an *absolute* and a *relative* one. The former is relevant with respect to an organisation's internal capacity to change, while the latter refers to competitiveness on the level of organisations.

From a strategic perspective, companies must consider their relative pace of change compared to their industry peers, and also the broader business environment (e.g., regarding legislation and stakeholder expectations). If a company is adapting, but not as fast as what has been coined the *clockspeed* of its industry, it falls behind its competitors (Fine 1998). An ever increasing speed of change has led some management scholars to argue for radical rather than incremental change to avoid a rat-race within the sector (Voelpel et al. 2005). Moreover, if a company does not adapt fast enough to the changing broader business environment – e.g. with respect to sustainability standards – it could lose its social licence to operate in the worst case (e.g., Sells 1994). On a larger scale, an entire industry sector that continuously lags behind the developments of the broader business environment risks coming under scrutiny or even becoming obsolete eventually (Hart & Milstein 1999). The introduction of one revolutionary green business model, however, might halt this development. Yet Green Business Model Transformations also carry the risk that a company gets too far *ahead* of its time. Companies that introduce too progressive business models too early or too quickly risk that they are rejected by bewildered customers, unconvinced business partners or even internal resisters. For example, unlike Toyota's hybrid car, the Prius, GM's more radical design of a fully-electric vehicle (the "EV1") was discontinued a few years after its initial release in 1996. The exact circumstances of the programme cancellation remain controversial, but one factor certainly was that the business environment for electric cars was not as favourable yet as it is today (see 7.8). Conversely, this is not to imply that firms should sit and wait – rather they need to actively prepare their business ecosystem and relevant stakeholders (see Schneidewind 1998).

From the internal perspective, a high pace of change puts a strain on the organisation. Sastry (1997) stresses that if organisations do not have sufficient time to rebuild competences after strategic reorientations, the performance level of the organisation will be consistently low. All organisational change is thereby subject to the respective time cycles of an organisation (Pettigrew et al. 2001, 700). To control the pace of change, Brown & Eisenhardt (1998, 162ff.) thus propose to introduce an artificial, calendar-based organisational rhythm of change which they refer to as *time pacing*. In contrast to reactive *event pacing* (which, for example, is triggered based on the industry clockspeed), time pacing "counteracts the tendency of most managers to wait too long, move too slowly, and lose momentum" (Brown & Eisenhardt 1998, 167). Conversely, in turbulent industry settings time pacing may be used to slow down change in firms that risk being too responsive to external events – which may turn out to be passing fads.

In conclusion, a company can make too large or too small steps, and change too fast or too slowly. In order to find the right balance, Kotter & Schlesinger (2008) identify four situational factors that they believe need to be considered for every change initiative: (1) the amount and type of expected resistance; (2) the balance of power between the initiators and resisters; (3) the sources of required information and support; and (4) the actual need and urgency for change. The authors argue that the stronger the anticipated resistance, the weaker the initiators, the more distributed the required information, and the lower the short-term pressure for change, the slower the change should be carried out. Furthermore, they state that a faster change requires more clear-cut plans, a more aggressive stance on resistance, and allows for less involvement of others (Kotter & Schlesinger 2008, 137ff.).

Resistance against Green Business Model Transformations should be expected to be strong and multifaceted, the resisters numerous and powerful, the relevant information widely dispersed and the impact deep but often rather long-term. All this suggests applying a very long timeframe for change. However, a slow transformation carries the risk that some basic business model assumptions become invalid before it is fully implemented: some opportunities can fade quickly (e.g., because competitors seize them first) and formerly invisible threats can materialise unexpectedly (e.g., in the form of new entrants or legal requirements). Hence, the timing cannot be freely chosen independent of external conditions. Because the success of Green Business Model Transformations is also heavily dependent on a strong internal commitment which conflicts with strong coercion, other means of accelerating change need to be employed. These may include a sophisticated communication strategy to create a strong sense of urgency and build momentum, intense involvement of the CEO and other board members to counterbalance powerful resisters, disciplined project management, and a suitable transformation organisation to make information flows across organisational boundaries as efficient as possible. The roles that individuals and groups play in change efforts are thus summarised next, the other aspects in later sections.

6.3.2 *Roles in Change*

During the discussions of change in the previous chapters organisations were mostly treated as abstract entities that undergo or resist change according to various theoretical models. However, eventually, change is implemented and received by human beings. Hence, the human factor, the roles they play as individuals and groups within organisations are critical to every change effort.

6.3.2.1 Change Agents

The dedicated group of people that are usually considered responsible for making change happen are referred to as change agents. Interpretations of the role of the agent range from rather technical to strategic tasks (Dunphy 1996, 543f.). With respect to episodic (transformative) change, Weick & Quinn (1999) consider the change agent to be the "prime mover" who creates (and not only redirects) change. He does so by concentrating on inertia and key leverage points, and thereby "changes meaning systems: speaks differently, communicates alternative schema, reinterprets revolutionary triggers, influences punctuation, builds coordination and commitment" (Weick & Quinn 1999, 366). Kanter et al. (1992, 376ff.) remark that a too narrow view on change agents is highly misleading; they identify three main groups that need to be distinguished in accordance with Lewin's unfreeze-change-refreeze model: *change strategists*, *change implementors*, and *change recipients*. In an earlier work, Ottaway (1983) proposed a taxonomy that is also compatible with Lewin's change model and even more granular, featuring ten types of change agents. Whether it is useful in practice to distinguish ten types of change agents is debatable; yet one important insight for the management framework in chapter 9 is that different individuals may be needed as change agents during different stages of Green Business Models because their respective roles differ substantially.

6.3.2.2 Leaders and Leadership Styles

Leaders have long been an important subject in management science. Although some scholars have come to question the influence that leaders actually have on the success and direction of "their" firm (e.g., Stacey et al. 2002), their role continues be considered pivotal in mainstream management literature (Nadler & Tushman 1990).

Change initiatives depend on the successful work and interplay of several types of leaders, namely *executive leaders* that "need to shape the overall environment for innovation and change" (Senge 2006, 320), *local line leaders* who help to establish changes in work routine, and *network leaders* who disseminate new ideas and practices throughout the organisation (Senge 2006, 319ff.). Green Business Model Transformations require strong leaders. Their success, however, not only depends on their skill, but also on the appropriate leadership style.

Over half a century ago, Tannenbaum & Schmidt (1973, originally published in 1958) presented a continuum of leadership behaviour and discussed under which circumstances the respective patterns should be applied. The continuum contrasts the authority used by the manager with the degree of freedom of their subordinates. The authors provided a framework consisting of three "forces" to determine the appropriate leadership style: forces in the manager (e.g., his confidence in the subordinate and own leadership inclinations), forces in the subordinate (e.g., the subordinate's needs for independence, tolerance of ambiguity, commitment, and expertise), and forces in the situation (e.g., organisational culture, team effectiveness, and the type of problem). The appropriate leadership style can thus vary even on group level depending on the particular setting.

In a retrospective commentary 25 years later, Tannenbaum and Schmidt acknowledge that they had ignored some factors related to the external environment that impact the appropriate leadership style and have grown in importance meanwhile. Among other things, these environmental factors include societal demands for considering stakeholder interests and sustainability issues (Tannenbaum & Schmidt 1973, 166ff.).

The importance of external factors may have increased even further. Nadler & Tushman (1990) discuss the requirements of leadership in the wake of intense global competition and the need to manage system-wide change of incremental as well as revolutionary nature at a rapid pace. They conclude that what is needed is a leadership style that goes beyond what the *charismatic leader* alone can offer: they further demand instrumental leadership and institutionalising the leadership for change.

The charismatic leader is described as someone who is capable of envisioning the future in a compelling way, can energise (or motivate) the organisation, and enable its members to take action. However, the above described prototype of a change leader also faces multiple problems in practice; these may include the creation of unrealistic expectations, strong dependency and the "need for continuing magic" (think of Steve Jobs from Apple), as well as disempowered lower level management with a reluctance to disagree (Nadler & Tushman 1990, 84). Hence, according to Nadler and Tushman, charismatic leadership should be complemented with *instrumental leadership*. Instrumental leadership encompasses structuring the change effort, controlling of behaviour and results, and rewarding desired behaviour systematically (Nadler & Tushman 1990, 85ff.). Obviously, a company cannot magically dig up gifted leaders as described above just for the occasion of transformation: However, it can at least assign leaders that match the requirements most closely and resist the temptation to keep all "indispensable" leaders in their line jobs.

6.3.2.3 Teams and Team Effectiveness

Teams (or groups, the terms are used interchangeably) represent one of the central elements of modern corporations (Cohen & Bailey 1997). Not surprisingly, the topic of team performance has received a large amount of attention (Hackman 1987; Mathieu et al. 2008). Based on Kozlowski & Bell (2003), Mathieu et al. (2008, 411) define *work teams* as

> *"collectives who exist to perform organizationally relevant tasks, share one or more common goals, interact socially, exhibit task interdependencies, maintain and manage boundaries, and are embedded in an organizational context that sets boundaries, constrains the team, and influences exchanges with other units in the broader entity."*

Various typologies of teams have been suggested, but the adequacy of a single distinguishing criterion is questionable. Possible criteria include *homogeneity of members*, *stability of their operating environment*, or the *duration of existence* (Mathieu et al. 2008, 411f.). Examples of types of teams other than work teams that are addressed in management literature include *project teams*, *management teams*, and *virtual teams* (Cohen & Bailey 1997).

One of the best-known models for team performance is Tuckman's framework of *forming, storming, norming, performing,* and – added later – *adjouring* (Tuckman 1965; Tuckman & Jensen 1977). According to this model, teams undergo a fixed sequence of stages before they start performing. Gersick (1988) challenges that view. Instead, she proposes that team development follows a pattern of punctuated equilibrium (see section 6.2.5): After the team is formed it enters a stable period of execution of the work procedures agreed during the team's initiation. At about half time towards the team's official deadline, a revolutionary period starts. During this transition, outsiders are consulted, old work procedures are questioned and new perspectives are adopted (Gersick 1988, 16). This revolutionary period shows dialectical characteristics and is "an opportunity for, not a guarantee of, progress" (Gersick 1988, 34). In summary, responsible managers and project leaders should pay special attention to the two transition periods when the team is formed and at midpoint, where the leverage to positively influence the prospects of the outcome is especially high.

Researches have been seeking to extract insights about the factors that influence team performance for decades. Mathieu et al. (2008) have reviewed research on this topic from 1997 to 2007. They use an input-mediator-output framework that is an extension of the traditional input-process-output framework. In the framework, performance is considered in terms of quantity and quality of output. Inputs that affect performance are "*individual team member characteristics* (e.g., competencies, personalities), *team-level factors* (e.g., task structure, external leader influences), and *organizational and contextual factors* (e.g., organizational design features, environmental complexity)" (Mathieu et al. 2008, 412). Mediators also influence performance; they include team processes and emergent states (e.g., team confidence, empowerment, team climate, cohesion, trust, and collective cognition) (Mathieu et al. 2008, 420ff.). The research on team performance is not easily summarised, and a comprehensive discussion of the results would go beyond the scope of this work. Hence, only selected results are presented that are specifically relevant for teams that can be expected to play an important role in Green Business Model Transformations. These include factors that can be influenced regarding project teams and (top) management teams.

The composition of *top management teams* influences their performance (Cohen & Bailey 1997, 269ff.), but is hardly considered a factor that can usually not be altered just for the occasion of a Green Business Model Transformation and is hence omitted in this overview.

Strategic consensus was found to be positively correlated with performance of top management teams (Mathieu et al. 2008, 430). However, empirical research conducted by Kilduff et al. (2000) reveals that high performing teams in fact show high interpretative ambiguity in the beginning and only reached strategic consensus towards the end of their life cycle (low performing teams show the opposite patterns). These results suggest opening up the opportunity space for the executive team at first and reducing ambiguity by means of a structured planning process afterwards.

In their review of team performance literature from 1990 to 1996, Cohen & Bailey (1997) found that in contrast to work teams, *project teams* do not necessarily benefit from higher autonomy (possibly due to cultural factors). Instead, leadership – both in terms of task- and person-based leadership – seems to play a pivotal role in the success of teams (Burke et al. 2006). This highlights the importance of assigning a gifted leader to Green Business Model Transformation projects. In addition to leadership skills as such, it is highly important to select someone who is respected in the line organisation (which makes the team's output more credible) and is well connected to provide the team with the necessary contacts and resources within the organisation. The organisational support is a central issue in strategic sustainability-related projects. In addition to a well-connected leader, project teams should have sufficient mandate to get access to respective resources within the organisation. Moreover, a motivating reward system is essential for strong team efforts (Hackman 1987, 324ff.). In its extreme form, project teams get the feeling that they can truly make a difference; paraphrasing Maslow's words, "the task [is] no longer separate from the self" (Senge 2006, 194).

With respect to team composition, Hackman (1987, 327) proposes to make them just large enough so that the work can still be accomplished, but to avoid any dysfunctions that large groups tend to show. For large transformations, this implies the necessity of a hierarchy of teams (that needs to be managed, too).

Functional diversity was found to have both positive and negative effects on project team performance (Cohen & Bailey 1997, 263). On the one hand, diversity allows covering a high share of the required expertise within the (core) team (Hackman 1987, 326f.) and the team can integrate different perspectives which can spur innovativeness of teams (Cohen & Bailey 1997, 263). On the other hand, diversity has been observed to negatively affect performance. This may be due to increased conflict within the team. And indeed, functional diversity has been found to be positively correlated with task disagreement. However, the ability of a team to be innovative and effective largely depends on the management and communication of task disagreement by members and leaders, and *not* the degree of functional diversity as such (Lovelace et al. 2001). It has also been suggested that teams with high diversity may seem less effective in the short-run, but that they in fact play out activities that are then prevented later in other parts of the organisation (Cohen & Bailey 1997, 263). That is, ideas are tested earlier and work products become more robust.

In order to achieve a higher degree of diversity in project teams of large, geographically dispersed organisations, teams may be designed as (partially) virtual teams. However, Gibson & Gibbs (2006) warn that the team's ability to produce innovation may suffer from its "virtuality"[13]. Virtuality is related to geographic dispersion, electronic dependence, structural dynamism (frequent changes in members, roles and relationships), and national diversity. All four aspects of virtuality were found to be negatively correlated with innovation. However, a "psychologically safe communication climate, defined as an atmosphere within a team characterized by open, supportive communication, speaking up, and risk taking" (Gibson & Gibbs 2006, 455) can protect a virtual team's power to innovate (even more so than for non-virtual teams). Team requirements shift depending on the transformation stage, and thus may necessitate reassignments.

Proponents of the paradigm of the learning organisation consider *team learning* as one of its five essential disciplines (Senge 2006, 216ff.). Senge stresses that teams need to be aligned in order to reduce the waste of energy and redirect it for efficient team effort. Without alignment, attempts to tap team member's full potential via measures of empowerment will fail to show positive results.

Teams are advised to enter a state of dialogue in addition to discussion. While discussion is understood as a competition of viewpoints leading to decision-making, dialogue is considered a process of suspending own assumptions to fundamentally reconsider issues and learn in a deeper sense. This way, teams are believed to become much more intelligent than the sum of their individual members can be. A prerequisite for fruitful dialogue is the absence of leadership during these phases; contributions of team members must weigh equally – independent of rank (Senge 2006, 221ff.). Together with a shared vision, team learning techniques can surface defensive routines and thus support the team's efforts with energy rather than generating inertia (Senge 2006, 239f.).

Dealing with the phenomenon of *inertia* is central to the success or failure of any change initiative; it needs to be investigated both on the level of whole organisations (organisational inertia), and with respect to managing behaviour of individuals or groups.

6.3.3 Organisational Inertia

Inertia is a key concept in all analytical frameworks for transformative change (Weick & Quinn 1999). Pfeffer (1997, 163) defines inertia not as the absence of change but as "the inability for organizations to change as rapidly as the environment".

[13] Innovation was measured in terms of perceptions of members, and "the percentage of innovation achieved as rated by customers downstream in the process", respectively, in two separate studies (Gibson & Gibbs 2006, 486).

6.3.3.1 An Evolutionary Change View on Inertia

Population ecology theorists actually consider organisations with high inertia (and not adaptability) to have a selection advantage. Hannan & Freeman (1984) regard structural inertia as a by-product of institutionalisation which in turn facilitates *reliability* and *accountability*. Structural inertia can be generated by internal factors (e.g., sunk costs, political coalitions, normative standards) and external factors (e.g., barriers to entry, relations with other organisations). Within organisations, their core aspects are the hardest to change (listed in decreasing order): stated goals (or general purpose), forms of authority, core technology, and marketing strategy in a broad sense, including target customers and the securing of resources (Hannan & Freeman 1984). Obviously, business model transformations can affect core aspects of organisations, at least the latter two. Hannan & Freeman (1984, 157ff.) further argue that structural inertia generally increases monotonically with age and size of organisations, while the death rate decreases ("liability of newness"). Following this line of argumentation, organisational renewal should thus increase the death rate of organisations, and more so for complex ones which require more time to change fundamentally. However, empirical research in the airline industry casts some doubt on the general validity of these conclusions (Kelly & Amburgey 1991).

Punctuated equilibrium theory as described by Tushman & Romanelli (1985) similarly puts inertia at the centre of the change process. However, punctuated change is considered the result of managerial decision-making, not random variation. Sastry (1997) has translated this change model (see 4.5.2.1) into a causal loop diagram:

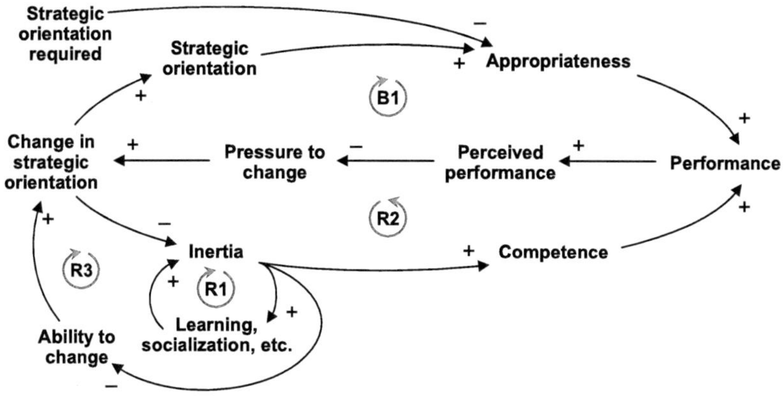

Fig. 6.2 Simplified causal loop diagram of the punctuated equilibrium change theory (Sastry 1997, 244)

Inertia is modelled as a stock that builds up or is reduced over time. Other state variables are the organisation's strategic orientation (i.e., deep structure), its perceived performance, and the pressure for change. The causal loop diagram contains four feedback loops:

- R1 reflects the building up of inertia as structures and routines become institutionalised
- R2, set in motion by R1, leads to increased competence, which improves (perceived) performance and reduces pressure to change.
- R3 describes how increasing inertia also reduces the ability to change.
- B1 describes the increasing misalignment of the strategic orientation with the changing environment as long as no fundamental organisational change happens.

Hence, R1 through R3 describe the reinforcing feedback loops that prevent major change during equilibrium periods. Only when the appropriateness of the strategic orientation becomes so low that it outweighs the achieved high competence levels, revolutionary change erupts. However, the simulations of Sastry (1997) revealed that the predicted change patterns did not always materialise. Sastry proposes to add two change management routines to the theory in order to get the desired patterns of change: monitoring for organisation-environment consistency and trial periods for change that prevent the unnecessary destruction of competencies.

In summary, evolutionary organisational theories view inertia as the flipside of competence and performance (punctuated equilibrium) or reliance and accountability (population ecology). Managers cannot influence inertia directly, rather its rate of change. As can be observed in figure 6.2, only previous fundamental changes are thought to reduce structural inertia "by disrupting existing networks, injecting new themes into the organizational culture, and bringing in new organizational members" (Sastry 1997, 249). However, Kelly & Amburgey (1991) remark that this may only be true for strategic change of the same kind. Moreover, recent empirical work on change questions whether at all prior change increases the likelihood of further change based on methodological issues of previous studies (Beck et al. 2008).

Furthermore, the evolutionary view seems to be conflicting with Peter Senge's systems view on organisational change. He states: "The harder you push, the harder the system pushes back." (Senge 2006, 58). However, both views might eventually describe the same in the sense that one cannot simply force change upon the system. Indirect measures like emphasising the misalignment with the market requirements and questioning the factual versus the perceived performance of the firm in order to generate a sense of urgency for change are much more elegant ways to affect inertia. Scholars have also argued that the arrival of a new CEO may make this task a lot easier (Kotter 1995, 60; Greiner 1998, 64). On the other hand, inertia is also seen as a protection from too much or unnecessary change. It does not protect, though, from distorted views on performance and strategic fit with the environment, or an unbalanced prioritisation of the same (Sastry 1997, 265f.).

6.3.3.2 A Competition-Based View on Inertia

Miller & Chen (1994, 2) define *competitive inertia* with respect to "the level of activity that a firm demonstrates in altering its competitive stand". It can be strategic or tactical in nature. According to Miller and Chen, the former might concern M&A or business model components like strategic alliances or important new products; the latter price changes, advertising, or minor product adjustments.

Competition- or adaptation-based perspectives on inertia and change management differ from the selection-based, evolutionary view in several ways: First, inertia is mainly considered a problem to be overcome, not an inevitable by-product of organisational virtues. Second, the managerial influence to make organisations "change-ready" is assumed to be much higher. Finally, resistance to change is considered more on an individual, behavioural level rather than in terms of structural inertia on the organisational level. The factors that influence competitive inertia also differ from structural inertia: They include *incentives to act*, the *awareness of action requirements and alternatives*, and the *constraints on managerial action* (Miller & Chen 1994).

As discussed before, past performance can become an *incentive to act*. However the relationship between past performance and change is contested and probably quite complex with respect to competitive inertia. While some scholars argue that poor performance provides an incentive to question current routines and scan the environment for new opportunities, others believe that managers may feel pressured to vindicate their past decisions and stay on their current course (Miller & Chen 1994, 3ff.). Miller and Chen propose that poor performance can pave the way for tactical rather than strategic changes. The latter is considered to be induced by market growth and opportunities. This line of argumentation fits with the rationale that companies who consider environmental sustainability a threat tend to focus on incremental measures like reducing energy use, waste or CO_2 emissions, while companies that view sustainability as an opportunity are eager to invest in new products and markets.

In contrast, success is generally associated with high inertia because it reinforces current practices and makes current leaders more powerful (Miller & Chen 1994). Moreover, organisational learning theorists (e.g., March & Olsen 1975; 1976) warn that learning under ambiguity can result in misinterpretations regarding the causality of success. Individuals then tend to perceive and interpret later events in a way that perpetuates their previous (potentially false or outdated) attitudes. Alternatives are frequently not even considered and little practical experience accumulates in areas that are not recognised as being important to the success of the firm. Consequently, such neglected areas are subjectively considered less promising and systematically underestimated (Berger & Bernhard-Mehlich 2002, 157). Sustainability advocates argue that this is exactly the case with many areas of opportunity in the sustainability domain (Berchicci & King 2007).

As a second relevant factor, the *awareness of alternatives* can reduce competitive inertia: Diversity of customers and competitors inspire companies to experiment on competitive practices. The same has been suggested for another factor, past experience with competitive change, but empirical evidence was weak (Miller & Chen 1994, 14). As discussed earlier, structural inertia is believed to increase with age and size of the firm (Hannan & Freeman 1984). However, Miller & Chen (1994) did not find empirical support that *managerial action is constrained* this way in the airline industry. Their study, however, supported the warnings of population ecologists that strategic change may actually hurt performance, at least in the short run (Miller & Chen 1994, 14f.).

The reflections on the phenomenon of organisational inertia presented above illustrate two things: First, the emergence of inertia is a complex issue, and second, its implications on an organisation's performance are ambivalent. Nevertheless, every change manager responsible for a major organisational change initiative needs to deal with it in management practice – ideally in a way that allows the change to happen without destroying valuable competencies.

6.3.4 Resistance to Change as a Change Management Issue

6.3.4.1 Phases of Resistance and Adoption

There are many models that describe the process of adoption – and resistance – of change. Carnall (2007) presents a simple five-stage model (see figure 6.3) derived based on works of Adams et al. (1976) and Kets de Vries & Miller (1984).

During the *denial* stage, employees highlight the effectiveness of existing procedures, self-esteem is reinforced and performance tends to remain unaffected (in the case of a disruptive change, however, it may suffer immediately). In the *defence* stage the change takes a clearer form, change recipients consequently start to devote much energy to keep the status quo, and as a result, performance drops. Resistance can take various forms: for example, it can be open or beneath the surface; it can be individual or collective (Tushman et al. 1986); and it can be destructive (to the point of sabotage) or constructive – and should then be considered valuable feedback to improve the change effort (Ford & Ford 2009). Once change recipients recognise that the change is inevitable and/or necessary they enter the *discarding* stage. Self-esteem starts to recover while performance approaches its minimum. During *adaptation*, performance is still low initially as new systems, procedures, structures, etc. have to be (re-)learned first. This learning process requires effort and can be rather painful. It is not until the *internalisation* phase that performance reaches or exceeds pre-change levels. At that point, the change effort has been fully integrated into the organisation (Carnall 2007, 240ff.).

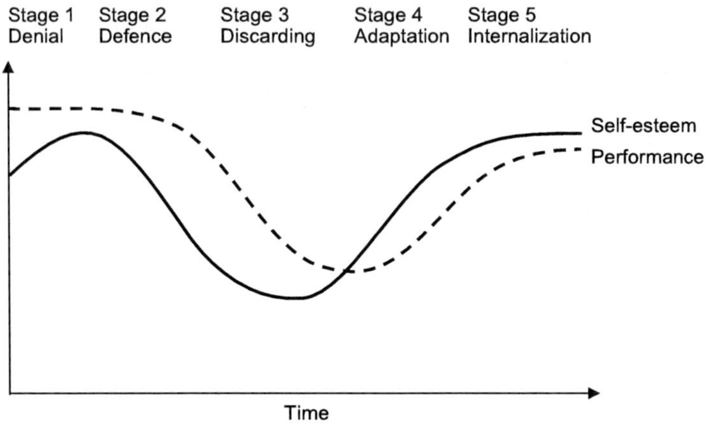

Fig. 6.3 The coping **cycle** (Carnall 2007, 241)

Research on smoking addiction illustrates how difficult it can be for humans to unfreeze their behavioural patterns (Prochaska et al. 1992). Progression through the stages is not linear and people fall back to the old habits three or four times before the change effort is finally successful. Hence, adoption on an individual level may actually follow a spiral rather than a linear pattern (Weick & Quinn 1999, 373).

Moreover, Senge (2006, 203f.) points out that there are qualitative differences in how committed people are to a cause: Most people just show variations of *compliance – grudging, formal*, or *genuine* – depending on how convinced they are. *Enrolment* means that someone will do what is possible within a given frame; an enrolled person intrinsically wants to make it happen. A *committed* person would go even further and is willing to change the relevant conditions for action if necessary.

Obviously, individuals that experience large scale change in organisations will not all go through the above described phases simultaneously. Some will adopt changes earlier than others, based on how the change affects them, but also based on the individuals' personality and preferences (if they are bold vs. cautious, etc.). With respect to innovations, people are often categorised according to a bell curve (Rogers 1962): a small number of *innovators* adopt the innovation first, followed by the larger group of *early adopters*. The *early* and *late majority* represent the largest groups followed by the *laggards* who are the last to adopt the innovation. A successful adoption would therefore follow an S-shaped pattern over time.

Whether or not that adoption pattern can be realised may depend upon whether or not a *tipping point* is reached (Gladwell 2000). That is, there is no guarantee that changes are not reversed again once the pressure for change is lowered – even though many people (but not a critical mass) have fully embraced the change (Sterman 2008). The more sources of resistance there are, the higher this tipping point will generally be. Hence, change management puts strong emphasis on tactics to identifying and neutralising such sources of resistance.

6.3.4.2 Sources of Resistance to Change

The first important step when tackling resistance to change in practice is to identify the underlying reasons for resistance. Jameson (1999) describes a framework that distinguishes three dimensions of conflict, each of which has different implications for conflict resolution strategies: First, there is the *content* dimension, i.e. the nature of resistance. For example, it can be based on objective or subjective grounds, task-based or driven by relational issues. Second, there is the *relational* dimension. That is, resisters can be more or less independent of the change initiators; they can have equal or unequal status, have high or low trust, and be a homogenous group (e.g. one department) or multi-party. Third, there are *situational* factors.

Del Val & Fuentes (2003) have identified two dozens of individual sources of resistance to change in literature that occur during the change formulation and implementation stages. Discussing the implications of each of the identified items is not expedient. Instead I will use four broad categories of sources of resistance adapted from Kotter & Schlesinger (2008, 132ff.):

1. *Self-interest*: One of the most dangerous sources of resistance is wilful resistance from powerful individuals. They may realise that the planned change would bring about a loss of power and status, or create new undesired interfaces and tasks. In this case, "politics" start. The special danger lies in the fact that this type of resistance is immune to rational arguments (the resister well understands the rationale for change) and will often be underneath the surface – individuals will usually refrain from arguing based on their own interests when they contradict the company's goals.
2. *Misunderstanding and lack of trust*: This source of resistance is usually based on poor communication and asymmetric information. Hence, simple misunderstandings can often be easily rectified. However, if the organisation has built a deep mistrust against management over time (possibly due to unfair behaviour in past initiatives), the situation is much more difficult to resolve.
3. *Divergent assessments*: When incomplete information exists either on the side of the initiators, or of the resisters, their assessments may diverge. It has been argued that this type of resistance is not only a threat, but also an opportunity (Ford & Ford 2009). Divergent assessment can also be based on rather subjective grounds like values or perceptions (Jameson 1999). Often, employees are truly interested in the success of their company and may have good reasons to object the change. Consequently, in any case the initiators should reconsider both content (as some assumptions may be wrong) and process (the degree and scope of involvement in decision-making) of the change initiative.
4. *General aversion to change*: Uncertainty and the fear of the unknown can be a powerful source of resistance that does not directly relate to the specific change initiative at hand. This type of resistance may even include the "winners" of the change who are unsure whether they can cope with the new situation. People might also feel pressured by their peers, or feel obliged to positions they endorsed in the past that stand in conflict with the current change initiative.

6.3.4.3 Mitigating Measures for Resistance to Change

There are various strategies and tactics to deal with resistance to change. They differ in how appropriate they are with respect to the type of resistance but also in which objectives they can meet satisfactorily (Jameson 1999). For example, management may want to stay fair in the process and maintain a good relationship with the resisters. In other cases, avoiding future resistance in similar situations may be a major concern. Moreover, resistance management can lead to more or less optimal solutions from a content perspective (Prein 1987). And finally, timeliness, costs and required resources always represent more or less important constraints. It is needless to say that the objectives listed above are not without trade-offs (Jameson 1999).

Kotter & Schlesinger (2008) list six practical measures to address resistance to change. They emphasise that it is the right mix depending on the reasons for resistance that determines their effectiveness. This fact often represents a major issue in practice as leaders find it difficult to adapt their personal style (which may be authoritative, cooperative, etc.) to the specific situation. The six measures are (summarised from Kotter & Schlesinger 2008, 134ff.):

- *Education and communication:* To clarify the logic behind the change effort is effective, but expensive, time-consuming, and requires a certain level of trust. Various channels should be considered.
- *Participation and involvement:* This measure is crucial when necessary information is missing or commitment (not only compliance) is needed; however, it is infeasible for time-critical change, and results may be poor if participation is not thoroughly managed.
- *Facilitation and support:* Measures may include training and various forms of support or rewards (also emotional) to cope with the change. This path is well-suited to reduce fear and anxiety, but also time consuming and expensive.
- *Negotiation and agreement:* Offering something in return for agreement can be a good option if resisters are powerful. However, it may harm content aspects, with a (future) risk of being blackmailed.
- *Manipulation and co-optation:* (Potential) resisters can be manipulated through the selective use of information. Co-opting means giving powerful resisters a special role in the change effort to gain their endorsement. This strategy is sometimes highly effective, but risky in terms of building and preserving trust.
- *Explicit and implicit coercion:* Playing it hard is fast but risky as compliance is reached at best; it is more attractive in cases when the change would be unpopular anyway.

The following table summarises the effectiveness (○ = low, + = medium, ++ = high) and risks (○ = low, - = medium, -- = high) associated with each measure related to the four reasons for resistance:

Table 6.3 Effectiveness and risks of measures against resistance to change by source of resistance

Measure \ Source	Self-interest	Misunderstanding, lack of trust	Divergent assessments	General aversion to change
Education and communication	o / -	+ / -	++ / o	+ / o
Participation and involvement	+ / --	++ / -	++ / -	+ / -
Facilitation and support	o / o	+ / o	o / o	++ / o
Negotiation and agreement	++ / --	o / -	+ / -	o / -
Manipulation and co-optation	+ / --	+ / --	+ / --	+ / --
Explicit and implicit coercion	+ / -	+ / -	+ / --	+ / -

Of course, the representation in table 6.3 is highly simplistic: Neither does it consider the plurality of goals mentioned above, nor are the individual ratings of effectiveness and risk to be considered definite for all types and contexts of change. Rather, they serve as a rough guideline to develop tailored strategies and tactics to deal with resistance to change:

Manipulation and coercion can be effective against all types of resistance; however they are also very risky as they jeopardise trust and cooperation between management and the larger organisation. Therefore they should be only employed if other measures prove ineffective or take too long (Kotter & Schlesinger 2008, 136f.).

Due to their deep impact, Green Business Model Transformations will likely trigger all four types of resistance. Many individuals in various areas will lose or gain power and status. However, negotiation must not be used to a degree that it endangers the logic of the new business model. Therefore, involvement of all critical areas of the company is crucial as it can ensure a broad commitment within the organisation. The commitment works as a social barrier to objecting change out of self-interest. Misunderstandings will also be a common issue as sustainability is a wide and complex subject that few people know and understand in all its facets. This may of course also lead to strong disagreement on the business case for the transformation. Again, this reinforces the need for involvement and emphasises the importance of education and communication. Third parties like external experts or consultants can also be used to educate the organisation and manage participation. As "neutral" mediators they can be effective in resolving diverging assessments. Moreover, consultants may serve as lightning rods; they can deliver unpopular change without putting too much strain on intra-organisational relations.

Due to the magnitude of change, strong reflexes based on a general aversion to change should also be anticipated. However, to the degree that the transformation is communicated in a credible way, employees may also develop a strong emotional feeling that the company is moving in the right direction in a moral sense. This may be a powerful source of momentum that can easily outweigh any unspecific aversion to change (Hage & Aiken 1970).

6.3.5 Practical Guides for Change Management

Numerous "cookbook" recipes for change are available to change agents and managers that hope to find a way to make executing change manageable (e.g., Beer et al. 1990; Galpin 1996; Kanter et al. 1992; Kotter 1995). These "n-step guides" for change ('n' being the number of sequential steps, phases or stages) usually claim to be applicable to a wide array of different change efforts and types of organisations. N-step guides for change have been criticised for being overly simplistic and mechanistic (Collins 1998, 82ff.). On the one hand, one cannot deny that these guides inevitably neglect the complex, non-linear aspects of change that put limitations on prediction and planning. Dimensions of change like social and political activity, corporate culture, external circumstances, or even the personality of important leaders tend to be ignored. On the other hand, at least they can keep change agents from making easily avoidable mistakes. Furthermore, management science has little to offer as an alternative: Change theorists have failed to-date to produce a comprehensive general theory of change, let alone one that is easy to apply in management practice. After all, using no guide for change whatsoever is unlikely to produce better results than a thought through n-step guide – at least if it is applied with careful consideration of its limitations and common sense.

One example for a comprehensive n-step guide to organisational change is Kotter's "Eight Steps to Transforming Your Organization" (Kotter 1995). This approach will be used to exemplify how theory on organisational change[14] can be translated into management practice.

Kotter presents eight steps that companies need to go through in order to make successful change happen, and explains the major pitfalls that exist in each phase. Kotter does thereby not address a particular type of change initiative. Like in other comparable approaches, making a big mistake in one of the phases, as well as trying to skip or shorten any of them is considered potentially devastating (Armenakis & Bedeian 1999, 303). Kotter's eight steps and respective pitfalls are summarised in the following table:

[14] Like for most prescriptive approaches to change, the underlying viewpoint focuses on the teleological nature of change. As I have argued before, insights from other process theories of change (see chapter 6.2) should not be discounted, even though they are harder to depict in an easy-to-digest format.

Table 6.4 Eight-step guide for organisational change (based on Kotter 1995, 60ff.)

Step	Description (verbatim from Kotter 1995, 60)	Major pitfalls (shortened from Kotter 1995)
1. Establishing a sense of urgency	▪ Examining market and competitive realities ▪ Identifying and discussing crises, potential crises, or major opportunities	▪ Underestimating the difficulty to create openness to change ▪ Overestimating the already achieved level of urgency ▪ Lack of patience (skip the phase) ▪ Lack of leadership (executives fear downside risks)
2. Forming a Powerful Guiding Coalition	▪ Assembling a group with enough power to lead the change effort ▪ Encouraging the group to work together as a team	▪ No history of teamwork (especially outside of hierarchies) ▪ Lack of line management involvement
3. Creating a Vision	▪ Creating a vision to help direct the change effort ▪ Developing strategies for achieving that vision	▪ Vision too complicated or blurry ▪ Without a clear vision, the initiative can degrade to a number of incompatible projects
4. Communicating the Vision	▪ Using every vehicle possible to communicate the new vision and strategies ▪ Teaching new behaviours by the example of the guiding coalition	▪ Too little communication: all available channels must be used repeatedly ▪ Behaviour and statements of some senior managers contradict communication messages
5. Empowering Others to Act on the Vision	▪ Getting rid of obstacles to change ▪ Changing systems or structures that seriously undermine the vision ▪ Encouraging risk taking and non-traditional ideas, activities, and actions	▪ Narrow job categories impede realisation of change ▪ Conflicting compensation / performance-appraisal systems ▪ Reluctance to change from superiors
6. Planning for and Creating Short-Term Wins	▪ Planning for visible performance improvements ▪ Creating those improvements ▪ Recognizing and rewarding employees involved in the improvements	▪ Losing momentum without wins after 12 to 24 months ▪ Lack of clear plan to achieve short-term goals reduces urgency levels
7. Consolidating Improvements and Producing Still More Change	▪ Using increased credibility to change systems, structures, and policies that don't fit the vision ▪ Hiring, promoting, and developing employees who can implement the vision ▪ Reinvigorating the process with new projects, themes, and change agents	▪ Declaring victory too soon may lead to reversal of change effort; remaining conflicts with vision are not resolved ▪ The root-cause of problems during this phase often lie in half-hearted efforts in previous phases
8. Institutionalizing New Approaches	▪ Articulating the connections between the new behaviours and corporate success ▪ Developing the means to ensure leadership development and succession	▪ Link between performance and the change initiative is not communicated clearly enough ▪ Change is personified so that it ends with the next generation of managers

Kotter emphasises the importance of creating a sense of urgency in the organisation: He believes that 75 percent of management need to be convinced of the need for change. This is easier to achieve during times of organisational crisis, although less resources are available in such cases to carry out the change effort. In both cases, externals (e.g., consultants, analysts) can be used as messengers to amplify the sense of urgency (Kotter 1995, 60ff.).

Some of the above-mentioned problems of n-step guides to change can be avoided if they are tailored to specific types of change initiatives. Not only can the individual steps be described in much more detail, but opportunities and potential pitfalls that are specific to the type of change can also be addressed. One example of such an approach is a six-step approach designed to develop and implement a green corporate strategy by The Boston Consulting Group (Rubel & Sommer 2007): The first three steps represent the *outside-in-view* and include assessing the general business context (step 1), conducting a stakeholder analysis (step 2), and evaluating the activities and positionings of relevant competitors and peers (step 3). The desired result of this phase is the optimal strategic environmental positioning of the firm. The strategic positioning integrates the theoretical optimum of the firm's positioning (based on its capabilities and customer/stakeholder needs) with the competitive realities considering competitors' market power and positionings and their potential future moves. The second part, the *inside-out view*, is concerned with filling the strategic positioning with life and making it happen. It comprises the development of an explicit green vision, mission and strategy (step 4), the development of concrete, actionable items (step 5), as well as effective internal and external communication of the green transformation (step 6) (Rubel & Sommer 2007).

Kotter's eight steps deal with the change aspect of transformations while the BCG approach also incorporates content-related tasks. It is important to realise that both are intertwined: content generation itself can be considered a change management activity, and the most sophisticated change management approach is useless (and potentially even dangerous) if the company falls short on the design of the core concept. Creativity is usually required to enthuse customers, surprise competition, and create competitive advantage. Hence, the last section of this chapter examines the question how creativity emerges in corporations and is successfully commercialised.

6.4 Theoretical Perspectives on Organisational Innovation

6.4.1 What Is Organisational Innovation?

Innovation has been studied by scholars of economics for a long time (e.g., Schumpeter 1912). In the context of organisations, the following definition has found broad acceptance:

> *"[Organisational innovation is] the adoption of an idea or behaviour that is new to the organisation."* (Hage 1999, 599). Hage continues, *"The innovation can either be a new product, a new service, a new technology, or a new administrative practice."*

This definition implies that an innovation does not need to be completely new to the world; it also includes ideas transferred from other industries or markets, for example. Furthermore, it is congruent with the elements of business models, although the elements beyond products (e.g., novel revenue models, target groups, processes, etc.) do not always receive due attention in academic discourse. Moreover, technical and administrative aspects of innovation cannot be seen in isolation (Van de Ven 1986, 592), just as the business model concept highlights.

Yet, business models as such have largely been neglected in the context of innovation both by organisation and strategic management literature (Teece 2010, 192). Works on the topic that have already been discussed (see 4.6) mainly deal with innovating new business model designs. However, general insights from innovation research are vital to manage the broader innovation process and to inform the management framework in chapter 9.

Innovation is often contrasted with *invention*. While both represent novelties, the latter has not yet been commercialised and widely adopted. An invention has merely proven to work in a laboratory or in simulations and might still take a long time before it becomes an innovation (if at all). One example is that of commercial aircrafts: After the invention of the aircraft (proven by a successful test flight in 1903), more than three decades passed until commercial viability could be demonstrated and aircrafts became an innovation (Senge 2006, 5f.). However, as Van de Ven (1986, 592) remarks, "[n]ew ideas that are not perceived as useful [...] are usually called mistakes." As a result, there is a certain bias in innovation research as "mistakes" are normally not included.

As a rare exception in the field of sociology, findings on innovation were largely consistent and have accumulated over decades (Hage 1999, 597), although some questions remain disputed. After a short presentation of the various types of innovation, important insights that help fostering innovativeness are laid out.

6.4.2 Typologies of Organisational Innovation

The three most common typologies contrast *administrative* vs. *technical* innovations, *product* vs. *process*, and *radical* vs. *incremental* (Damanpour 1991, 560). Incremental innovations are the only ones that are out of scope in this work (as long as they are not part of a larger, non-incremental business model innovation).

Henderson & Clark (1990) have proposed a more granular typology for product innovations: They distinguish between innovations that (a) reinforce or overturn core concepts, and (b) change *linkages* between core concepts and components – or not. The resulting four types of product innovations are (1) *incremental innovation* (reinforcing / unchanged), (2) *modular innovation* (overturned / unchanged), (3) *architectural innovation* (reinforced / changed), and (4) *radical innovation* (overturned / changed). Within this framework, incremental innovation would typically be out of scope. However, the most interesting point that arises from this typology is the following: Incremental innovations (like minor efficiency improvements of the engine of a car) – even when introduced by competitors – work in favour of currently dominating players as they can make full use of their existing competencies. In the case of modular innovation (e.g., involving more significant engine redesigns like incorporating new injection systems or turbo chargers, Magnusson & Berggren 2011, 319), most competencies are still useful, although some new ones are needed and others may become less important. The remarkable insight that Henderson & Clark (1990) bring forward is that in the case of architectural innovation (and also radical innovation), existing competencies can in fact be a *handicap* for incumbents, because unquestioned assumptions – what they refer to as *dominant design* features – become invalid (e.g., fully-electric power trains allow for very different car designs, see 7.8). Unlike most radical innovations, however, architectural innovations initially may not even be recognised as such.

Many business model innovations will be of an architectural or even radical nature. Hence, there is a very real threat for established companies that their long-standing competencies in executing their existing business model disadvantage them vis-à-vis newer competitors when undertaking Green Business Model Transformations.

With his bestseller "The innovator's dilemma" (Christensen 1997), Clayton M. Christensen popularised the concept of *disruptive technology* (also see Bower & Christensen 1995), later more accurately referred to as *disruptive innovation* (Christensen & Raynor 2003; Christensen et al. 2004). In contrast to Henderson and Clark, Christensen argues that distinguishing between incremental and revolutionary innovations based on their attributes is not very meaningful. Instead, he proposes to view them as either *sustaining* or *disruptive*. The former means to sell better products for more money to attractive customers – circumstances under which incumbents were found to be greatly advantaged. In contrast, *disruptive innovation* targets new or unattractive customer segments with simpler products for less money. This type of innovation is supposed to be better suited for new entrants. The great risk for incumbents is that the formerly inferior products evolve and increasingly fulfil the needs of more demanding customer segments, eventually capturing the entire market (Christensen & Raynor 2003, 31ff.).

Looking at low value segments may also be rewarding for green business models. Simple, low-tech solutions embedded in smart business models often save energy and other resources, come at lower cost, and are still sufficient for most customers. Especially base of the pyramid (BoP) markets have to be considered a fertile ground for respective experiments (WEF 2009). Increasingly, large multinational companies seek to tap into these new growth markets (Simanis & Hart 2006, 44). As large and established companies are at the centre of consideration in this work, respective specifics relating to innovation are discussed next.

6.4.3 Innovation in Large and Established Companies

The relationship between the size of an organisation and its rate of adoption of innovation is complex. While Damanpour (1992) found a positive relationship, Hage (1999, 606) even pleads that the variable of organisational size should just be ignored by researchers. He claims that irresolvable methodological problems are associated with it – for example, the relationship greatly varies by industry (Freeman & Soete 1997, 234). Generic statements concerning this matter should thus be met with caution. Meta-analyses of innovation research (e.g., Damanpour 1991; Hage 1999) have revealed significant correlations with other variables, though. Most notably, specialisation (typically measured by the number of different occupational types) is positively correlated with all kinds of innovation types of organisations (Damanpour 1991, 576). Other positive associations were found for "functional differentiation, professionalism, managerial attitude toward change, technical knowledge resources, administrative intensity, slack resources, and external and internal communication" (Damanpour 1991, 569). In contrast, centralisation was found to have a significant negative correlation with innovation (Damanpour 1991).

The case of radical innovations

One claim related to organisational age made by some researchers is that companies become less and less likely to create radical innovations the older they get (Anderson 1999, 227; Van de Ven 1986, 596). Others argue the same to be true for large companies, suggesting that their inertia hinders them in realising radical innovation (Schaltegger & Wagner 2008, 40ff.). Ettlie et al. (1984) argue that size is correlated with higher structural complexity, formalisation and decentralisation – factors which in turn rather promote incremental innovations than radical ones. However, Damanpour (1996) found that complexity and size are in fact *positively* associated with radical innovation. Slack resources, the availability of technical knowledge, and an attenuated financial risk of failure are among the potential causes he names (Damanpour 1996, 699).

As researchers seem to disagree whether or not size and complexity favour radical innovations, perhaps, the chosen strategic orientation of an organisation provides a more fruitful explanation (Miles et al. 1978; Damanpour 1996). At least for the companies that strategically seek to strive through radical innovations, it is essential to put due effort into overcoming respective disadvantages and exerting own strengths for radical innovation. This includes forgetting "old certainties" from the established business model that are no longer valid, borrowing key resources that are invaluable, and being ready for explorative learning (Govindarajan & Trimble 2005).

Establishing a focus on innovations

As noted above, the challenges for large corporations to remain innovative are manifold: They are highly complex, fragmented, and often very "political" organisations – a characteristic that goes against the common notion of an entrepreneurial and innovative firm. Kanter (2008) proposes to overcome these disadvantages by implementing standardised processes to free capacity for high-value activities and establishing shared values that guide decisions without formal leadership mechanisms. This way, the organisation is not occupied with continuous fire fighting and stays flexible yet focused to pursue innovative ideas. Senge (2006) goes beyond the concept of shared values and calls for establishing a shared *vision* within the company: According to Senge, through a shared vision an organisation can hold the creative tension necessary to produce significant innovation – even during times of crisis. Dougherty (1992) remarks that innovators who belong to different functional areas within large firms may not disagree about general goals, but produce "interpretive barriers" that need to be dealt with for effective innovating.

Innovation strategies for large and established companies

Damanpour & Wischnevsky (2006) suggest distinguishing between innovation-generating and innovation-adopting companies. Established companies in many ways are incentivised *not* to search for disruptive innovations in their own industry sector. Not only is the risk of failure relatively high, but even success could jeopardise profits from current products and processes. Hence, it may be reasonable to wait for proof of concept from a small competitor and then conquer the market as a fast second mover through superior competencies and resources (Markides & Geroski 2005). However, incumbents need to be careful not to kill such innovations through ill-conceived financial analysis: Christensen et al. (2008) point out that traditional financial analysis can lead to non-competitive costs and performance erosion in the long-run: First, when fixed and sunk costs are ignored when evaluating an investment in a new technology, there is a heavy bias towards the current technology. However, new entrants who do not have this bias can gain a long-term competitive edge if the incumbent considers the *usable* rather than the *competitive* lifetime of its assets. Second, methods of discounted cash flow (DCF) and net present value (NPV) usually assume that doing nothing (as the alternative to investment in innovation) leads to constant performance. However, as new entrants and competitors do innovate constantly, the base case should be performance erosion rather than stagnation (Christensen et al. 2008). Established companies can circumvent many of the problems mentioned above if they innovate at the periphery instead. Yet, venturing away from core business also bears risks, and many companies have failed in doing so (Zook 2007). Once companies have found their strategic view towards innovation, they need to make the most of the sources of innovation that are available to the firm.

6.4.4 The Sources of Innovation

Von Hippel (1988) has pointed out that in contrast to common knowledge it is not always product manufacturers themselves who conceive innovations, but also users (i.e., customers) or business partners. The distribution of innovations by source depends on the expected profits from the innovation. In contrast to a manufacturer's innovation, user innovation only has to be useful to a very limited number of users (including the innovator himself). The innovators among users tend to be a highly concentrated group termed *lead users*. Hence, instead of relying exclusively on in-house R&D, a company may also follow the strategy to redesign products in order to facilitate innovation by lead users and consequently commercialise their ideas.

Moreover, Von Hippel (1988) remarks that firms' engineers engage in informal trading of proprietary knowhow with other firms (and even competitors). In more general terms, the ability "to recognize the value of new, external information, assimilate it, and apply it to commercial ends is critical to [a firm's] innovative capabilities" (Cohen & Levinthal 1990, 128). This ability is known as *absorptive capacity*. Cohen & Levinthal (1990) argue that absorptive capacity is path dependent, i.e. prior investments in knowledge further promote the acquisition and application of valuable knowledge in the same domain. Therefore, firms with own internal R&D departments do not only generate more original innovations, but should also find it easier to tap external sources of knowledge as a by-product.

Many firms have sought to compensate for their lack of competency with strategic alliances (see 4.4.3) to create and commercialise new business models (Kale & Singh 2009). For example, energy giant BP has entered several alliances with chemical firms as it lacked the required expertise to develop and commercialise biofuels quickly[15] (BP 2009). Furthermore, alliances can be designed "to generate innovations for an entire industry or to impose a new global standard" (Hage 1999, 611). However, it should be noted that managing alliances itself is challenging, with many alliances delivering little value (Kale & Singh 2009).

Chesbrough (2003a; 2003b) has popularised the term *open innovation* that contrasts with the old paradigm of closed (exclusively internal) innovation. Open innovation does not discard the concept of internal R&D; rather it proposes a blended innovation process equally relying on external knowledge as it has become more readily available in recent years. Chesbrough (2006; 2007) further propagates to move to *open business models* as a next step. He argues that increasing internal development costs and shorter product life cycles (i.e., reduced market revenue) can thereby be compensated with cheap external development and revenue from external commercialisation of intellectual property through licence fees, spin-offs, or divestments (Chesbrough 2007).

6.4.5 Innovativeness and Absorptive Capacity

The innovativeness of a firm heavily relies on the interfaces both between subunits and with external sources of innovation-relevant knowledge. If these interfaces do not possess either the required expertise or the capacity to effectively capture and redirect this knowledge, the absorptive capacity of the firm – and hence its innovativeness – will be limited (Cohen & Levinthal 1990). Diversity of background is essential as a broad range of expertise is necessary to relate to diverse emerging fields of knowledge (Cohen & Levinthal 1990; Hage 1999, 605). Cohen & Levinthal remark that gatekeepers can be used to "translate" knowledge domains if internal and external actors have little overlaps in expertise. In the case of rapid and uncertain technical change, an *organic structure* (Burns & Stalker 1994) of the organisation is believed to be preferential (Cohen & Levinthal 1990, 132). This means that work is conducted in the absence of strong formal hierarchies, methods, and duties. Communication tends to be more lateral than

---

[15] This includes a partnership with chemical firm DuPont for the development of the advanced fuel molecule biobutanol, and a joint venture with Verenium, a biotechnology firm, to produce ethanol from lignocellulosic feedstocks. In addition, BP is investing US$ 500 million in its Energy Biosciences Institute (EBI) to explore applications of biotechnology to energy (BP 2009).

top-down compared to centralised, *mechanistic structures* (Burns & Stalker 1994, 5f.).

One problematic implication of the concept of absorptive capacity is that it cannot be built within a short period of time. Prior technological knowledge and complementing market knowledge are essential to recognise relevant external information as such and to be able to capitalise on it commercially (Lichtenthaler 2009). This issue is explored further based on practical examples in chapter 7.8.

Furthermore, absorptive capacity follows dynamics similar to some of the systems archetypes (see 4.5.3). Absorptive capacity can be viewed as a double self-reinforcing loop (Cohen & Levinthal 1990, 135ff.): On the one hand, the higher the level of absorptive capacity, the more efficient the accumulation of additional knowledge becomes. On the other hand, absorptive capacity increases the sensitivity to commercial opportunities and resulting expectations. Both factors increase the attractiveness of further investments in absorptive capacity. These virtuous circles can be further strengthened by organisational structures that foster learning. At the same time, inactivity results in an erosion of absorptive capacity, depending on the rate of change in the external environment.

Of course, the described dynamics also work in reverse. That is, if a company fails to build-up knowledge in a promising new field initially, catching up requires disproportionally high effort; the firm may become "locked out" (Cohen & Levinthal 1990, 136). Due to a lack of prior knowledge, missed opportunities may not even be recognised anymore. This effect may be prevalent in many companies that view sustainability as a public relations issue, but lack the technological and market-related knowledge to appreciate its true potential (Berns et al. 2009b, 5). The lock-out effect is further intensified when low absorptive capacity leads the firm to rely more and more on its existing knowledge; over time, it will develop high levels of competency in potentially inferior or outdated fields of knowledge. This will not only increase inertia but also result in a loss of diversity of backgrounds, thus speeding up the erosion of absorptive capacity. Cohen & Levinthal (1990, 135) remark that buying absorptive capacity via external hires, consultants, or acquisitions can be an option, but is limited to the extent that integration with the firm's activities is important for innovation and also takes time.

In conclusion, firms with high absorptive capacity will find it relatively easy to maintain a high level, but it is difficult to significantly improve absorptive capacity from a low level. It should be noted that absorptive capacity is also subject to limits to growth. Once absorptive capacity approaches its maximum, the marginal return from further improvement will become negative – despite the high efficiency in accumulating knowledge: there will just not be much valuable information left that has not been captured yet. Because absorptive capacity is usually restricted to specific knowledge domains (Cohen & Levinthal 1990, 148), a firm may face tradeoffs between research spending on the one hand, and the level of absorptive capacity and the number of active knowledge domains on the other hand. Next, the actual process of how innovation is created in practice is laid out.

6.5 Innovation Management

6.5.1 Antecedents and Challenges of Successful Innovation Management

Van de Ven (1986, 591) asserts that "From a managerial viewpoint, to understand the process of innovation is to understand the factors that facilitate and inhibit the development of innovations". These factors can either impact the innovativeness of the firm in general terms (see chapter 6.4.5), or only apply to specific innovations. For example, the failure of single potential innovations can be caused by many factors, but most fall into one of two broad categories: (1) the innovation is flawed (i.e., has an insufficient value proposition), or (2) it is implemented poorly. This suggests that high quality innovations which are implemented properly always get widely adopted and low quality ones do not. Unfortunately, reality does not quite confirm this train of thought as the example of the failed, but superior Sony Betamax video format, and the still prevalent "QWERTY" keyboard[16] illustrate (Sterman 2008). Explanations of such seemingly unjustified cases of success and failure, respectively, usually include factors like hidden costs, agency problems, or externalities. However, Sterman (2008) concludes that all-too often the managerial problem originates from mental models that neglect feedback processes, delays and other system dynamics elements, which also happen to play an important role in sustainability-related innovations. Systems thinking (see 4.5) can help understand these causal relationships.

McGrath et al. (1996) propose that, in addition to causal understanding, there are three further antecedents necessary to capture rents from innovations: The four antecedents are (1) causal understanding; (2) innovation team proficiency; (3) emergence and mobilization of new competences; and (4) creation of competitive advantages.

The antecedents build upon each other. McGrath and colleagues thus argue that lower-order antecedents (e.g., causal understanding) precede higher-order ones (e.g., competitive advantage), although they can still develop further in parallel with higher-order competencies. Hence, during more mature phases of the development process, the rent potential grows faster than in the beginning (McGrath et al. 1996, 393f.).

To fulfil the four antecedents mentioned above, innovation management must overcome four central kinds of problems as formulated by Van de Ven (1986): (1) "the *human* problem of managing attention", i.e. getting the attention focused on fostering new ideas instead of protecting existing practices; (2) "the *process* problem is managing new ideas into good currency", (3) "the *structural* problem of managing part-whole relationships" as multiple, heterogeneous (and sometimes loosely

[16] QWERTY" refers to the order of keys on English language keyboards. This order is a remainder of the age of early typewriting in the 19th century. It was chosen in order to prevent typebar clashes. The Dvorak keyboard, an alternative designed to improve efficiency and reduce fatigue, has not been adopted widely despite its obvious superiority in the absence of the initial mechanical problems (David 1985).

coordinated) parts of the organisation are needed to implement the innovation, and (4) "the *strategic* problem of institutional leadership", thus "creating an infrastructure that is conducive to innovation" (Van de Ven 1986, 591, emphasis altered).

6.5.2 The Process of Innovation

Utterback & Abernathy (1975) have proposed a macro model of the innovation process according to which both the development of production processes and the development of products each go through three sequential stages: In the first stage the market is still immature, and new production processes tend to be uncoordinated and associated with frequent changes (that are often *not* innovations). For products, the focus lies on performance-maximising accompanied by frequent changes and high margins. As the market matures in stage two, the rate of process innovations reaches its peak and segmental leaps in process efficiency take place. Companies now try to maximise sales, and product innovations tend to be more geared towards differentiation. In the third and last stage, the whole production process becomes highly integrated and every change is costly due to systemic dependencies. Contingent on the context, this may concern single divisions or even several firms of a vertically fragmented industry. Products now tend to become standardised in order to minimise costs.

Utterback & Abernathy (1975) also propose that most innovations during the first stage are stimulated by market needs, by new technology in the second, and by production-related factors (for minimising cost) during the last stage. In contrast to process innovation, product innovation tends to fall continuously the more mature the market becomes. However, the conclusion that large companies should therefore produce relatively more process than product innovations has not been confirmed empirically – in fact the opposite seems to be the case (Damanpour 1996).

Johnson (2010, 55ff.) argues that there are several sequential shifts in the basis of competition: from performance to reliability, to convenience (may be skipped), to cost. Johnson also states that product, and later process innovations dominate the first two stages, yet that shifts to convenience and cost are often the result of business model innovations. However, neither of the authors seems to acknowledge that a shift *back* from cost is possible; yet empirical examples show that a shift towards sustainability as the basis of competition may facilitate just that (see chapter 7).

More often than not, individual product innovation is considered the output of R&D processes. The classic stages of the linear product innovation process in a manufacturing context have been identified as follows: (1) basic science; (2) design and engineering; (3) manufacturing; (4) marketing; and (5) sales. This *technology-push* view is contrasted with the *market-pull* model for more mature markets: (1) market need; (2) development (3) manufacturing; and (4) sales (Rothwell 1994, 8f.). However, as discussed above, innovation can have many different origins within and outside of organisations. Hence, the process stages listed above may also be performed or supported by external parties including suppliers, users, or partners (Von Hippel 1988). In addition to integrating external parties into the innovation process, its stages are often conducted in parallel. The innovation process is increasingly geared towards speed and efficiency (Rothwell 1994, 15ff.).

Just because innovations are – by definition – new to the organisation, they are not necessarily conceived with big eureka. Instead, more or less structured processes exist to uncover potential starting points for innovation (Anthony et al. 2006). As a concrete example, Shapiro (2001, 33ff.) proposes seven dimensions of potential changes for process innovation. These changes can not only improve the economic, but also the environmental performance of a process:

1. *Rethink*: Always question why processes are the way they are.
2. *Reconfigure*: Consider ensuring quality early in the process and look at best practices from other industries; eliminate duplicated and low-value activities – as expressed concisely by Peter F. Drucker: "There is surely nothing quite so useless as doing with great efficiency what should not be done at all."(Drucker 1963, 54)
3. *Resequence*: Improve the sequence of tasks by doing them earlier (when better prediction is possible), by postponing them to become more flexible, by performing them in parallel, or by reshuffling them to reduce bottlenecks and dependencies.
4. *Relocate*: Consider relocating activities to customers, suppliers, closer together, or making them virtual.
5. *Reduce*: Use better information or simplify in order to reduce the frequency of tasks or to make better use of critical resources.
6. *Reassign*: Consider in- and outsourcing as well as shifting activities to suppliers or customers.
7. *Retool*: Consider new technologies, automation, leveraging additional competencies, and re-skilling of staff.

Specific considerations for managing the innovation process are laid out in the following. For this purpose, I distinguish the following four generic stages (that cover all kinds of innovations): idea generation, validation, realisation, and institutionalisation of innovation practices.

6.5.2.1 Idea Generation

An innovative firm enjoys a constant flow of ideas, with the promising ones being selected efficiently for further exploration and eventually commercialisation. Internal idea generation, trends from within the firm's industry as well as from other external sources can stimulate creativity. In fact, many disruptive innovations historically came from the periphery – with Apple (iPod, iTunes, iPhone, etc.) being just one of the most well-known examples of companies that repeatedly disrupted adjacent markets.

However, most companies are neither innovative in the sense described above, nor do their ideas lead to radical innovations very often. Three techniques briefly described in the following can be used to stipulate idea generation within a relatively short period of time: scenario planning, idea generation workshops, and idea collection. However, long-term development of innovation capabilities (see section 6.5.2.4) should not be neglected, even if short-term idea generation methods are applied successfully.

Scenario planning

Pierre Wack, who pioneered scenario planning at Royal Dutch/Shell, circumscribes the idea behind scenario planning as follows: "Scenario planning aims to rediscover the original entrepreneurial power of foresight in contexts of change, complexity, and uncertainty." (Wack 1985a, 150) For this purpose, scenario planning addresses two major problems that many long-established companies have in common: overconfidence and tunnel vision (Schoemaker 1995). That is, "[s]cenarios address blind spots by challenging assumptions, expanding vision and combining information from many different disciplines" (Shell 2008, 16). Scenario planning workshops usually include executives, but need to be prepared thoroughly. A detailed explanation of how to use scenario planning to pave the way for new breakthrough innovations, as well as for the application of the other two techniques that follow, will be presented in chapter 9.

Idea generation workshops

Idea generation workshops depend both on the creativity and the expertise of its participants to generate a large amount of ideas within one or two days (approximately one idea per person and hour is realistic). During the workshops it is important to find the right balance between structure and freedom: People need some guidance in order to produce meaningful ideas, but also need to be stimulated to take new perspectives of thinking (Coyne et al. 2007). This can be achieved by providing mental bridges, e.g. by thinking in terms of how a highly successful company like Google would approach an issue (see Jarvis 2009). Another technique involves carefully chosen probing questions (Coyne et al. 2007, 73ff.). Like scenario planning, idea generation workshops require significant pre and post work to be effective.

Idea collection

Idea collection is based on the assumption that there are many good ideas floating around in the organisation that have just not been captured and combined in a meaningful manner. GE has successfully conducted energy "treasure hunts", a process developed by Toyota, to find ways to save energy throughout the organisation. These involved 181 locations, 2,600 employees and led to projects that reduced CO_2 emissions by 250,000 tons in 2006/07 (General Electric 2006, 31). The collecting of ideas does not need to be restricted to employees; suppliers or customers can also be a valuable source of innovation. GE also uses so called "dreaming sessions" that include key customers who can freely express their wishes in order to capture the next "big thing". GE's highly successful Ecomagination campaign, inter alia, originated from one of them (Fisher 2005).

Some of the world's largest *Crowd Ideation Events*, as this technique will be referred to in the following, are IBM's "Innovation Jams" (Bjelland & Wood

2008; IBM 2008a; IBM 2010c). In the 2006 event, 150,000 people participated in two 72-hour online sessions and posted more than 46,000 ideas in web forums and wikis. The first session was structured along 25 clusters of technologies in six broad groupings. After the first session, some 31 "big ideas" were selected through text mining software and substantial management time. Not surprisingly for such a gigantic online brainstorming session, discussions were hard to moderate and many ideas were completely useless; however, many ideas helped to realise incremental improvements to existing businesses, and some even led to the creation of new businesses.

Participants had already discussed many of the ideas brought forward before the Innovation Jam with their managers. The event gave them the opportunity to expose these ideas to a wider audience with different (and potentially critical) perspectives. Moreover, many small ideas could be combined with others so that they had an impact. However, as Edward Bevan, vice president at IBM Research and one of the organisers, remarks: "Idea generation is in some ways the 'easy' part [...] of innovation, whereas advancing, refining and building support for those ideas is the really tough part." (Bjelland & Wood 2008, 40).

6.5.2.2 Idea Validation

With its Innovation Jam 2006, IBM did not only try to capture ideas, but also sought to speed up their commercialisation significantly. It was found that the second Jam session – which was supposed to refine selected business ideas – was ineffective in that format and hence was not repeated at later events. Apparently, this work is better done by managers, not by the crowd. Nevertheless, ten businesses funded with 100 million dollars have been created from the Innovation Jam 2006. The most remarkable one, not surprisingly, belongs to the environmental sustainability domain: "Big Green" started during the Innovation Jam as intelligent combinations of many different ideas and insights, and eventually became "the largest single initiative in IBM history: a billion dollar program to change radically how IBM and its customers use energy and other resources for computing." (Bjelland & Wood 2008, 36). Amazingly, the Jam revealed that IBM could make significant contributions to water management, one of the most significant environmental challenges of the future that no senior manager at IBM had considered so far. Nevertheless, after the Jam senior executives founded a new business unit. Many of its later members had contributed environmental ideas during the Innovation Jam (Bjelland & Wood 2008, 39). The Jam in 2010 even was dedicated to eco-efficiency (IBM 2010c).

However, not all of the initial ideas went on a path to quick commercialisation. Some ideas were transferred to R&D, and one of the ten big ideas was shelved after market demand turned out to be insufficient (Bjelland & Wood 2008, 36). A crucial part of innovation management is thus to abandon misjudged ideas early enough and refocus on more promising ones (Anthony et al. 2006, 111f.).

Challenges for engineering-driven firms

Even IBM, despite its reputation as a great innovator, has difficulties in commercialising its inventions. IBM and other technology-savvy companies can find it difficult to identify practical uses for their leading-edge technology. Or, members of engineering-driven firms wrongly assume that highly challenging technologies guarantee good profits if they can be tamed. Customers buy product and service functions – and not the scientific achievement behind them. Hence, evaluating how well a proposed innovation meets the market demand (potentially in reference to competitors' offerings), can be enlightening (McGrath et al. 1996). If a product, however, is designed to fulfil customer needs in a novel way (e.g., requires an unfamiliar usage pattern) or addresses previously unmet customer needs (rather than simply replacing a substitute product), even the most sophisticated market research may fail to accurately determine the prospects of the new value proposition (Von Hippel 1988, 102ff.). Thus, such benchmarks need to be applied very carefully with radical innovations – if at all. Von Hippel (1988, 106ff.) proposes to use lead users instead of typical users to explore novel and unfamiliar value propositions. Although lead users are also constrained by their past experiences, they are more familiar with future-oriented or experimental product uses. In contrast, Shapiro (2008) points out that the laggards who do not use a company's products yet may be an even more valuable source of information. After all, they are the target segment for game changing, discontinuous innovations (Christensen 1997).

Challenges for financially-driven firms

More financially-driven companies instead tend to be vulnerable to the pitfalls of financial analysis tools. Predicting the economic success of particular innovations is generally considered to be very difficult (McGrath et al. 1996, 401). Investment decisions for innovations based on DCF and NPV are problematic (see 6.4.3); in addition, most project teams know the required thresholds and "make the numbers work" by playing with the assumptions. However, it is in fact the assumptions that should matter most to executives, and not the largely speculative revenues or profits in five or ten year's time (Christensen et al. 2008).

For this reason, *discovery driven planning* (McGrath & MacMillan 1995) has been developed. Discovery driven planning turns the whole game around: the minimal financial requirements in the form of a reverse income statement are determined first. Based on this, assumptions that suffice minimum requirements are backtracked and tested for their viability. If the feasibility of the requirements can be credibly demonstrated (including implicit assumptions!), the project gets approved; if not, further study has to be done or it gets shelved. Analogously, pilots need to confirm the tracked assumptions and not necessarily meet financial measures. Especially with new business models that involve non-linear dynamics (e.g., word of mouth) small changes to underlying assumptions (e.g., contacts per user or conversion rates) can have huge effects on resulting financials that are thus prone for manipulation.

Whether a company is engineering- or financially-driven, the insights described above all underline how much value a systematic check of assumptions along the business model elements can add to the validation process – especially when the innovation is radical in nature.

Challenges related to strong uncertainty

Scenario planning is a useful tool to make decision makers aware of uncertainty and its potential implications. However, scenario planning is silent about which scenarios should be the basis for action, or to what extent. A solution to this problem may be applying *real options* to business strategy (Luehrman 1998a; 1998b). That is, a firm invests just to have the *option* to carry out a course of action at a later time if a certain future materialises (Amram & Kulatilaka 1999). However, Mintzberg et al. (2009, 65) remark that such a use of real options is difficult in practice: For one, the technical methods are not as well-established as in their original finance context. Moreover, in highly uncertain environments, i.e. when forecasting is impossible (a major reason for doing scenario planning in the first place), judging options can become very unreliable. Large option values with hidden underlying assumptions can thereby change the recommendation without deciders realising it (Bierman and Smidt 1993, 486). And lastly, the real options approach does not work well for any Green Business Model Transformation that requires all-or-nothing decisions. For example, GE could not sensibly have decided to implement only part of Ecomagination as the majority of benefits stemmed from the consistent whole of the initiative. In fact, due to reinforcing feedbacks, many business models should only be implemented fully-fledged or not at all. In other cases, though, making a big entrepreneurial bet can be avoided through business model pilots. For business models that *might* become dominant in the future, but whose time has not yet come, a company may opt for a precautious course of action and build competencies that are needed. Yet, due to the limited general applicability to Green Business Model Transformations, the real options approach is not pursued further in the management framework of chapter 9.

6.5.2.3 Idea Realisation

Even when a firm has correctly validated the assumptions of an innovation investment, success is not guaranteed. In some cases, assumptions about the innovation idea are correct, but the assumptions about the own implementation capabilities are too optimistic (McGrath & MacMillan 1995). McGrath et al. (1996, 400) remark that innovation managers tend to focus on technical challenges and frequently neglect critical economic success factors like team proficiency and new required competencies. The process of building competencies is difficult to measure and therefore also challenging to manage. Even if managed well, building significant new competencies takes time; a partnership may be the best way to fill the gap when time is critical.

In reference to Schön (1971), Van de Ven (1986, 592f.) emphasises the socio-political aspects of innovations: Ideas do not induce lasting change unless they activate political debate and become a means to gain influence and resources. Before an idea reaches legitimisation and becomes taken for granted, it can fall victim to short-term focus and problem evasion as described in the garbage can model (see chapter 6.2.6). Project management tools can effectively be applied to address these issues (Huber 1984, 938; Van de Ven 1986, 593f.).

The innovator's job, however, is not even done when the innovation is implemented and working according to the project plan. Especially with respect to process and administrative innovations, the termination of now obsolete procedures need to be actively managed in order to avoid "shadow" structures and activities that undermine the success of the innovation (Van de Ven 1986, 600f.).

Another common pitfall is ignoring the fact that the environment is changing, too. Assumptions may become invalid as competitors and the broader business environment evolve. Continuous monitoring combined with adaptability is critical. Speed is important, too, as opportunity windows may close quickly. With reference to the great speed at which Japanese manufacturing firms used to convert ideas into commercial products, Simon (1993, 140) proposes to involve other relevant functions early in the development process. This way, R&D is less likely to come up with products that are expensive to manufacture or miss the customers' needs. However, R&D must be careful not to sacrifice the new product's distinctiveness in order to comply with everyone's requests.

One important domain to mention here is marketing and sales. For example, an ill-conceived launch of a revolutionary new green product can break the whole business model. The respective growth dynamics that marketing and sales can activate have been laid out in chapter 4.5.4.

As a last caveat, innovation managers need to be aware of the fact that commitment – often praised as a pivotal enabler for true breakthroughs in organisations – has a flip-side: Innovators can become so enthusiastic that they are viewed as fanatics by people outside their inner circle (Senge et al. 1999). This may materialise in the form of the use of special language (e.g., project-specific terms) and ignorance with respect to the interests of those affected by the change. These in turn will react with more or less directly provoked resistance. It is thus important for innovators to stay in constant contact with the wider organisation and remain open for feedback in order to ensure a smooth diffusion of the innovation (Senge 2006, 296ff.).

6.5.2.4 Institutionalisation of Innovation Practices

Firms should generally strive to foster innovation practices that overcome the human tendency to shirk from decisions under high complexity and uncertainty. Failure to do so can result in gradual performance erosion. The famous frog analogy applies here: If a frog is thrown into boiling water, it immediately jumps back out. Apparently, if the frog is put into cold water which is then heated slowly, it stays in the water until it cannot escape anymore and eventually dies after the water starts boiling. In such situations, innovation efforts equal crisis management, which is argued to frequently produce "mistakes" instead of innovations (Van de Ven 1986, 595f.). Even if the "crisis innovations" are successful, a continuously well-performing innovation practice is obviously better than the recurring need for a serious crisis to spur innovation.

Institutionalising innovation practices is also relevant for Green Business Model Transformations. Usually these entail the strategic decision to venture into a domain which the company is not fully familiar with and that requires significant learning. Moreover, new business models often require substantial optimisation, even if the introduction was successful.

Continuous environmental scanning

Simon (1993) emphasises the importance of institutionalised intelligence activities that connect the firm with its environment in order to anticipate the future. Firms in dynamic industries tend to have internalised the fact that they have to generate new ideas well before the old ones lose their traction. In a study of UK-based companies from various sectors (automobiles, book publishing, investment banking, and life insurance), Pettigrew & Whipp (1991, 104ff.) found that environmental assessment is one of the central distinguishing factors of high-performing firms. Instead of technical, one-time efforts to assess the environment, these companies are open and continuously learn on all levels.

Therefore, R&D and customer-facing functions should routinely be involved with the external environment in order to capture promising new trends. Other functions like marketing – although maybe not as critical in this respect – should foster two-way communications as well. Moreover, environmental scanning activities need to be differentiated enough to recognise the different needs and interests of various stakeholder groups (Van de Ven 1986, 594): If relevant distinctions are "averaged out", important opportunities as well as potential sources of resistance may be overlooked.

Organisational structures and innovation

Anthony et al. (2008) present four organisational structures that are suited to support innovation in different contexts (but can be used in parallel): First, *training units* complement the infrastructure for innovation and ensure that the required skills are available. Second, *funding or oversight mechanisms* can be used to champion ideas, fund them, and remove internal obstacles. They may as well oversee the discovery-driven planning process. Third, *incubator groups* may be used to rapidly develop innovations from a first idea to a business concept which is then reintegrated into the core organisation. Fourth, *autonomous growth groups*, composed of innovation generalists and high-potential leaders, can be used to advance innovations that are incompatible or even conflicting with the organisation's core (Anthony et al. 2008, 49). As radical (and also architectural) innovations entail managing an unfamiliar, complex recombination of business model elements, the innovation unit needs to be able to take a holistic view. According to Van de Ven (1986, 598ff.) this may be best achieved if the innovation unit is not only autonomous, but if its members represent the rich variety of the relevant business environment and if everybody understands the overall concept of the innovation beyond his or her functional specialty.

However, due to economic considerations the degree of conflict between the old and the new business model should not be the only relevant criterion whether to keep the innovation effort separate or integrated: The *strategic relatedness* of the business models, i.e. whether they address similar markets, determines the potential for synergies between them (Markides & Charitou 2004). Markides and Charitou argue that a separate unit as described above will usually be best if conflicts are serious and markets do not overlap. Even then, senior management should coordinate the business models in order to realise at least some synergies (Markides & Charitou 2004, 26; O'Reilly III & Tushman 2004). If the markets are similar, however, synergies may justify re-integration of the new unit once the business model has matured. In this case, the firm should try to avoid that the two organisations diverge from each other more than necessary (e.g., by aligning IT systems, transferring managers or conducting company-wide events) (Markides & Charitou 2004, 30f.). On the other hand, if the new business model does not threaten the core business and no markedly different mindset or steering mechanisms are required, it will usually be expedient to leverage existing resources as much as possible and integrate the new business model from the beginning (although interferences from the old business model still need to be managed). Only if the strategic overlap is low it makes sense to separate them later to increase autonomy in order to compete more effectively (Markides & Charitou 2004, 31f.).

Steering innovativeness within the firm

Some firms successfully manage decentralised innovation through challenging targets and smart incentivising for divisions. For example, GE has set stretching revenue targets for its Ecomagination portfolio (products internally certified to be ecologically friendly) and provided funding to spur innovation with green products (General Electric 2008). Setting some focus can actually spur innovation rather than dampen creativity – as long as it is not too narrow regarding the means to achieve it. Firms should generally attempt to maintain a balanced portfolio of innovations that improve the core business, extend it, and enable growth in novel areas (Anthony et al. 2008). Moreover, the traditional "big bets" approach to innovation, in which a small number of ideas receive significant funding based on forecasts and business case analysis, can be complemented with an evolutionary learning approach (Shapiro 2008): a large amount of ideas each receive limited funding initially. Then, once early signals become apparent for which ideas show good prospects, less promising ideas can be eliminated. During the next phases, the remaining ideas receive more funding until a few high-potential ideas emerge from the selection process. Due to the large amount of considered opportunities, informed selection and the cumulative funding process, the likelihood of success is high despite moderate financial investment.

It is important for senior management to withstand the temptations of neglecting long-term performance in order to boost short-term results as the earnings-per-share thinking prevalent in many established companies suggests (Christensen et al. 2008). This also means establishing a positive (but sensible) attitude towards risk; otherwise only incremental, low-risk innovations will be supported. Eventually, radical innovations may not even be suggested anymore – they become known to be rejected anyway.

Moreover, management should seek to cultivate innovators within the organisation (Cohn et al. 2008): Once they have been identified (e.g., via talent management programmes), they need to be developed through gradually more challenging innovation assignments. Mentoring, peer networks and placement in a high-impact position maximises innovators' effectiveness. For that reason some companies place them as an "innovation hub" outside of the normal line organisation.

6.5.3 Using Sustainability as a Catalyst for Innovation

Fussler (1996) was among the first to use the term *eco-innovation* in literature, suggesting that firms should focus on innovations fuelled by environmental sustainability issues. Nidumolu et al. (2009) argue that sustainability has meanwhile become the key driver for innovation. They claim that innovation management and sustainability management may increasingly become synonymous. This suggests that institutionalising corporate sustainability in organisations (see Jennings & Zandbergen 1995) should not only improve the environmental performance of a firm, but also its innovativeness. Nidumolu et al. (2009) propose to manage such innovation based on a five-stage model that involves building cumulating competences. Each stage offers unique innovation opportunities. In the first stage, firms start to view compliance as an opportunity. They can then move on to making their value chains sustainable (stage 2), designing sustainable products and services (stage 3), developing new business models (stage 4), and finally arrive at what the authors call "creating next-practice platforms" (Nidumolu et al. 2009, 64). This last stage corresponds to revolutionary green business models that impact more than just a single market.

The sequence and implied ranking order of types of sustainability innovation proposed by Nidumolu et al., however, is overly simplistic. For example, green innovations in value chains can be much more challenging and significant than certain green product innovations. Apart from that, they are certainly correct to infer that green business model innovations of various types require significant competences – competencies that many companies do not possess today (see also Berns et al. 2009b, 18f.).

Eco-innovations – at least if understood broadly – cannot be reduced to their direct environmental impact (see Esders 2008, 23ff. for a discussion of definitions). In fact, some eco-innovations have far-reaching systemic effects (Andersen 2008). For firms that undergo a Green Business Model Transformation this means that they should try to understand, and possibly steer, these effects in their favour. As concrete examples greatly facilitate exploring this complex topic, however, it will be part of chapters 7 and 8.

This chapter provided the theoretical background for understanding organisations, change, and innovation. Respective practical implications and tools described in management literature have also been discussed. Whenever appropriate, the relevance of sustainability for these topics has been highlighted. In addition, environmental sustainability issues and various means to improve them have been laid out in chapter 3.3, business model change in 4.6, and a variety of business model prototypes that exploit these means in chapter 5.5. Combined with the insights on innovation and change presented in this chapter, all theoretical and conceptual groundwork necessary to create, implement, and operate innovative green business models is therefore provided, and ready to be synthesised into a concise management framework (chapter 9). However, to ensure that this synthesis does not suffer from abstract inferences that do not conform with reality, a broad survey on Green Business Model Transformations has been conducted. The remainder of this book thus offers novel insights on opportunities and challenges arising from Green Business Model Transformations, derived from management practice.

PART III

TOWARDS A PRACTICAL MANAGEMENT APPROACH

Chapter 7
A Survey on Green Business Model Transformations

A central motivation for this publication is the fact that Green Business Model Transformations have been rarely seen in the business world to date. Although sustainability reports and myriads of related corporate press releases suggest otherwise, most activities in the domain have been non-substantial or limited to isolated aspects of corporate practices. Hence, it comes as no surprise that textbook examples of companies that successfully changed their business models based on environmental considerations are difficult to find. Moreover, of the few well-known examples that exist, critics may question motives, environmental benefits, and the genuine contribution to economic success that can be attributed to the green transformation itself.

Nevertheless, a comparison of theoretical and conceptual considerations with economic reality is an essential step to develop a credible management framework. For this purpose, not only recognised successful transformations need to be examined, but also companies that operate in industries which have produced no Green Business Model Transformations so far, like the airline industry. In addition, companies from industries that are in the process of restructuring are examined; these companies are currently trying to seize opportunities for hitherto unknown green business models. Covered transformation stages thus include the anticipation of green business models yet to come, experimentation with novel designs, scaling up a tested business model, and monitoring and refinement of established green business models. Table 7.1 provides an overview of the case vignettes presented in this chapter:

A. Sommer: Managing Green Business Model Transformations, SPLCEM, pp. 171–219.
springerlink.com

Table 7.1 Overview of case vignettes

Industry sector / topics	Stage of transformation	Covered business model transformation aspects	Firms covered (i=interviewed)
1) Food, packaged consumer goods • Organic food • Sustainability in agriculture	• Monitoring & refinement • Anticipation	• "Green" turnaround • System dynamics in business models and design refinements • Green consumers and product-level transparency of sustainability • Sustainability-based competition • Ambiguity of sustainability issues	• Frosta (i) • Nestlé (i) • Danone • P&G (i) • Wal-Mart
2) Lighting • Off-grid lighting in Africa	• Experimentation	• Base of the pyramid markets • Business model design and launch • Partnering	• Osram (i) • Philips (i)
3) Building technology • Performance contracting	• Scaled implementation	• Non-technical business model innovation • Challenges of scaling up, replicating a business model in different markets	• Siemens (i) • ABB (i)
4) Cross-industry • Platforms for green growth • Example: GE Ecomagination	• Monitoring & refinement	• Focusing innovation power • Creating synergies on corporate level • Green brand building • Mobilisation of employees • Alternative platforms to foster green growth	• GE (i) • Siemens (i) • IBM (i) • Google
5) Chemicals • Sustainability as a service • Green chemicals	• Scaled implementation	• Complexity of environmental sustainability • Building a business on sustainability expertise • Challenges from legacy issues	• BASF (i) • Bayer (i) • Cognis (i)
6) Transport (airlines, shipping) • Green airlines • Low carbon shipping and logistics	• Anticipation • Experimentation • Scaled implementation	• Role of industry rules & myths • Credibility and economics of alternative paths towards sustainability • Shifting basis of competition (towards sustainability-based competition)	• Lufthansa (i) • Ryanair • Maersk Line (i) • Damco (i)
7) Automotive • Electric cars	• Anticipation • Experimentation	• Revolutionary vs. incremental change • System-dependent innovation • Customer perceptions of "green" • Industry dynamics: strategies of incumbents and challengers; value chain disruption; competing technology paths; alliance building • Competition on different levels: within the firm, between firms, between industries, between countries	• Toyota (i) • BMW (i) • BYD (i) • Daimler (i) • Better Place • Renault • RWE (i)

After a brief recapitulation of the rationale for Green Business Model Transformations, seven case vignettes are presented. Contemporary examples of transformations are contrasted with firms that have (so far) rejected the idea. Differences in firm motivations, starting points, and approaches are highlighted.

7.1 A Plea for Green Business Model Transformations

In this brief recapitulation of the rationale for Green Business Model Transformations two principal motivations are compared: normative and economic ones.

7.1.1 Normative Justification

For decades, critics have warned that the present development path of mankind will lead to disaster (e.g., Carson 1962; Meadows et al. 1972; von Ditfurth 1985; WCED 1987; von Weizsäcker 1988; Senge et al. 2008). Some dire warnings turned out to be inaccurate (like the predictions on resources by the Club of Rome), and some problems have been addressed effectively (like the depletion of the ozone layer). However, other significant problems – most notably climate change – have newly appeared. And despite some successes in fighting environmental degradation, the overall ability of modern societies to solve global environmental problems remains questionable (e.g., Diamond 2005; UNEP 2010a). Furthermore, the failure of governments to act decisively on climate change (Economist 2009; UNEP 2010b) highlights that sustainable development cannot be left within the responsibility of governments alone. As Schaltegger & Burritt (2005, 185) remark, "Many large corporations control more resources than many sovereign nations." Yet, the prevalent approach towards sustainability based on incremental improvements may be too slow to overcome the looming environmental crisis (Jackson 2009, 67ff.). In the worst case these measures may even reduce the sense of urgency and hinder the pursuit of more fundamental solutions. A cascade of Green Business Model Transformations in various industry sectors could deliver just that by profitably using market forces to support environmental quantum leaps.

Numerous CEOs, managers and other individuals within corporations meanwhile embrace this obligation towards environmental sustainability (Senge et al. 2008). If nothing else, leadership on sustainability is a convenient way for CEOs to make the headlines without a corporate scandal (e.g., Gunther 2006). Moral convictions and vanity will help to advance a sustainability agenda within the firm, but is not sufficient. Ultimately, a Green Business Model Transformation needs to be supported by a business case, a frequent barrier for companies that want to tackle sustainability issues (Berns et al. 2009a). Not least, a convincing economic justification reduces the immense personal risk associated with large-scale transformations for responsible managers. The economic benefits of green value creation levers (see 3.2.3) that can be identified for the examples from the vignettes are summarised in the following.

7.1.2 Economic Justification

Setting aside ethical motivations, firms pursue ecological activities to gain legitimacy (e.g., license to operate, risk reduction, employee satisfaction), and competitiveness (e.g., lower cost, higher market share, differentiation, or valuable resources) (Bansal & Roth 2000). The vignettes illustrate that Green Business Model Transformations can offer economic benefits in all business model components and elements (see 4.4).

7.1.2.1 Green Value Propositions and Target Groups

The case vignettes offer a rich spectrum of ways to establish successful value propositions based on sustainability. With regard to Porter's three generic strategies of cost leadership, differentiation and focus (see 4.5.4), most people have traditionally associated sustainability with companies that follow the focus strategy. Nevertheless, a strong green value proposition can be consistent with all three generic approaches. The cases also highlight how important a differentiated consideration of target groups can be for the success of a green business model.

Wal-Mart – one of the most obvious examples of the cost leadership strategy – tries to lower its cost base through strategic sustainability initiatives related to its supply chain, transportation, and energy efficiency (Wal-Mart 2007). Among other things, its savings result from Wal-Mart imposing standards on its suppliers, and reaping the cost benefits.

A very different example of sustainable and cost efficient offerings provide the off-grid lighting business models by Osram and Philips in Africa. Yet, the special target group characteristics of base-of-the-pyramid markets pose additional challenges for the design of a suitable value proposition.

In contrast, green airlines serve as an (hypothetical) example how firm can compete strategically on cost by disproportionately *increasing* the cost base of low-cost competitors though sustainability (see Reinhardt 1999, 152ff.).

Two prime examples for successful differentiation through sustainability are Toyota with its hybrid car segment and Frosta with its sustainably redesigned food products. Both companies were able to establish a price premium, although especially Frosta had to reassess its original growth objectives because its target group did not uniformly tolerate the price increase.

Ultimately, green differentiation means helping your customers to become green. The chemical company BASF has even created an entire business model on this basis and offers its expertise to customers as a service. It thereby not only generates additional revenue, but drives volume of conventional offerings through deeper customer relationships. Logistics company Damco has designed a similar offering for greening logistics chains of customers.

Shipping company Maersk Line also hopes to grow its revenue through higher volumes (and prevent price-based competition that hurts freight rates). However, it does so by resetting the criteria that are most relevant to the customer within the whole industry through sustainable operating practices (i.e., reliability instead of speed). Another example in which a company redefines competition is the service-based performance contracting of Siemens Building Technologies. In this business model, the company takes on the risk, including financing, and interlinks its profits with the energy savings at the customer site.

The emerging markets for electric cars offer rich insights on appropriate value propositions and relevant target groups. Different approaches of incumbent car manufacturers like Toyota, Daimler, and BMW are contrasted with that of new challengers like Chinese BYD Auto, and players from other industries (e.g., utilities) that compete for influence.

7.1.2.2 Key Resources and Processes

Corporate environmental sustainability can generate resources that contribute to sustained competitive advantage in numerous ways (Hart 1995). The struggle to acquire such resources is most visibly demonstrated by the vignette on electric cars. While some firms merely try to adapt their current technology with modular changes (i.e. electric drivetrains), the example of BMW and its e-mobility programme *project i* shows that much more ambitious approaches are possible; BMW aims to reinvent the entire value creation side.

Principally all aspects of value creation are affected by or can be improved through environmental sustainability (Esty & Winston 2009). The case vignettes cover all business model elements related to key resources and processes (see 4.4):

General Electric improved its brand as an environmental leader, and thereby – among other things – manages to better motivate its employees. Toyota still benefits from its technological advances in hybrid cars, and many automotive firms have engaged in strategic partnerships to stay competitive in the electric car market. BASF secures access to raw materials and acquired a firm specialised in green chemistry. As mentioned above, Wal-Mart currently works hard to make its supply chain more sustainable and efficient. It also creates a large database on sustainability information on the products it sells. Siemens Building Technologies makes a virtue out of scant investment budgets of its customers and sells energy efficiency as a service.

Furthermore, Frosta changed key production and sourcing processes in order to deliver on its new green value proposition. In addition, it adapted its communications to be aligned with its new attitude towards transparency about product ingredients and business practices. Siemens Building Technologies needed to adjust its recruiting policies in order to support its service-based business model, and GE strongly increased its R&D spending to deliver on its promise for green innovation. Furthermore, some companies like BMW and Maersk Line modified management processes or defined exceptions (e.g. related to rules and metrics) that accommodate the needs of their new green business models.

7.1.2.3 Financial Logic

Some green business models do not differ from conventional alternatives in their financial logic. However, there are various examples for which it plays an essential role. The most obvious examples are servicising models (see Rothenberg 2007) like that of Siemens Building Technologies and Osram: Revenues are no longer generated from product sales, but instead come from a steady stream of service charges.

However, the financial logic is the Achilles heel of many green business models. Frosta, for example, swapped revenue for a higher margin. In this case the gain in margin per product sold was higher than the loss in volume, but the example nevertheless demonstrates that firms need to pay close attention whether their profitability assumptions are robust.

The opposite logic is usually applied in markets at the base of the pyramid. Osram thus needed to apply a different cost model in Africa than it would have in a developed market. For instance, it could not set up an expensive distribution network and now relies on deliveries by independent individuals with donkey carts.

Another instructive example is the electric car. Because of the high cost of batteries and low cost for electricity compared to fuel, companies need to think of innovative ways to offer electric cars at attractive prices and take advantage of their lower operating cost.

7.1.2.4 The External Environment – Competitors and Non-targeted Stakeholders

The business model concept already integrates certain stakeholders that were traditionally considered "external to the business" as their crucial role in value creation is recognised (see 4.4). This includes customers as well as key suppliers and business partners. For green business models, also the remainder of the external environment (e.g., competitors, regulators, media, NGOs, the public at large) often plays a vital role as they tend to be strongly interlinked: Firstly, some business models require joint actions by numerous industry players (Astley & Fombrun 1983) – the shipping industry serves as an example. Moreover, many green business models are strongly dependent on the goodwill of its broader business environment. Green business models will usually generate stronger acceptance within affected stakeholder groups (e.g., communities at production facilities). At the same time, they rely on the perception that they indeed help protect the environment, much more than conventional business models do. Respective reputational risks tend to be higher the larger the company, and the closer it is to the consumer – such business models are simply attractive targets for NGOs. For instance, if Lufthansa tried to become green by using first-generation (i.e., unsustainable) biofuels, NGOs would most certainly criticise their questionable effect on the environment and their interference with human food supply.

However stakeholders are not only a threat, they can also facilitate the success of a business model (Harrison et al. 2010). Osram and Philips benefit from the local knowledge of NGOs they cooperate with in Africa, and governments are willing to subsidise some green business models, e.g. for electric cars. As explained earlier, if this support is central to the success of the business model, the respective groups should thus be considered part of the *target groups* component rather than an external factor.

7.1.2.5 The Consistent Whole – More Than the Sum of its Parts

All successful business models have in common that its elements are consistent, and ideally even reinforce each other. In addition, business models need to be designed flexibly enough to account for external changes (Demil & Lecocq 2010). This managerial challenge tends to be exacerbated in the case of green business models as environmental sustainability often requires dealing with diverse sets of technologies and approaches. It also encompasses complex sociological and political elements (Porter & van der Linde 1995a; Coenen & Díaz López 2010). This also means that the necessary expertise for managing Green Business Model Transformations is frequently scattered within a company (in functional and potentially also geographical terms). Combined with the ambiguity and uncertainty that usually accompanies new business models this leads to major organisational challenges and puts special emphasis on the importance of sound change management to bring a planned green business model to life.

Furthermore, plausible business model design and competent management of the implementation process may not be enough. The aesthetical appeal – or "sexiness" – of the new green business model concept may also be an important factor for its adoption (see Gersick 1991, 27). Even a catchy slogan that has emotional appeal and mobilises people can make a big difference (Van de Ven 1986, 592). There is no doubt that the label "Ecomagination" contributed to the success of GE's initiative.

In the following, the above mentioned cases are critically examined. The opportunity at hand is explained, as well as related risks and barriers to seizing them. Relevant business model elements are highlighted, and rivalling concepts are contrasted with each other where appropriate. The case vignettes are meant to serve as illustrative pieces of thought, interpretation, and – in some cases – even speculation. They are not, however, designed to provide a comprehensive documentation of the history of Green Business Model Transformations.

The first case vignette is a comparatively clear-cut example of a Green Business Model Transformation in the food sector. Off-grid lighting in Africa, a servicising model in a base of the pyramid context is presented next. After that, the process of scaling up a servicising model in the building modernisation sector is examined. Then, the well-known case of GE Ecomagination is discussed, followed by an investigation of environmental concepts in the chemical industry. The remainder of the case vignettes is located in the transport sector. First, opportunities and barriers in the airline industry are contrasted with recent developments in container shipping. Finally, various aspects of the emerging market for electric cars are explored.

7.2 Vignette 1: Green Food – Niche or Inevitable Transformation?

Food security and safety is at the very centre of sustainability efforts (WFP 2010). Food producers and their supply chains are therefore highly affected by recent sustainability trends.

Historically, through the application of modern farming techniques like synthetic fertilizers, pesticides, high-yield crops and irrigation infrastructure in developing countries as well as productivity gains in industrialised countries, world-wide output in agriculture has increased dramatically during the last decades (Leetmaa et al. 2004; Wikipedia 2010a; USDA 2010). In fact, developed nations now face problems related to abundance, not shortage of food (WHO 2003). However, hunger is still prevalent in many poorer nations (WFP 2010), and obesity and malnutrition even frequently coexist (WHO 2003).

At the same time, productivity measures in agriculture and food production may have been pushed too far – or in a wrong direction. Food safety debacles, like the mad cow disease (BSE) or the Chinese milk scandal to name just two, have caused countless victims and severely undermined the trust in food producers that were directly or indirectly associated with these incidents (Buzby 2003; Branigan 2008; Ramzy 2008). The long-term effects on the environment and human health of practices related to food chemistry, genetic engineering of crops, animal feed, and so on are also critically viewed by NGOs (e.g., Foodwatch 2010) and raise concerns among citizens (TNS Opinion & Social 2006). Moreover, agriculture is a major source of greenhouse gas emissions. Livestock alone is responsible for more GHG emissions than all transportation combined (Steinfeld et al. 2006, xxi). In addition, ethical issues related to animal treatment and the social conditions of farmers in developing nations are controversial.

As a result, organic food production has grown strongly since the 1990s in Western countries (Dimitri & Greene 2002; Warner 2005; European Commission 2005) and now constitutes an established niche (Smith 2007).

Against the background described above one might conclude that green business models are a sure-fire success for food producers. Reality, however – both in terms achieving sustainability and economic success – turns out to be rather challenging. The topic is examined based on two company examples: the frozen food producer Frosta AG – which actually attempted to transform its business model – and Nestlé, the largest food producer in the world.

7.2.1 Frosta's Turbulent Quest for Profitability through Greener Products

In January 2003, the German frozen food producer Frosta AG had launched a Green Business Model Transformation, fundamentally changing the whole product range of its main brand (also called "Frosta"), which then represented 25 percent of total revenue. Ten months later, Frosta AG issued a press release stating that its CEO at the time would step down (Frosta AG 2003b) – the Green Business Model Transformation was widely considered a failure (Willenbrock 2003).

However, the privately-held company did not revert to its previous business model and instead corrected initial design flaws of the new, green business model during 2003. Eventually, the willingness to take the risk has paid off. Since 2004, the company continues to enjoy profits that are higher than before the transformation – in relative *and* absolute terms. Figure 7.1 shows the financial history of Frosta AG two years before until five years after market launch:

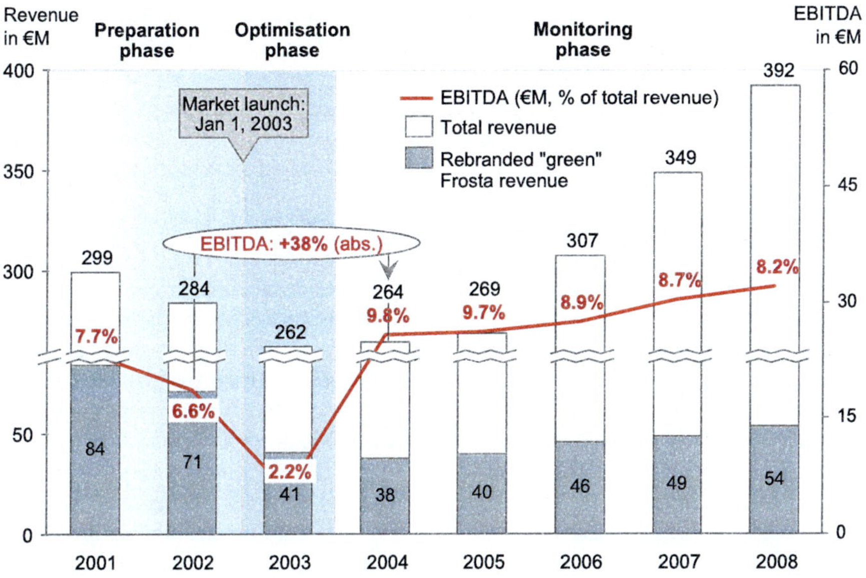

Fig. 7.1 Revenue and EBITDA of Frosta AG 2001 to 2008[17] (data source: Frosta AG 2005; 2008)

So why did Frosta almost go bankrupt although in retrospect the concept was viable?

Frosta is an example of a radical, turnaround-type of Green Business Model Transformation. As it is often the case in when seeking a radical, fundamental solution, a worse-before-better pattern materialised (see chapter 4.5). In this case, Frosta's branded revenue had plummeted by a staggering 42 percent in 2003 and the company as a whole went into the red (Frosta AG 2003a, 8). One may speculate whether or not the management of a public company would have maintained the strategy given the intense financial pressure which the company faced in 2003 (or even attempted the transformation in the first place). In any case, Frosta started its transformation as a response to intense competitive pressure.

The company had already faced a steady decline in branded revenues for a number of years before its Green Business Model Transformation in 2003 (Frosta AG 2001). Maybe more worrisome, margins with retailers continuously eroded and Frosta was caught in a race to the bottom: Low prices of discount competitors on the one hand and big marketing budgets of premium segment competitors including Iglo and Knorr (two brands then owned by Unilever) on the other hand severely constrained Frosta's room to manoeuvre (Ahlers 2008). Consequently, Frosta decided to make a bold strategic move and rebranded Frosta based on the idea of a purity command as it has been applied to German beer for centuries. The "FRoSTA Purity Command" implies that all meals sold under the Frosta brand are

[17] Due to slight distortions resulting from changed accounting standards, figures from 2004 onwards are not always fully comparable to previous ones (Frosta AG 2005, 10)

produced without the use of additives such as flavour enhancers, aromas, artificial colouring, stabilisers, and emulsifiers (Frosta AG 2008, 9). Although Frosta emphasised quality and taste rather than environmental sustainability as such, the latter was the means of achieving the former – instead of applying food chemistry. Moreover, pork was sourced from selected brand meat schemes that forbid to feed ground animal products and growth stimulants. Also, medicines may not be used preventatively. Fish was purchased exclusively from sustainable sources certified by the Marine Stewardship Council (MSC) (Frosta AG 2002, 11).

Frosta invested 25,000 work hours in implementing the new business model. Ensuring that raw materials were free of additives was a major challenge since over 240 new specifications had to be developed and respective suppliers to be found. Recipes were redesigned with the help of renowned cooks in order to create superior taste (Hamprecht 2005, 114f.). In the end, it turned out that half of the product range had to be discontinued altogether as the meals were no longer feasible under the purity command requirements (Ahlers 2008, 3). A number of operational processes were also affected adversely. For example, storage capacities for chilled materials had to be increased as milk powder was replaced with fresh milk; the use of salt without anti-caking agents required the development of a special grinding machine for production. As a result of all the changes, the average price had to be increased by 30c to 60c to over € 3 per unit (Hamprecht 2005, 115f.).

Frosta AG's stated goal was nothing less than to "reinvent frozen foods", thereby doubling its market share within five years (Willenbrock 2003). However, the German consumer apparently was not ready for Frosta's food revolution.

Frosta may have achieved a unique selling proposition with its purity command label, full listing of ingredients, and advertising efforts. However, with unit prices crossing the psychologically important €3-threshold, many of the traditional Frosta customers were not willing to pay the premium. At the same time, the group of eco-minded customers who were gained in return was far too small to make up for the loss of price-sensitive consumers (Frosta AG 2003a, 10). Given the price consciousness of the average consumer and the reduced product variety, the initial market share goal seems totally unrealistic in hindsight. Indeed, consumer research shows that only a small share of consumers (anywhere in the world) is willing to sacrifice price for sustainability aspects (Manget et al. 2009; P&G 2010). Hence, true mass market success with green business models seems unattainable in the food sector, at least when a static view is employed (this aspect will be enlarged upon later).

In the case of Frosta, it was not the idea of a green business model itself that was ill-conceived, but rather the goal of increasing market share by these means. Instead, the focus should have been profitability from the beginning (and volume only in the very long term). And indeed, today's CEO Felix Ahlers assures that retail margins for the Frosta brand have improved significantly due to the transformation (although this cannot be observed directly from the numbers in figure 7.1, which shows EBITDA margins related to the total revenue).

However, Frosta first had to implement some corrective measures to make the business model economically successful. Measures included an ordinary restructuring programme, i.e. Frosta reduced marketing and consultancy expenditure and cut jobs (Frosta AG 2003a, 10ff.). While these measures were necessary to adapt to the lower volumes, Frosta did not fall trap to the classic "fixes that fail" temptation (see 4.5.3) and cut on its expensive, but differentiating product characteristics. Instead, management decided to reduce the unit size in order to push prices under the €3-threshold. This measure, of course, was not received without criticism by customers. Yet, Frosta decided to strike a new path in terms of communications and engage with its critics. Instead of the traditional closed, one-way corporate communication controlled by functional experts, Frosta decided to create the "FRoSTA-Blog" in 2005. Since management was confident that it had eliminated most potential causes for eligible criticism, completely open communication was the logical next step. Unlike most other corporate blogs, Frosta decided to run its own unedited and post all user comments without any time delay. Furthermore, corporate bloggers were not limited to communications professionals, but included dozens of employees from various functions – including the CEO (Arnhold 2010, 192ff.). Consequently, Frosta's Felix Ahlers (then board member for marketing and sales) proactively started a discussion about the reduction of unit size, explaining the company's rationale. Interestingly, even critics of the decision gave Frosta credit for the openness, and some consumers even defended the company against criticism by others (Ahlers 2007). Meanwhile, the effectiveness of Frosta's blog as a branding tool has also been confirmed scientifically, along with a positive effect on employee commitment towards the company (Arnhold 2010).

Frosta's initial attempt to achieve a fast, radical rebranding had failed and recognised advertisement themes had to be revived. Presumably, the company simply lacked the necessary marketing muscle. Nevertheless, the company found an effective way to address its customers through user generated branding – an option that most big-brand firms still shy away from.

The example of Frosta shows how entrepreneurial spirit can lead to lasting business success through Green Business Model Transformations. However, it is also a warning that such radical transformations can be risky and thus lend themselves to applying a systematic approach which prevents major mistakes from happening, or at least allows the company to counteract quickly and effectively. This is of course at the centre of consideration of the management framework presented in chapter 9.

However, the comparatively simple example of Frosta – incorporating a homogenous product offering and limited geographic footprint – cannot credibly be used as the sole reference if the management framework claims to be applicable for diversified multinationals as well. Hence, Frosta's case is contrasted with another food company which encompasses as much complexity as one can possibly find in this industry: Nestlé, the largest food producer in the world (Nestlé 2010b, 5).

7.2.2 Nestlé and its Delicate Path towards Sustainability

When Nestlé introduces a new product, it needs to satisfy Nestlé's so-called "60-40-plus" rule. This means that the new product has to be rated better in blind testings by at least 60 percent of probands compared to competitor products. The "plus" criterion has been added to incorporate quality and health criteria beyond taste. That is, more fat, salt or sugar (or other food chemicals) will make a product more tasty, but less healthy.

However, it is easy to see that Nestlé's approach is quite different from that of Frosta. Frosta starts with rigorous standards and tries to create products that are as tasty as possible under these constraints, while Nestlé's food engineers probably do it the other way round. To be fair, frozen foods are easier to produce without food chemistry than many of Nestlé's products. Even Frosta does not fulfil all criteria for organic (or "bio") food due to cost reasons (some of Frosta's own standards are stricter than those for organic food, hence adhering both standards would be very challenging). On the other hand, Nestlé has also achieved improvements in the sustainability of its purchasing practices (Hamprecht 2005, 122ff.). Furthermore, through *ecologies of scale* (Schlich 2008), Nestlé with its enormous size is principally well-positioned to produce food with a low environmental impact per unit.

Ultimately, the true reason why Nestlé shies away from going beyond its current eco-practices is less technical than market-driven. Until now, it is very hard to compete on sustainability (and not only taste or price). The sustainability performance of mainstream food products is simply not transparent – not yet. Although attempts to introduce simplified, but consumer-friendly measures of transparency like the proposed traffic light system in the EU have been rejected (in accordance with Nestlé's official position) (European Parliament 2010), web 2.0 applications will increasingly provide this kind of information – with or without the consent of big corporations. While user-generated branding efforts like the FRoSTA-Blog remain at least partially under the control of corporate marketers, some emerging web 2.0 phenomena completely defy control of companies.

One effect of the web is that NGOs and the civil society have gained power. NGOs have significantly increased their reach through social media and no longer rely on formal members alone for their campaigns, as Gerd Leipold, former Executive Director of Greenpeace, explained during an interview. Nestlé itself felt the force of a recent social media-based Greenpeace campaign against Nestlé's Kit Kat (chocolate bars). It included an internet-film featuring very figurative pictures of a man eating the bloody finger of an Orang-Utan instead of a chocolate bar (the habitat of Oran-Utans was implied to be endangered due to Nestlé's sourcing practices). The prompted removal of the clip from the video platform YouTube actually greatly increased the traffic to alternative sites that were created quickly. After the campaign had picked up momentum, Nestlé felt impelled to reconsider its initial defensive stance on the issue (Greenpeace 2010).

The best way for a company to immunise itself against unforeseen scandals, campaigns or longer-term shifts in ethical beliefs is to know its supply chain and the life cycle of its products very well. This way, the company can quickly inform the public or decide to change its current business practices. And it can preventively improve the sustainability of its business. For a company like Nestlé, this transparency can mean to know exactly which cows the milk comes from and how they are kept – rather than buying milk anonymously from wholesalers. Considerations at Nestlé a few years back to cut costs by giving up its control over the supply chain would have been devastating during the Chinese milk scandal, had they been implemented. Ideally, this transparency is extended to the consumer, so that eventually, sustainable practices become a source of competitive advantage rather than mere risk management.

Therefore, real-time product-based information could potentially become even more impactful than social media campaigning (or dedicated sustainability websites): Consumers can use smart phones already today to scan a product bar code in a store and obtain related information. Such mobile applications affect retailers as they allow instant price comparisons in a store. But the truly disruptive novelty with respect to sustainability is that, for the first time, consumers are conveniently empowered to vote on products with their wallet based on real-time sustainability information (e.g., GoodGuide 2010). So far there are many problems remaining, notably generating and aggregating huge amounts of product and company information in a meaningful way. However, should this practice become commonplace and studies confirm a significant influence on the consumer buying decision, even powerful firms like Nestlé would have no choice but to compete on "green" on a product-basis.

This description above may sound futuristic, but Claus Conzelmann, Head of Safety, Health & Environmental Sustainability at Nestlé, believes that a sustainable vision of the future food industry would necessarily include standards to make sustainability characteristics transparent on a product level. This way sustainability can be directly integrated into the value proposition and greatly enlarge the group of potential target customers. An influential company like Nestlé should give serious thought to whether it wants to actively participate in shaping such a future, or whether it prefers to oppose it as long as possible and let other parties define the standards (see Schneidewind & Rehm 2010). In the US, the world's largest retailer Wal-Mart is already working on a sustainability index due for launch in 2013 (Rockwood 2010). The goal is to provide consumers with simple and convenient product information on sustainability. It will be based on supplier questionnaires and informed by a global product lifecycle database created by a consortium of universities together with suppliers, retailers, NGOs, and government representatives (Wal-Mart 2009a). Wal-Mart is one of the few companies world-wide that are powerful enough to initiate such a project on its own (Rosenbloom 2009). It will be interesting to see whether this project produces credible and reliable results. After all, even when considering the supply chain of the food industry alone, sustainability is a matter of great complexity. In addition, a holistic approach to sustainability would need to include consumer behaviour and its effects. This would, among other things, require a shift in marketing activity from selling products to promoting sustainable consumption. Such a shift, however, contradicts the financial logic of many current business models.

Since coming under scrutiny in the wake of the infant formula controversy (Time Magazine 1984; Sethi 1994), Nestlé tries to keep sustainability issues rather low-key. And indeed, relying on ecological food production and agriculture is not even unambiguous in ethical terms: As the world population is estimated to grow to nine billion people by 2050 (UN 2004), the goal of feeding the masses may seem contradictory to ecological agriculture with its lower productivity and higher food prices – at least in the short run. While Frosta is only active in Europe, a global corporate giant like Nestlé may actually be held accountable for contributing to food security of the poor (Shrivastava 1995b, 939f.). Consequently, Nestlé propagates the "Creating Shared Value" concept (Nestlé 2010c) which amends value creation for shareholders by aiming to improve the economic and social conditions of other involved parties of the value chain, too – thus expanding the total value created (Porter & Kramer 2011).

Some companies already view feeding the poor as a business opportunity. For example, Danone actively promotes its self-declared mission to "bring health through food to as many people as possible" and sells affordable yoghurt in Bangladesh (Knaut 2010, 1). However, morally-charged claims that link for-profit activities with basic human need satisfaction can be a two-edged sword. Danone may get credit for improving people's lives by many, but others may find the ethical pretension to be insincere or reject making a profit from it altogether. Nestlé, for example, is being criticised for earning money with bottled water – water being argued to be a public good that is not to be exploited for profit, especially where water is scarce (Nestlé 2010a).

The question what ethical conduct means in the context of food and beverage production will likely change and remain controversial in the foreseeable future. The list of topics is long (Maloni & Brown 2006), both regarding developed and developing countries. In any case, low income consumers in general are increasingly seen as a business opportunity by companies from developed economies. However, as the next example will show, challenges and success factors can be quite different from what these companies experience in their traditional markets.

7.3 Vignette 2: Off-Grid Solar-Powered Lighting for the Base of the Pyramid

In 2008, lighting manufacturer Osram launched a small pilot project at Lake Victoria in Kenya to provide local fishermen with solar-powered lanterns that offer an alternative to kerosene lanterns hitherto used to attract fish at night. The global potential market for off-grid lighting is gigantic: an estimated € 30 to 50 billion are spent annually to fuel kerosene lamps – by the quarter of the planet's population that has no grid access (Rybak 2009). Therefore, it comes at no surprise that Osram's great rival Philips (and over 100 other, mainly very small competitors) has also begun to target the largely untapped off-grid lighting market (Philips 2008; Philips 2010a; Lighting Africa 2010, 34f.).

Kerosene lanterns are not only hazardous to health, they also harm the environment through spilling, and produce significant amounts of CO_2 – 50,000 tons per year in the Lake Victoria area alone (Osram 2010), and at least an estimated 20 million tons in the whole of Africa (Lighting Africa 2010, 14). However, the main appeal of the value proposition that Osram developed was supposed to be the savings for the customer of about 20 percent vis-à-vis kerosene lanterns – a substantial amount considering that up to 70 percent of local fishermen's income is spent on kerosene (Zeug 2009; Osram 2010). Hence, the concept behind the new business model sounds phenomenal – a flawless triple win for Osram, the local fishermen, and the environment. However, it turned out not to be all that simple: Despite a year-long planning period to get the details of the business model right, immediate success did not materialise and Osram had various painful lessons to learn.

The new business model originated from an internal innovation award at Osram and the cooperation with an NGO, Global Nature Fund (GNF), that had been experimenting with alternative lighting solutions at Lake Victoria for some years (Osram 2010). This cooperation was valuable for Osram in several ways: GNF and its partners had proven through tests that electric lanterns worked just as good to attract fish as kerosene lamps (Sauer 2009). Maybe more importantly, it gave Osram access to knowledge about cultural characteristics of its new customers and business environment – which Osram was completely unfamiliar with.

Osram did a lot of things right and applied many of the known success factors for BoP business models (see Seelos & Mair 2007; WEF 2009): The product has been designed to meet the needs of the target customers, i.e. the lanterns are robust, water-resistant, and economical (GNF 2010). They can also be used for other purposes than fishing, e.g. to run a radio, charge mobile phones (a critical feature) or simply provide light in homes. Moreover, Osram focuses on the target group of potential *users* – not buyers. This is a smart thing to do, given that fishermen (like most BoP customers) have low and fluctuating incomes. That is, unattractively large investments or instalments are avoided through a servicising model. In addition, the process of recharging the lantern is sufficiently quick and easy as batteries are simply exchanged in dedicated solar-powered charging stations (so-called "O-Hubs") (Osram 2010). This newly created channel of O-Hubs is probably the most innovative part of Osram's business model, and enables the servicising model in the first place. While it requires high initial investment, the O-Hub concept effectively deals with the prevalent problem of equipment reliability in Africa: Cheap, low-quality lighting products threaten to spoil the market (Lighting Africa 2010, 69f.). As the knowledge necessary for repairs is scarce, and even minor spare parts are often difficult to get, equipment is frequently broken and useless. This circumstance makes private investments like a solar-powered lantern even more unattractive. However, because Osram and its partners keep the responsibility for the functionality of the equipment and effectively sell portions of electric light – and not lanterns – the O-Hub concept greatly reduces the risk on the side of the consumers.

However, Osram's business model also had various flaws initially: For example, some people would have liked to rent a lantern, but without any means of transportation, the walking distance to the O-Hub was too far for them. Because setting-up a delivery service would have been too expensive, an independent local delivery network based on donkey carts has emerged meanwhile. Also, rental charges were reduced and the deposit of 2,000 Kenyan Shillings (about € 20) turned out to be too high and consequently was cut in half. Microfinancing options that were initially thought to make the deposit affordable were hardly used, presumably due to the complicated process and bad past experiences. At the end of the day, local fishers have not acknowledged the compelling "business case" as anticipated – at least not immediately. Osram may have underestimated the general scepticism towards the new product that first needs to be overcome (see Lighting Africa 2010, 65). Moreover, bonds with kerosene dealers were stronger than initially thought. Kerosene dealers can rely on long-term business relationships with their customers and can also be more flexible regarding the payment, i.e. fishermen can sometimes pay with their catch or even put the kerosene on the slate. This flexibility naturally results in a certain dependency on the side of the customer. Osram's efforts once more demonstrate that many characteristics of the business environment are very different to what companies of developed countries are used to. BoP business models often require a more holistic assessment of the economic and social effects on sellers, buyers, and the community (see London 2009 for a framework to assess respective impacts).

According to Jochen Berner, Project Manager Off-Grid at Osram, the three Kenyan O-Hubs are making money now. After one year of planning and more than one and a half years since the opening of the first O-Hub, Osram will now move forward and build more of them. Berner still considers the relatively long planning period justified to get the technical specification right before launch (changes become much more expensive afterwards). However, the progress of the project could have been accelerated by scrutinising the business model design earlier in the planning phase with people who are knowledgeable about local conditions.

Philips follows a different approach. For its current project, Sustainable Energy Solutions for Africa (SESA), the Dutch competitor of Osram has entered a public private partnership (PPP) with the Dutch government and, like Osram, cooperates with a number of local NGOs. For Philips SESA is just a first step in its global off-grid lighting programme (Philips 2008; Philips 2010a). However, Philips relies on the conventional ownership model and struggles with amortisation times of 6 to 12 months for its products, a number it aims to cut in half eventually (see also Lighting Africa 2010, 28f.).

Like Osram, Philips is still occupied with learning how to best serve the market and scale up. Although there is a social responsibility element to Philips efforts – for instance, it has given away some solar-powered lanterns in Ghana and Rwanda for free (Philips 2010b; Philips 2010a) – Philips eventually aims to create a large for-profit business, too. Initial activities thus function as trials, e.g. to get the target specifications and price point right before a full-fledged market entry.

One problem that both Philips and Osram will likely be facing in the future is that market insights gained in one region or country do not necessarily apply to other local BoP markets. In fact, attempting to centrally plan and monitor business model operations – as multinationals tend to do – can backfire. Experiences from previous BoP market entries show that traditional business rules do not always apply: Instead of scaling up through centralised operations, *scaling out* is usually a more appropriate strategy. That is, while seeking scale in terms of standardised off-grid lighting products can still make good business sense, companies should at the same time allow for localisation of their offerings and channels by means of a high degree of local autonomy in decision-making (WEF 2009, 19ff.). Traditional means of distribution are neither available nor cost-effective (see Unilever's Project Shakti for an innovative solution, Porter & Kramer 2011, 70f.). Philips and Osram may therefore try to engage in entrepreneurial partnerships with locals who can choose the means of selling off-grid lighting that work best in their local market. For example, such a business partner may know that Osram's required deposit significantly inhibits the adoption of its offering, yet that there would be no substantial risk of equipment loss without it in his community, and thus decide to decrease or abolish it altogether. There is no way to make informed decisions of this kind centrally for thousands of O-Hubs. In contrast to an authoritative model, an entrepreneurial model can also significantly reduce principal-agent problems without excessive (and expensive) measures of control, because both parties are inherently interested in the striving of the business. A suitable model to share financial risks with business partners and a basic set of governance rules still needs to be provided and enforced, of course (see WEF 2009, 21ff.). But apart from that Osram could limit its local activities to selected support tasks like educating its business partners about the products – and rely on their entrepreneurial spirit to work out the rest.

Another seemingly obvious long-term threat to off-grid lighting business models is the advancement of the electric grid (Sauer 2009; Zeug 2009). However, Africa's non-electrified population is actually projected to *grow* in the next few years and even decades as electrification does not keep pace with population growth (Lighting Africa 2010). Historic experiences in developed countries, though, suggest that dropping prices for grid access will eventually make it the dominant solution for customers. Yet, in African countries like Kenya, price is not the only relevant variable due to a lack of availability and reliability. Hence, this outcome cannot be taken for granted (Steel 2007). In fact, in her ethnographic study that is based on a system dynamics model (see chapter 4.5), Steel (2008, 1) found that there are indeed "realistic scenarios under which off-grid generation will become the dominant supply source". Moreover, it is even imaginable for Osram to run some off its O-Hubs with electricity from the grid if that is cheaper and if many local families can still not afford necessary installations for access to the grid.

Although the off-grid lighting business models presented above arguably belong to the category of green business models – their contribution to reducing environmental impact is significant – it is worthwhile noting that this effect does not play an obvious role for their economic success. Adverse health effects of kerosene lamps cannot be presumed to be an overly important decision criterion for the poor. Avoiding CO_2 emissions likely plays no role at all. The protection of Lake Victoria can be argued to preserve (and increase) the potential customer base of fishermen in the long run. However, an analogous argument hardly applies to other potential off-grid customers in Africa and Asia. Yet, the stakeholders who *do* care about environmental and social impacts of these business models are the NGOs who help to put them into reality. It is hard to quantify the value of the resources and competencies that these partners offer to companies operating in BoP markets, but building these internally without external help most likely takes longer and is more expensive (Dahan et al. 2010). In this sense, ecologically (or socially) desirable characteristics of BoP business models can work as a motivator for NGOs (and government organisations) to help facilitate their economic success, even if green aspects are rather circumstantial with respect their significance for the customer value proposition.

Whether or not Osram and Philips will succeed with their off-grid lighting initiatives on a large scale remains to be seen, as is the eventual design of the winning business model. It is unlikely, though, that one single business model will dominate the global market considering the important differences in local characteristics (Lighting Africa 2010, 35ff.). At least, the transformations of Osram and Philips are pilot-based and thus carry low risk. Although the new green business model represents a radical change to the existing one, it can be launched in parallel, starting at a small scale. Both Osram and Philips are in the comfortable position to compete for *additional* revenue, without being under pressure regarding their current business model.

These kinds of transformations require early experimentation and rapid learning to get the details of the business model right (McGrath 2010) – otherwise they tend to have high failure rates. But due to the limited investment (in Osram's and Philips' case a few million Euros so far), they do not threaten the company's existence. If successful, however, the new business model can eclipse the old business model over time, and possibly even replace it entirely eventually – although the latter seems highly unlikely in the particular case under consideration.

7.4 Vignette 3: Energy Savings in Buildings – Performance Contracting Services

Service-based business models do not only offer advantages in BoP markets (see Rothenberg 2007). With the energy-saving performance contracting concept Siemens Building Technologies is pursuing a business model that is almost an antithesis to Siemens' original mindset to engineer innovative products, but leave responsibility for their optimal application primarily to the customer. In contrast, performance contracting already encompasses identifying the opportunity on behalf of the customer, then taking over responsibility for seizing it. Performance contracting thereby interlinks Siemens' financial return with the need satisfaction of the customer. Specifically, Siemens Building Technologies offers its customers complementary energetic analyses of its real estate if it is allowed to carry out the identified measures afterwards. Because Siemens only claims a share of the saved energy cost and requires no upfront investment from the customer, the value proposition is virtually risk-free (Siemens 2007). Siemens still realises enjoyable profitability because the savings tend to be substantial, i.e. a reduction of 20 to 30 percent in energy consumption can usually be achieved (Siemens 2010a). For many customers it actually makes a huge difference that they can use operating expenses to cover the project cost and do not need to tap notoriously scarce investment budgets.

Although Siemens already captures revenues of several hundred million Euros per year with performance contracting, the addressable market ranges in the billions. Maybe surprisingly, the main impediments that hindered Siemens to fully capture that opportunity since the inception of the business model in the Unites States 15 years ago do not primarily originate in difficult market conditions, but have to be searched for internally. According to Gunnar Liehr, Global Head Energy Services & Performance Contracting at Siemens, the ongoing transformation effort remains challenging.

As mentioned earlier, the business model is not compatible with a traditional product-based mindset (see Rothenberg 2007). Cost allocations that presume much higher shares of value added – which are typical for product-based business models – can severely strain servicising businesses, even if they deliver solid profitability. Internal competition for resources and the reluctance to spend premiums for highly sought after experts jeopardise sales and execution capabilities. Large, bureaucratic companies like Siemens often spend months until they acknowledge the need to deviate from standard recruiting practices and take respective actions.

Another reason why performance contracting has not grown larger so far is that it is difficult to extend the business model into the industrial domain. In addition to the more difficult task of establishing a baseline for production sites compared to commercial buildings (primarily due to fluctuating utilisation), the needed expertise of customer processes is difficult to build up coming from a product-based business model.

In order to avoid disturbances from the established part of the business, it turned out to be critical to separate the new performance contracting unit (see also Govindarajan & Trimble 2005). Depending on the minimal organisation necessary to support the business model and the size of the existing business, the new unit should grow at least to the size of a few millions of revenue before it can be reabsorbed safely and remain intact afterwards, says Liehr. Siemens Building Technologies has lost significant time and money in regional markets in the past where it broke this rule and failed to establish a sufficient local footprint quickly. Hence, Siemens' overall strategy to grow its business model relies on a staged transformation: the initial focus is to make a local presence self-sustaining and only then expand further geographically within a country or region. Regions are consequently distinguished by their level of maturity, which in turn is associated with different degrees of freedom, the portfolio selection, and targeted customer groups.

As the learning curve during the transformation has been considerable, overall growth is believed to be exponential currently. Only the economic crisis in 2008 has disturbed this pattern – ironically because public institutions had *too much* money to invest from the stimulus packages, which made the financing model completely unattractive to this important customer group. Apart from the resulting two-year delay, Siemens' expansion plans are largely on track. One key success factor is thereby to develop sound business plans for each local market. A structured go-to-market approach is something which Siemens Building Technologies has been struggling with in the past. During the initial incubation phase, excessive, too detailed planning is futile, even counterproductive. Planning is nevertheless crucial for learning and helps scaling up the business model (Govindarajan & Trimble 2005, 66ff.). Accordingly, business plans have turned out to be essential for Siemens Building Technologies as they allow identifying and reviewing essential market assumptions, and monitoring the progress of the business model with adequate KPIs. Because the new business model follows a different financial and operational logic compared to its alternative, KPIs need to be adapted in order to accurately reflect its progress (Johnson et al. 2008, 59). Without this, the business model risks being questioned fundamentally each time that results seem unsatisfactory.

Obviously, it is the energy savings that represent the green aspect of this business model. Of course, simply selling the products would eventually lead to a similar result for the environment if all involved players completed their tasks adequately (which cannot be taken for granted, though). However, performance contracting has made the energetic optimisation of buildings attractive to customers who would not have invested otherwise. By now, the sustainability megatrend has even created another customer group that did not exist ten years ago: multinational companies that seek a reliable partner to help them execute on their global sustainability initiatives. To serve this customer group effectively, Siemens has created a central unit that coordinates such projects and takes advantage of the execution capabilities of its regional offices (which could not handle such a client otherwise). As the next vignette will show, coordinating sustainability-related business activities is not only rewarding across regions, but can be done successfully across (heterogeneous) divisions as well.

7.5 Vignette 4: Green Corporate Platforms for Growth – GE Ecomagination

General Electric is one of the largest, oldest and most diversified companies in the world. Usually one would expect such a company to be plagued with extreme inertia. Nevertheless, with "Ecomagination", GE has performed one of the most remarkable green transformations to-date.

GE launched the Ecomagination initiative in 2005 at the urging of its CEO Jeff Immelt. The initiative was inspired by so-called customer dreaming sessions (Fisher 2005) and aimed at providing new opportunities for growth. General Electric reported revenues of US$ 134 billion (circa € 108 billion) in 2004; hence any company-wide growth initiative is necessarily very large in magnitude. At the time, GE had six major divisions – GE Industrial, GE Healthcare, GE Infrastructure, NBC Universal, GE Commercial Finance, and GE Money (General Electric 2006, 37), each with sometimes very different business models.

Instead of pursuing green opportunities in an uncoordinated way at the discretion of individual business unit managers, GE thought to exploit the megatrend of environmental sustainability in a systematic way for the company as a whole. GE was thereby following the mantra that "green is green", i.e. that customers demand high environmental performance and are willing to pay for green products if (and only if) they offer a good overall value proposition and operating performance (General Electric 2008, 6).

GE Ecomagination consisted of several components (General Electric 2008):

- a multi-hundred-million advertisement campaign (including TV, internet, magazine, and billboard ads)
- commitments to reduce greenhouse gas emissions, energy, and water use and to be transparent about its environmental performance
- a target of selling US$ 20 billion (later updated to US$ 25 billion) of so-called Ecomagination products in 2010 – up from US$ 10 billion in 2005
- the doubling of research investment in green technologies within five years

The most interesting part of Ecomagination from a business model perspective is the stretch goal for green revenues. It is somewhat relativised by the fact that many Ecomagination products have already existed or been in the pipeline before the launch of the initiative. Moreover, GE's certification procedure is rather intransparent and allowed products to be qualified whose environmental impact is not undisputed among environmentalists, like nuclear power technology or efficient, but fossil-fuel powered turbines.

Despite the reservations expressed above, GE's public stance on promoting a greener economy was a revolution in corporate America. Not surprisingly, reactions of some customers were harsh – especially from not-so-green customers like operators of coal-fired power plants who feared their business models could come under additional public pressure. GE's CEO Jeff Immelt paraphrased the reaction of some large customers as follows: "Can't you just shut up and sell us stuff?" (Kranhold 2007). In addition, support from managers internally was rather reluctant at first (MacMillan 2008).

The transformation seems even more remarkable since GE used to be considered an environmental villain in the U.S. Among other things, GE had – then legally – poisoned the Hudson River for decades and later was reluctant to clean up the hazard it had caused (Fisher 2005). But since 2005 the tide has turned: European competitors like Siemens, Philips or ABB that all have a long track record of environmental management, disgruntledly have to accept that GE is now seen as a global leader in addressing sustainability (SustainAbility 2008, 25; Berns et al. 2009a, 69; Haanæs et al. 2011, 10).

A key question is how a company as large as GE was able to align its leadership team and mobilise the organisation for such a seemingly risky, complex undertaking. According to Carlos Haertel, Director of GE Global Research Europe, the answer is culture. GE has a long tradition of executing large transformations successfully (Lowder 2006). GE's flat hierarchy and practice of accountability ensured that corporate leaders supported Ecomagination, not by coercion, but by engagement. GE's combination of central direction setting and delegated responsibility for implementation defanged the great complexity of the initiative. Of course, it was clear to managers that inactivity or even resistance would not be tolerated and negatively affect their careers. Moreover, GE managers knew that good quarterly results alone would not be appreciated if they were not based on the new strategic direction – which is believed to be crucial for long-term success. As a result, even divisions that are not affected by environmental sustainability at first glance found ways to contribute. For example, GE Money created a credit card for which up to one percent of turnover was donated to climate protection projects conducted by Greenhouse Gas Services, a joint venture between GE and AES Corporation (General Electric 2007, 21). Not only did consumers pay for greenhouse gas abatement that GE could use for communication purposes, the projects are also supposed to make use of GE's featured Ecomagination products (Greenhouse Gas Services 2007).[18]

Whether or not Ecomagination can be considered to be a green (meta) business model in its own right seems to be a difficult question to answer. It has certainly influenced the future evolution of multiple business models within the conglomerate, and also changed GE's perception in the marketplace. According to Haertel, some successful products would not have been developed had there not been the Ecomagination platform that gives GE's green value propositions additional credibility and reduces the perceived risk of failure internally. Moreover, Ecomagination has opened doors – of customers and political institutions alike – and can thus be considered to have positively influenced relationships with target groups. In addition, the search for ways to save energy and greenhouse gas emissions internally has saved the company millions on the value creation side of the business model – something that had not been expected a priori. And last but not least, Ecomagination has a positive influence on employee motivation and talent attraction (General Electric 2006, 5).

On the other hand, Ecomagination has not yet evolved into an "innovation engine" that produces big ideas itself. In the future, for instance, it could be conceivable that Ecomagination will generate new cross-business solutions for customers or entire new business models which each individual unit would not be able to create itself. So far, it "only" works as a catalyst to facilitate and accelerate green value creation within the various parts of the conglomerate.

[18] As remarkable as this example of synergy and innovation sounds, the credit card offering in the U.S. has apparently been discontinued. Only a similar offering in Australia is still featured on GE's Ecomagination website (GE Ecomagination 2010).

One step in this direction, however, was GE's recent clean energy challenge (General Electric 2010b). In July 2010, GE invited researches to participate in an open innovation contest (see Chesbrough 2003a) for future clean energy solutions. GE and its partners provided US$ 200 million (circa € 160 million) investment capital for the winning proposals. GE thereby tries to tap into the innovative power of small firms and creative individuals and combine it with its own financial strength and capabilities in commercialisation. Consequently, this use of the Ecomagination platform can help GE to overcome some of the impediments to radical innovation usually attributed to large, established firms (e.g., Damanpour 1992; Damanpour & Wischnevsky 2006).

Other firms take different approaches. GE's rival Siemens, for example, has initially been described by commentators as slow to capitalise on the smart grid wave (e.g., Berst 2009; Smartmeters.com 2010). Meanwhile, the company aggressively communicates its target of € 6 billion in revenue until 2014 (up from € 1 billion in 2009). To bring together the distributed resources and expertise in the field, Siemens has therefore created a dedicated unit for smart grids headed by the former CEO of Siemens China (Siemens 2010c). By dedicating significant resources and senior executives to this field of innovation, Siemens hopes to successfully incubate a number of green business models (Günther 2009).

A very different approach is the one pursued by Google: Google has founded the philanthropic initiative Google.org based on the founders' commitment to dedicate financial resources as well as employee time towards solving some of the world's most urgent problems. Founder Larry Page is quoted as follows (Google 2010b): "We hope someday [Google.org] may eclipse Google itself in terms of overall world impact by ambitiously applying innovation and significant resources to the largest of the world's problems."

Meanwhile, Google.org has shifted its focus from grant making to technical contributions, including green initiatives like the Google "PowerMeter" (smart grid) and "RE<C" ("renewable energy cheaper than coal") (Google 2010a). Google believes that it can have a higher impact with contributions based on its own engineering talent (see McKinsey 2008). Although Google.org certainly manages the effectiveness of its projects tightly, the philanthropic nature of the organisation eliminates any business case constraints. Most daring projects can be launched that would otherwise fall prey to investment rules (see Christensen et al. 2008). As even not-for-profit goals require wide-spread adoption in order to make a difference, Google may at some point decide to build commercial business models around some of its successful ideas. For example, it could design a "free" business model that charges for premium services or relies on advertisement (like its search engine). This kind of growth platform may thus generate some unique innovations. However, as soon as the pure philanthropic mindset is questioned and some hidden commercial agenda creeps into the project approval process, this effect is likely to disappear. Firms should not expect a positive overall return from such a green platform for growth.

Google and other firms in the information technology sector face limited threats from sustainability compared to many resource-intensive and polluting industrial firms. GE and Siemens even benefit greatly from the push towards clean energy because they produce many of the essential technology products. In contrast, the chemical industry has come under scrutiny by environmentalists long before climate change has entered mainstream debate. The following case vignette demonstrates, however, that the response of chemical firms need not be a defensive one.

7.6 Vignette 5: The Chemical Industry – Sustainability Services and Green Chemistry

The chemical industry has once been hailed as the solution to many of the most important problems of mankind. But over time, large-scale accidents and health hazards from products led the public to become rather suspicious of the chemical industry as a whole (Aftalion 2001). For instance, Europeans perceive man-made disasters and chemicals in everyday products as some of the most severe environmental threats (TNS Opinion & Social 2008, 8). This condition has not been without effect on the management of environmental issues: Today, many chemical companies are among the most advanced in this field (Hoffman 1999). One firm that is certainly part of this group of most advanced companies is BASF (SAM & PwC 2010), the world's largest chemical company. During an interview with Andreas Kicherer, Head of Marketing & Sales Sustainability Services at BASF, he emphasised that environmental sustainability within the chemical industry is a complicated matter. "There is no such thing as a green business model, they can only be more or less green", he says. Even when comparing products with renewable versus oil-based feedstocks, it is not at all clear without detailed assessment which of the two alternatives has superior environmental characteristics over the full life cycle of the product. With its eco-efficiency analysis, BASF provides a scientific basis for strategic product development decisions. Environmental and economic aspects are thereby considered jointly; after all, an environmentally-friendly, but very expensive product is as unfit for the marketplace as a cheap, but very "dirty" product.

BASF has used eco-efficiency analysis to determine the characteristics of at least one key product for each of its 100 product categories, thereby gaining a good overview of its circa 12,000 products. However, internal transparency of the environmental characteristics of a product does not resolve the widespread problem of distorted consumer perceptions of the same ("green myths"). Nevertheless, the methodology was so successful that BASF started to sell it exclusively to its customers about 10 years ago. Under the label "BASF Success", the company today provides eco-efficiency and other environmental services related to products, processes, and strategy (BASF 2007). This eco-services business model is profitable by itself, although BASF is silent about the revenue and profit it derives from it. However, Kicherer emphasises that the benefits from having a deep and trustful relationship with its customers by far exceeds direct financial returns.

The business model meanwhile has been lifted to the next level, when BASF started not only to offer its environmental expertise to customers, but also to governments in Northern Africa and the Middle East. In fact, this type of business model is a very special breed as it requires no direct flow of money anymore. In exchange for its services, BASF receives something potentially much more valuable instead: access to critical raw materials. Hence, the business model can be said to be based on the *currency of sustainability*. This new business model type is significant as it could also be transferable to other industries. A closely related model would be its adoption by mining companies. These have significant experience in large-scale projects in remote areas. This expertise could be very valuable if combined with world-class environmental expertise. This way, the process of getting the necessary permissions and concessions to do business in respective countries could be smoothed and accelerated.

Building a competence-based green business model like that of BASF cannot be done without previous long-term engagement. It takes years, potentially even decades, to develop the necessary competencies. General Electric, after having intensified its environmental efforts for the last five years, may now be ready to take that step and extend Ecomagination in a similar way. GE certainly was not ready for that five years ago.

Despite the widespread scepticism towards the chemical industry, it can offer key contributions to make other industries greener. Yet, this fact is not sufficiently recognised, argues Daniel Steiners from the chemical and pharma company Bayer AG. Among other things, Bayer's contributions encompass the already discussed food and construction industries (Steiners 2010). Its CropScience division develops products that improve the stress tolerance of plants against biological (e.g., diseases, pests) and other factors (e.g., weather, salt), thereby increasing the yield. It also provides insulation materials for buildings and currently establishes an "EcoCommercial Building program" that aims at "bringing together the best materials, systems, and technologies" for the construction of energy-optimised buildings (Bayer 2010, 44).

The innovative sustainability services business model of BASF and the mentioned contributions to the greening of other industries by Bayer should not hide the fact that the greening of their *core* businesses remains a challenging task for the chemical industry. Related technical and economic challenges are complex and greatly differ by subsector; they can thus not be addressed in detail as part of this short vignette. Schneidewind (1995) provides a respective analysis for important subsectors of the Swiss chemical industry.

On the level of the chemical industry as a whole, a new core business model paradigm (see 8.2.1) may be offered by *green chemistry*, the "design of chemical products and processes that reduce or eliminate the use or generation of hazardous substances" (US EPA 2011). Although the term has grown significantly in popularity (Linthorst 2009), green chemistry is still far away from being established mainstream. Due to the longevity of assets in the chemical industry a far-reaching shift would still take several decades, even if favourable regulatory frameworks were put in place and technological challenges in the development of green substitutes were solved rather quickly (see Poliakoff et al. 2002). Traditional strategies based on simple process and product improvements, however, are unlikely to allow firms to comprehensively green their core businesses. Additional options include portfolio shifts and function-oriented approaches that fundamentally reconsider how customer needs can be fulfilled (Schneidewind 1995, 238ff.). With respect to the former, BASF has acquired the specialty chemistry company Cognis in 2010. Among other reasons, this acquisition advances BASF's development towards green chemistry (Cognis 2010a). Cognis heavily emphasises its dedication to green chemistry and has implemented the so called *twelve principles of green chemistry* (e.g., Anastas & Warner 2000). Cognis reported sales of about € 2.6 billion in 2009 and currently half of its processed raw materials are from renewable feedstocks (Cognis 2010b).

The chemical industry demonstrates how threats stemming from environmental sustainability can be turned into a virtue. However, it also revealed the ambiguity and inertia that can impede the quick realisation of green business models. The following examples from the transport industry – a sector that is also increasingly affected by sustainability concerns – offer further evidence of factual and imaginary barriers for green business models and how they can be overcome.

7.7 Vignette 6: Sustainable Transport Industries – Airlines and Shipping

In 2007, transportation produced about 23 percent of the 28 Gt of global CO_2 emissions (IEA 2009a, 13). Furthermore, the IPCC believes that up to 2 Gt of yearly CO_2 emissions from transport could be saved at a cost of less than US$ 20 per ton CO_2 in 2030 (IPCC 2007, 11). A majority of this potential will be in road traffic, which also represents three quarters of transport emissions (Stern 2007, 172). Together with other associated environmental issues (air and noise pollution, resource usage, etc.), this makes transport one of the key fields of concern for the sustainability of human activity. Due to the heterogeneity of related business model aspects, the case of individual mobility (i.e., electric cars) is discussed separately from freight and passenger transportation. With respect to the latter topic, the industry survey focuses on potential business models for airlines and the shipping industry.

7.7.1 Green Airlines – An Oxymoron?

Air traffic causes less than two percent of global CO_2 emissions (IMO 2009, 3) – much less than road traffic. Yet, the airplane has become a symbol for the climate change problem. Although airlines put significant efforts into reducing their impact incrementally (IATA 2008), no truly green airline business model has been created so far. The following section explores why this is, and if it could be changed in the foreseeable future.

The single most important environmental impact in the airline industry stems from burning aviation fuel, which produces various emissions, most notably CO_2. Other problems include noise and local air pollution (for a comprehensive discussion of environmental impacts see Daley 2010). However, due to the prominence of the climate change problem the following discussion will focus on greenhouse gases emissions from burning fuel.

Historically, relative improvements in fuel efficiency have been substantial – 75 percent over the last 40 years. However, improvements are going to reach physical limits and would still need to be twice as high if they were just to stabilise absolute emissions at current levels given the annual increase in air traffic of about 5 percent (IATA 2008, 2). Currently applied measures include fleet renewal, the improvement of aerodynamics (e.g. through winglets), maintenance measures (e.g. enhanced turbine washing processes), route planning optimisation, and so on (Lufthansa 2009). Seat configuration and load factors theoretically offer substantial efficiency gains (atmosfair 2011), but the highly competitive airline industry has long concentrated on these measures – hence the practical potential is probably rather limited.

In contrast, the use of biofuels has only been tried on several test flights so far, but could almost eliminate the climate change impact of air traffic. However, scaling up production in a sustainable yet economical way represents a formidable challenge: In order to be sustainable, crop land must not be previously occupied by food crops or major sinks of CO_2 (e.g., forests). In addition, land requirements – depending on the plant – for producing the volume needed by only one of the larger airlines would be in the order of a small country (ATAG 2009, 18f.). Advanced biofuels (such as that derived from algae) promise to solve these problems but are still in an early stage of development. Their future economic viability is contested (Waltz 2009; Stephens et al. 2010). In any case, investment costs (e.g., land development, building refineries) for supplying one major airline would be many the billions of Euros. Considering that these investments may not even be acknowledged to help the environment – some media reports have been very critical after the first biofuel boom (e.g., Grunwald 2008b) – these investments seem rather risky. It will thus be crucial for the airline industry to prevent any deviation from sustainable business practices by their suppliers.

Unless some revolutionary green air plane design is invented, the path towards green airline business models thus includes the following options: Incremental efficiency improvements contribute a necessary, but not sufficient share of lowering environmental impacts. Sustainable biofuels promise to offer a solution for the mid- and long-term once they become available for widespread commercial use. Until then only *carbon offsets* (e.g., see Goodward & Kelly 2010) represent a realistic solution to achieve a discontinuous drop in net environmental impact. This way, own emissions can be neutralised by reducing emissions elsewhere, e.g. through reforestation or renewable energy projects in developing countries. For example, Lufthansa offers its passengers to offset the generated CO_2 from flying for 20 € per t CO_2. A one-way business class flight from New York City to Tokyo produces about 1.1 t CO_2 (myclimate 2010). As the fifth largest airline in the world by number of passengers (Wikipedia 2010d), Lufthansa emitted 24.2 Mt CO_2 in

2008, mostly for passenger transportation (Lufthansa 2009, ii). Hence, naïve extrapolation suggests that Lufthansa could become CO_2-neutral for a little less than € 500 million, or 2.5 percent of its traffic revenue. Scaling up these compensation efforts would be highly challenging, and potentially incur much higher cost than specified above. In addition, carbon offsets are contested means of reducing a company's environmental footprint, even when based on a recognised verification scheme (Revkin 2007; Davies 2007).

"Even if one assumes that major airlines [like Lufthansa] could become carbon-neutral at an average cost increase of only a few percentage points per ticket, the extreme price sensitivity of airline passengers would significantly dampen demand and thus reduce worldwide mobility", says Martin Koehler, member of the Supervisory Board at Lufthansa and former head of BCG's Aviation practice. In addition, the criterion of environmental friendliness is insignificant for most travellers as a study of the Boston Consulting Group shows (Ulrich et al. 2009, 2). Low interest of passengers for voluntary carbon offsets in the UK seem to confirm these claims, especially when considering that many of the surveyed passengers were aware that such offerings exist (Webb 2010). One plausible explanation is that consumers do not gain status from emission-neutral flying (unlike when driving a Toyota Prius, for example). Nor do they associate green travelling with higher quality (e.g., in contrast to organic food). Hence, the question is whether a well-designed green offering could change this, and eventually even make non-green air travel socially discouraged like fuel-guzzling SUVs already are in some countries. A Green Business Model Transformation in the airline industry would thus need to address not only technical and financial aspects, but alter the – possibly latent – *needs* of its customer base. Changing the rules of the game for the airline industry is certainly difficult, but relates to the core of what the business model approach is about.

Low-cost airlines have achieved just that: Many customers turned out to be eager to accept "no frills" flights in exchange for low prices. With respect to establishing widespread carbon offsets or the use of biofuels, however, large full-service players will likely be in the better strategic position: Not only would the resulting absolute increase in cost per passenger improve their relative cost position vis-à-vis low-cost carriers, but their customers are also less price-sensitive (Frontier Economics 2006, 71). This argument is less straight-forward than it might seem at first glance: Due to high seat density and load factors, low-cost carriers tend to achieve high fuel efficiencies. For instance, in 2008 Ryanair on average burned only 3.5 litres per 100 passenger-kilometres (Ryanair 2008) versus 4.34 for Lufthansa (Lufthansa 2009, ii). Thus, full-service carriers would need to cope with higher abatement costs per passenger-kilometre. The use of connecting flights further adds to this effect. In fact, Ryanair claims to be "Europe's greenest airline" (Ryanair 2008, 1) (many environmentalists would object to this claim with reference to the additionally created demand, though). Yet low-cost carriers are still unlikely to welcome any industry-wide cost increases due to carbon offsets or biofuels: Ryanair claims to have had average fares of €35 last year versus more than €200 for Lufthansa (Ryanair 2010). Even when considering higher fuel efficiency and shorter average distances per trip (about one third of the distance), the resulting *relative* price increase could still be highly damaging for a low-cost business model (Frontier Economics 2006, 69ff.). Ryanair, for that matter, is anyway unlikely to do more than necessary to protect the climate any time soon: During a recent interview Ryanair's CEO Michael O'Leary allegedly stated: "Do I believe there is global warming? No, I believe it's all a load of bullshit." (Hickman 2010).

Various measures of making green flying desirable for consumers are conceivable. Special green boarding passes may be a simple way to demonstrate status (it works for first-class tickets). However, paying extra for a green ticket is not very attractive if the customer then sits next to an ordinary non-green customer and in essence gets the same performance. A more promising measure to raise the environmental product characteristics in the customers' mind may be the introduction of special flights, e.g. with planes that are painted in green with a distinct logo. However, individual flights cannot be directed at green customer groups as simply as other products can. For instance, an airplane flying from Sweden to India will usually have to carry both Swedes and Indians (and others). Swedes, however, may be much more willing than Indians to pay a premium for green air travel. Nevertheless, carefully selected routes may be used to test green offerings and compare their success to the previous offering and other comparable flights. In fact, Lufthansa has recently announced as the first airline to offer commercial passenger flights between Hamburg and Munich with a 50 percent biofuel mix for one engine (Lufthansa 2010).

In other industries, founding a green subsidiary would be an appropriate way to then test the new value proposition on a grander scale at moderate financial risks. However, airlines are heavily relying on scale and flexibility, e.g. as they switch airplanes on routes based on bookings. Therefore, such large pilot projects could harm the cost base of both the new and the conventional part of the company – a circumstance that would be very unfavourable in the highly competitive airline industry. In conclusion, green airline business models are highly challenging with respect to all business model components.

The purpose of the brief analysis above was not to present a fully developed green business model for the airline industry – this would require much deeper analysis of many of the mentioned business model aspects. The purpose was to highlight available options and demonstrate that the economics of a green airline business model heavily depend on the question whether the rules of the game can be changed or not. Maybe the most critical aspect in this regard is whether the environment can be made a relevant factor in the consumer's buying decision.

In Europe, one first change to the rules of the game is already getting imposed on the industry: Airlines will be subject to the European Union Emissions Trading Scheme (EU ETS) from 2012 onwards (IATA 2008). While European airlines fear the associated distortion of competition with non-European competitors, the regulation also offers new opportunities as airlines with certified climate protection projects can expect to recover some of their cost through emissions trading or by preventing the need to purchase additional certificates to cover growth (EU 2008, 8ff.). Thus, politics may eventually turn the balance for green airline business models – whether any large player is willing to take the risk of pioneering such a business model at scale remains to be seen.

7.7.2 Maersk – Challenging Conventional Wisdom within the Shipping Industry

Shipping is the backbone of the globalised economy. Of all modes of transportation it is generally the most efficient per ton and kilometre of goods moved. Many companies currently realise substantial savings in their carbon footprint by shifting freight from air or road to sea (examples include Nike 2010, 113ff.; Otto Group 2009, 44; and IKEA 2008, 113ff., to name just a few). However, due to the massive volumes of international shipping, the industry currently produces nonetheless about 2.7 percent (or 870 Mt) of all man-made CO_2 emissions. Moreover, in its study from 2009 the International Maritime Organization (IMO) estimates that CO_2 emissions from shipping will double or triple until 2050 without effective policies in place. This would dispatch 12 to 18 percent of the total allowed CO_2 emissions in the 2°-stabilisation scenario of the IPCC (IMO 2009).

However, in the same report the IMO identifies a total reduction potential ranging from 25 percent up too a staggering 75 percent. The IMO even claims that many considered measures would be cost-effective, but that non-financial barriers may prevent industry actors from realising them (IMO 2009, 44ff.). Meanwhile, this study has partly been overtaken by reality as the following elaborations will show.

Most measures proposed by the IMO have the potential to contribute only well below 10 percent of the envisioned overall reduction. Analogous to the airline industry, measures include improving efficiency and reducing water and air resistance of the hull, superstructure, propellers, and other design parameters. Voyage management, energy management and the use of renewable energy (biofuels, wind energy) are also considered to promise limited contributions (IMO 2009, 44ff.). The use of very large ships could also increase efficiency: A ship that carries 16,000 twenty-foot equivalent unit (TEU) containers is estimated to have 20 percent lower capital cost and 40 percent lower fuel cost per TEU than one with half of that capacity. However, factors like utilisation, loading infrastructure, and flexibility (e.g. whether a ship can pass the Panama Canal and the Bayonne Bridge) limit the use of such ships in practice (Døhlie 2009).

One measure that has a comparatively large potential – but is considered unlikely to be realised in the study due to industry incentives – is *slow steaming*. By reducing the speed of the ship, the energy needed per distance travelled can be reduced significantly as fuel consumption increases roughly with the cube of the speed (Longva 2009, 44). However, slow steaming means that more ships are needed to transport the same volume of goods in a given period of time. Moreover, demurrage rates in ports can be so high that ship operators try to arrive as soon as possible (IMO 2009, 47). Also, running an engine on low loads over longer periods of time may damage it and necessitate additional maintenance efforts (Tampi 2009). And, maybe most importantly, fast delivery of goods is an important competitive factor in the eye of the customer. Hence, while being a nice idea to save energy in theory, slow steaming is one example for a measure that makes sense for the planet, but simply does not happen under the constraints of the economic system. At least, that was the generally accepted truth until quite recently.

Ironically, it was not sustainability considerations that led to the overthrow of these industry myths, but extreme economic pressure. During the financial crisis, industry giant A.P. Moller-Maersk had reported the first loss since its founding in 1904. Maersk Line, the company's 30-billion-dollar container shipping arm, had been hit by a dramatic drop in freight rates and low demand (Miller 2010). However, already in 2007 the company had started to experiment with slow steaming and proven to engine manufacturers that it is safe to reduce engine load to as little as 10 percent instead of the traditional 40 to 60 percent (Maersk Line 2009). While this change to Maersk Line's key operating process may not seem revolutionary at first glance, it in fact questioned the overall value proposition of the container shipping industry: Instead of as-fast-as-possible delivery, Maersk now offers customers something that might be much more valuable: punctuality.

According to Maersk Line's Søren Stig Nielsen, the final verdict of the customer is still pending. Yet when considering how common delays currently are – they concern almost every second shipment – the buffer that slow steaming offers becomes very valuable, especially if customers can be made aware of the hidden costs of such delays. To be even more attractive to customer, Maersk Line tries to compensate the slower voyage speed by reducing the huge inefficiencies in ports – a goal that depends on a reliable schedule on the part of the shipping companies, too.

For Maersk Line, slow steaming has turned out to be very attractive both in environmental and economic terms. Maersk Line saved 7 percent of CO_2 emissions and corresponding fuel cost per container moved (or 2 Mt CO_2 per year) in the first 18 months of its implementation of slow steaming and sees the potential to save even more (Maersk 2010). As of 2010, three quarters of the Maersk Line fleet steams slowly. Following the success of Maersk, the industry has started to follow its example (Rosenthal 2010).

Critics may argue that once the oil price drops or trade volumes increase and capacities get tight, shipping companies will revert to old practices. However, Maersk Line has decided that slow steaming will remain a cornerstone of its future business model. It is believed to be economically beneficial even in good economic times due to the advantages mentioned above (Maersk 2010). And indeed, Maersk Line has apparently started to sell relatively new, super-fast ships that have still been ordered under the old business logic and started to replace them with freighters optimised for slow steaming (Aston 2010; Marine Log 2010). In fact, slow steaming is only one aspect of Maersk Line's new overall environmental strategy that, according to Søren Stig Nielsen, encompasses a profound change to Maersk Line's business model.

The new strategy of Maersk Line envisions competing directly on environmental sustainability. Sustainability in shipping (like in most business-to-business industries) does not translate into a premium or play a key role in winning new contracts. Yet, Maersk Line believes it can drive volume vis-à-vis competitors by gaining or maintaining a preferred supplier status (Broberg 2009). This endeavour is further supported by trying to become more customer-centric as opposed to the inflexible, complicated business practices prevalent in the industry today.

So far, unsustainable competitors have not gone out of business due to poor environmental performance. However, sustainability-minded customers are expected to exclude these from negotiations in the future. To speed up this development and create transparency, Maersk Line is already providing its customers with monthly scorecards which also show their respective carbon footprint. Hence, a key assumption is that demand for low-carbon supply chains has to be *actively* generated. And Maersk Line is determined to lead rather than letting competitors (or regulators) set the pace. Through this proactive approach, Maersk Line not only hopes to secure volumes from other shippers and shape regulations in its favour, but also to promote shipping vis-à-vis other modes of transportation, particularly air and road. The company hopes to steer future discussions away from reducing absolute CO_2 emissions from sea freight to reducing overall transport emissions dramatically by modestly increasing those from shipping. As an integrated transportation company, A.P. Moller - Maersk has all necessary resources to optimise logistics chains holistically in this spirit.

As explained above, slow steaming and extensive optimisation of logistics chains can have a significant environmental impact, even on a global scale. The shipping industry has not been under much public scrutiny so far, and efforts of industry leader Maersk reduce the risk of coming so in the future. However, should the industry be required to reduce emissions even more drastically, for instance in the form of a very strict cap-and-trade policy, the industry would need to rethink again.

Erling Johns Nielsen, Head of Global Supply Chain Development at Damco (a logistics subsidiary of Maersk), believes that the introduction of a "grey fleet" could be the result. A grey fleet means that shipping companies share unbranded assets like containers, infrastructure, or even vessels. This would allow for a significantly greater overall efficiency, for example as many containers currently have to be shipped empty to their port of embarkation. Like slow steaming, the idea of a grey fleet is not new. In fact, there is a related industry that has successfully applied a similar concept for years: the airline industry. Airline alliances like Star Alliance, One World, and others allow individual airlines to offer more destinations to customers and share infrastructure cost. However, a leading company like Maersk Line would need to give up some of its differentiation in order to partner with competitors. Competition law and different service profiles within the industry further complicate matters. Moreover, there are many sources of inefficiency yet to be eliminated in the current system; the introduction of a grey fleet may thus be considered an option for the distant future only.

Although this strategic option does not seem to be attractive to Maersk in the short- or medium term, it may be more appealing to smaller competitors who could gain significant scale by forming an alliance. In this case, Maersk may be forced to follow.

Maersk Line, and with it the entire corporation A.P. Moller - Maersk, has changed its strategic direction based on environmental sustainability. Difficult economic conditions have fostered the eventual success of experiments with slow steaming. In comparison, Lufthansa has announced a trial with biofuel flights, but a far-reaching shift like in container shipping has yet to occur in the airline industry.

Whether environmental sustainability represents an opportunity of similar magnitude for the airline industry cannot be determined conclusively based on the analysis presented above. Yet, against the background of the now obsolete industry myths in the shipping industry, questions arise whether presumed consumer preferences of airline passengers are carved in stone – or, whether a convincing green airline business model could revolutionise the industry.

This vignette demonstrated that it can be fruitful to search for solutions for sustainability challenges in other industries. Maybe the airline industry can gain insights from how the shipping industry has started to shift the basis of competition in favour of sustainability. Conversely, shipping companies may benefit from the examples of alliances between airlines.

Yet, the idea that industry players largely exhibit uniform strategic behaviour (and occasionally change it collectively) is not valid for all industries at all times. The automotive industry demonstrates this fact in the wake of the emerging market for electric cars. In this situation of great ambiguity and upcoming upheaval, strategic responses differ substantially. Competition – and cooperation – thereby facilitates creating numerous variations of new green business models, as will be explained in the following.

7.8 Vignette 7: Electric Cars – An Inconvenient Revolution

In the public debate on climate change and automobiles, electric cars (also known as electric vehicles, or EVs) are currently praised as the saviour. Not only do fully electric cars emit no CO_2, but they emit no air pollution at all, and additionally have very low noise levels. Of course, this is only part of the story as electric cars require electricity to drive – which needs to be generated, e.g. by a wind turbine or coal-fired power plant. Depending on the source of energy, CO_2 emissions of an EV can be virtually zero or even be worse than that of efficient internal-combustion engine (ICE) cars (Horst et al. 2009; Daimler 2010b).

Despite the enormous public attention that pure EVs currently get, they are still an obscurity in terms of the number of actual vehicles on the road. Nevertheless, many industry observers now believe that the widespread adoption of the electric car is inevitable, although it may still take decades rather than years (Hensley et al. 2009). However, as Ford's Executive Chairman, Bill Ford, cautioned in a recent interview: Although he, too, thinks that electric is the most promising technology for automotive propulsion of the future, he had thought the same about biofuels 18 months earlier and about hydrogen 18 months before that (Bonini & Kaas 2010, 8). BCG forecasts that 2.7 percent of global new-car sales in 2020 will be fully electric cars, and another 20 percent hybrid electric. However, the authors of the study acknowledge that these figures vary widely under different sets of assumptions (Book et al. 2009, 7). All statements in the following on future cost, adoption rates, technology paths and other market factors have therefore to be considered speculative in nature – just as it will be the case for most Green Business Model Transformations that decision makers need to assess under uncertainty.

Although the story of mass adoption of the electric car is thus one that is still to be written, it likely started with a product that was launched in Japan back in 1997: the Toyota Prius (Toyota 2010b).

7.8.1 *First Mass Market Success – The Toyota Prius Hybrid*

The Toyota Prius is a hybrid, i.e. it has an electric as well as an internal combustion engine. Although the Prius was far from the first (partially) electric car, it was the first to be sold in large volumes. In fact, electric cars were once more widespread at the beginning of the automotive age than gasoline-powered cars (DCTI 2010, 14ff.).

The Prius was launched in the U.S. in 2000 with a premium of a few thousand US dollars to cover the higher production cost compared to ICE cars. Extra cost for hybrid technology is estimated at about US$ 7,000 (circa € 5,000) per vehicle (Book et al. 2009, 3). There is no comparable non-hybrid Prius, but Toyota sells other hybrid models at circa 35 percent higher price points (Toyota Motor Sales, U.S.A. 2009). Unlike any other car with a (partly) electric propulsion, the Prius managed to attract a significant customer base and continues to hold a commanding market share (see figure 7.2) in this fast growing market segment (hybrid sales have stalled during the economic crisis but are expected to grow to over 5 million vehicles per year by 2020 in North America alone (Book et al. 2009, 8)):

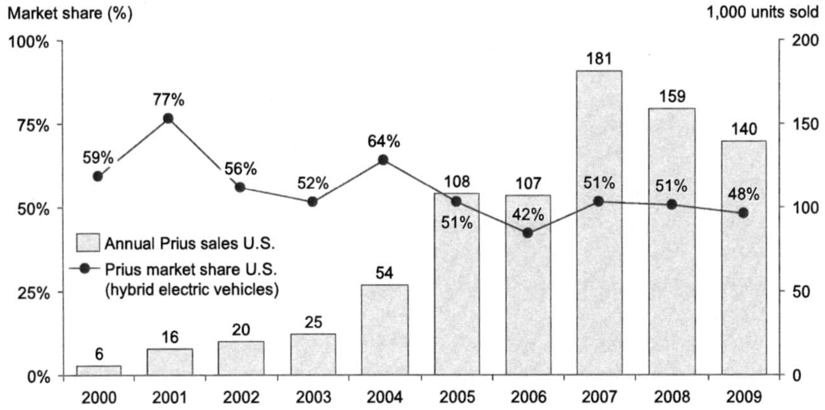

Fig. 7.2 Prius sales and market share in the U.S. 2000 to 2009 (based on AFDC 2010)

According to Dirk Breuer, Advisor Advanced Technology at Toyota, the Prius is the result of a systematic review of possible drivetrain solutions, from which the hybrid concept emerged as the most promising of all practically feasible alternatives. In addition, many basic features of conventional cars were reconsidered. Insights gained from that process have been used for many of Toyota's other models – thus even if the Prius had not succeeded in the market, not all efforts would have been in vain. On the other hand, Toyota's engineers tried to design a product that would not alienate customers because it is too different from what they are used to. For example, despite the superior acceleration of electric powertrains at low speeds, engineers emulated the feel of an ICE car. The low noise level of electric driving is already unusual enough for many customers who testdrive a hybrid for the first time. For an informative, more detailed account on the birth of the Prius and respective business model choices see Taylor (2006).

It would be misleading, however, to believe that most customers buy the expensive Prius because of its technical features or superior fuel economy (which is not revolutionary compared to highly efficient conventional cars). While technology serves as the foundation for success, Toyota has achieved true differentiation because it managed to attach an emotional value to the product: In a CNW Marketing Research survey, Prius buyers named "Makes a statement about me" as their most important reason to buy a Prius[19] (Maynard 2007). Along with the electric vehicle in general, the Prius in particular had become a symbol for a socially constructed "green" technology (see Pinch & Bijker 1989), largely independent of its actual contribution to the solution of the climate change problem that is associated with the automotive sector.

Speaking in evolutionary terms, the Toyota Prius is the missing link between the ICE and the fully-electric car. For further discussion, it is necessary to distinguish between different types of electric vehicles: In its popular meaning, the term *electric car* usually refers to fully-electric cars that are *battery electric vehicles* (BEVs), i.e. that are powered exclusively by electricity stored in their onboard batteries. These batteries need to be recharged, either with common household electricity or at dedicated charging stations. In contrast, *hybrid electric vehicles (HEVs)* feature an internal combustion engine in addition; they can be further distinguished into (at least) two variants, *full* and *mild*. *Full hybrids* like the Toyota Prius can drive using either electric or combustion propulsion, or both. *Mild hybrids* only feature a small electric engine that is not strong enough to propel a car on its own. Both types of hybrids use the electric engine and battery to improve vehicle efficiency, but are still reliant on fossil fuel as their batteries cannot be charged externally. A further step towards the fully electric car is thus a hybrid that includes the capability to plug it into the electricity grid. These cars are referred to as *plug-in hybrid electric vehicles (PHEVs)*. Toyota has announced a new PHEV model of the Prius for commercial launch in 2012 (Toyota 2010b). In essence, a PHEV is a BEV with an additional combustion engine. However, contemporary designs of PHEVs can only travel short distances in electric mode – a maximum of 13 miles in the case of the upcoming Prius PHEV (Toyota 2010a) – and thus still rely on conventional fuel for longer trips. A transitional technology to overcome this problem (until battery technology has advanced far enough) are

[19] Top reasons cited for buying a Prius (second quarter 2007): "Makes a statement about me" (57%), "Other (incl. incentives, business, etc.)" (42%), "Higher fuel economy" (36%), "Distinctive styling" (33%), "Lower emissions" (25%), "New technology" (7%) (Maynard 2007).

so-called *range extenders*. These vehicles (example: the Chevrolet Volt) drive electric but feature an additional onboard combustion engine to recharge the battery while driving (GM 2010). The different vehicle technologies are summarised below (see figure 7.3):

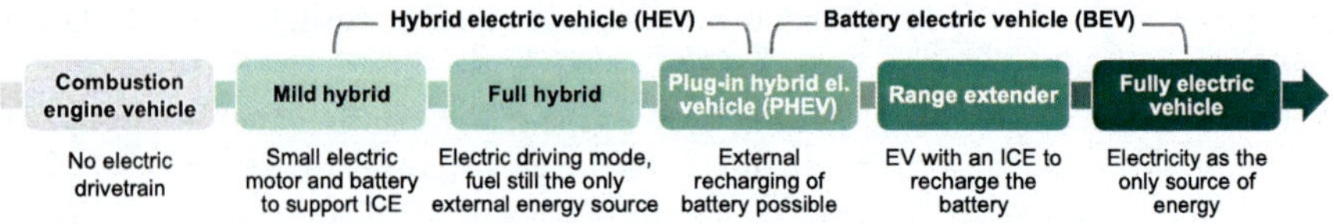

Fig. 7.3 Electrification path of power train technologies (based on Book et al. 2009, 2)

7.8.2 The First Wave of Fully-Electric Vehicles for the Mass Market

Automotive original equipment manufacturers (OEMs) have enjoyed a century of relative stability during which incremental changes, but no revolutions have taken place (Bonini & Kaas 2010, 7). While HEVs represent a modular innovation related to a critical component, a broad adoption of fully-electric vehicles would be far more radical: In contrast to HEVs, all-electric BEVs allow for a very different construction of the car (e.g., wheel hub motors without transmission), and might induce a range of completely new business models in the context of battery issues and services. Unfortunately for established OEMs, radical innovations tend to put incumbents out of business (Henderson & Clark 1990). Not only does their long-standing component knowledge (i.e., ICE drivetrains) get de-valued, but their legacy architectural knowledge may actually prevent them from competing effectively in the electric car game. Another challenge is that the *dominant design* (e.g., Suarez & Utterback 1995) has not emerged yet, not even in the broadest sense: alternatives like hydrogen cars or alternative fuels could well turn out to be more economical and scalable solutions.

Despite its pioneering role with the Prius, Toyota thus still pursues a range of alternative technologies. According to Breuer, the risk that the current enthusiasm towards EVs is followed by a phase of disenchantment just seems too great. In addition, OEMs need to invest in advancing ICE technology as it will continue to dominate new car sales for at least the next few product generations. Hence, OEMs are currently facing a big stretch in terms of R&D budgets. Respective decisions today may have a big impact on a company's future success (Book et al. 2009, 8).

Most OEMs have been quite reluctant to pursue the development of EVs. There are even conspiracy theories around the discontinuation of past electric car programmes, most notably GM's EV1 that was built from 1996 until 1999. The documentary "Who Killed the Electric Car?" accuses – among others – the oil industry and car manufacturers to have (self-)sabotaged the project (Paine 2006).

Considering the business case and strategic implications of electric cars, OEMs would at least have a motive (and oil companies even more so):

Apart from the general danger of radical innovations for incumbents mentioned above, the electric car severely disrupts the established automotive value chain. OEMs will lose a significant part of their competitive differentiation through the replacement of their sophisticated ICE technology with electric drivetrains. Moreover, the battery of an electric car will represent a major share of the total cost of electric cars, even when considering future economies of scale from large production volumes. Estimates suggest that a modern 15-kWh battery pack (equivalent to about 100 kilometres of range) will still cost the OEM US$ 6,000 in 2020 (Dinger et al. 2010, 7f.). This high production cost – many EVs will have batteries larger than 15 kWh – also makes EVs difficult to sell to consumers with the traditional, ownership-based business model. Yet, because the electricity needed for propulsion costs less than the equivalent amount of diesel or petrol, the total cost of ownership (TCO) vis-à-vis ICE cars is less problematic than the car price differential alone.

However, amortisation times of only a few years (for US consumers) will still not be realistic without a multiplication of fuel prices or government incentives of several thousand dollars. The conditions are better in Germany, France, the UK and China, but without government incentives, electric cars will not be an economical choice for the majority of car owners in these countries either (Dinger et al. 2010, 9f.).

A start-up that challenges OEMs with a novel business model to overcome the cognitive problem of high purchasing prices (see Kaenzig & Wüstenhagen 2010) is the company *Better Place*. Better Place is promoting a business model similar to that of a mobile phone provider: The EV is subsidised with the revenue from the electricity that needs to be bought exclusively from Better Place in return (TED.com 2009). Better Place has focused its initial efforts on "transportation islands" without frequent long-distance drives like Israel and Denmark. For the latter it has committed itself to a volume of at least 100,000 EVs together with its OEM partner Renault (Renault 2009). Australia, the US (California, Hawaii), Canada (Ontario), Japan (Tokyo), and now China have also been addressed meanwhile (Better Place 2010b).

To cope with the problem of low range and comparatively long charging times, Better Place intends to provide battery switch stations for its customers. That is, depleted batteries can be automatically exchanged for charged ones within about one minute (Better Place 2010a). Yet, many industry experts doubt that the battery switch model will find broad adoption. As the battery is a very costly part of an EV, stocking extra batteries has a very negative impact on the overall business case. If future standardisation of batteries and plugs remains low, Better Place's business case would be further strained as it intends to offer its charging stations to non-subscribers as well (Better Place 2010c). Whether or not Better Place will become profitable remains to be seen, but it certainly has accelerated overall market development for EVs.

A public charging infrastructure has a very problematic business case in general, not just for Better Place. OEMs are not incentivised to invest as they are unlikely to earn more by selling an EV compared to an ICE car. They could instead opt to target mainly commuters who can charge at home or at work.

In contrast, utilities would seem to be the most natural provider of a public electric-charging infrastructure: Electric cars promise to create additional demand, to provide energy storage capacities that can absorb fluctuating generation from renewables, and to replace peaking power plant capacity by using plugged-in EVs' stored electricity instead (vehicle-to-grid) (Arnold et al. 2010, 47f.).

And indeed, some activity from utilities can be observed currently. For example, German utility RWE has initiated an e-mobility programme and started to build a public electric-charging infrastructure in some major German cities (RWE 2010). In the future, RWE might try to earn money by selling its charging stations to government authorities, or to benefit from a general shift from oil to electricity. However, it seems unlikely that RWE can be earning much money from *operating* a public charging infrastructure: According to BCG, even with an amortisation time of 15 years, utilities would need to more than double the price of electricity compared to normal household electricity (Book et al. 2009, 9).

Yet, even the favourable effect on overall demand will be limited: Assuming efficient EVs which consume 15kWh per 100km, the electricity needed to propel *all* passenger cars in the USA would be equivalent to only 10 percent of the total US electricity demand in 2008 (based on data from FHWA 2009; EIA 2010). In addition, integrating large amounts of EVs into the grid will require major technical upgrades towards a smarter grid (ETP 2010, 7). Technological progress related to battery technology (e.g., charging cycle times) are also a prerequisite for advanced applications as described above.

In conclusion, developing a viable value proposition for the mass market will be challenging for OEMs, even when applying novel business model designs like that of Better Place. Without government intervention, the TCO is simply too high in the medium term, and a public charging infrastructure may not become widely available. Most OEMs have little incentive to push the EV, and consumers cannot be expected to buy it purely out of ethical reasons (although some will). Hence, government interventions will likely play an important role in setting the rate of adoption of the electric vehicle. As a result, there is not only competition between companies, but also between countries (Book et al. 2009; Bradsher 2009; Bundesregierung 2009).

7.8.3 The Electric Car Game Is a Business Model Game

No compelling customer value proposition; formidable technological challenges related to key resources and processes; problematic economics – given all these issues the obvious question is: Why do OEMs spend billions of Euros on the development of electric cars?

The answer is that incumbent OEMs simply cannot risk falling behind in case the market for electric cars starts soaring. The EV market directly cannibalises that for ICE cars. Because EVs have never been produced in large volumes, a head start can bring about enormous advantages in unit cost through experience and scale (see Mintzberg et al. 2009, 101f.; Sterman 2000, 368ff.). Moreover, low-cost competitors from emerging economies – especially China – will try to seize this unique opportunity of a level playing field and capitalise on their growing home markets and government support (Bradsher 2009; Gao et al. 2008). For example, the Chinese battery and car manufacturer BYD has just signed an agreement with German utility RWE to cooperate on EVs in Europe (BYD Auto 2010). While Chinese car makers were used to be seen as impudent imitators by established manufacturers (Landler 2007), the electric car challenge seems to have turned them into valuable potential partners. A recent cooperation between Daimler and BYD illustrates this fact (Daimler 2010a). Consequently, the bustling activity of incumbent OEMs can be interpreted as "hedging" (see also Avadikyan & Llerena 2010 for the role of HEVs in hedging strategies).

This hedging can be observed in all parts of the value chain. Toyota, through material supplier Toyota Tsusho, has even acquired a stake in a lithium mine to secure the supply of this important raw material for batteries (lithium itself is rather abundant, but current extraction capacities are not) (Krauss 2010). Recent strategic alliances between OEMs and battery manufacturers (e.g., Daimler 2008; Toyota 2009; Renault 2010) serve not only the purpose to secure access to suppliers, but also to gain internal knowledge about a critical technology quickly. In the long run, however, most OEMs can be expected source batteries like many other components, and not produce them in-house or in joint-ventures. It is unlikely that batteries will become a branded part of the EV that has importance beyond its specifications. In any case, the strategic importance of electric car technology is immense for many OEMs as the recent espionage scandal at Renault highlights (Pearson et al. 2011); to date, the Renault-Nissan alliance alone has invested € 4 billion in its EV programme.

In contrast to the majority of established OEMs, premium manufacturer BMW has chosen to strike a completely new path. BMW believes that building cars in particular and individual mobility in general has to be reinvented. Most remarkably, this insight was not generated under intense economic pressure (like many other Green Business Model Transformations), but was the deliberate conclusion of broadly applied market research between 2006 and 2007.

In accordance with increasing urbanisation, BMW focused its investigations on metropolitan areas. The results were sobering: Not only does consumers' affinity for cars decrease, but owning a car is not more important than owning a smart phone in some countries. Three relevant customer groups surfaced from the research (Blanco 2010; Holloway 2010):

- the *lifestyle customer*, who cares mostly about brand and is willing to spend heavily on optional equipment
- the *economy customer*, who calculates and chooses based on total cost of ownership
- the *responsibility customer*, who wants to demonstrate that he or she cares about environmental sustainability

Based on the understanding that the OEM business model of the past may not be successful in the future, BMW initiated *project i*, "a future mobility think tank" (BMW Group 2010c). It is sponsored directly by the board, was exempt from cost cutting during the financial crisis (Katzensteiner 2009), and is provided with far-reaching competencies (including the choice to look for solutions externally). In sharp contrast to the many alliances that can currently be observed, *project i* is strictly sealed off from competitors. Even during a personal, non-recorded interview with BMW, many aspects remained vague. However, what has been revealed is that BMW will launch a fully-electric car for urban environments called *i3* in 2013.

The i3 was developed under the project name *Megacity Vehicle* and is supposed to revolutionise the building of cars. After the general customer requirements were understood (e.g., fully-electric car with four seats, no compromises), engineers were handed over the task not to build a car that satisfies a specific value proposition, but to completely rethink the value creation-side – the production of the car. BMW believes that the current system is largely optimised, and that the switch to electric cars is the right time to go for a quantum leap with an entirely novel production system. The result – as far as it has been revealed – is a flexible, modular car design that includes a "life" and a "drive" module, referred to as the "LifeDrive concept". The drive module combines the battery and drive system in an aluminium chassis that provides basic crash functions. The life module is attached on top of the drive module and offers superior interior space and freedom of configuration. The module consists of a specifically developed Carbon Fibre Re-Inforced Plastic (CFRP). CFRP offers high crash safety and is very lightweight. It is 50 percent lighter than steel and 30 percent lighter than aluminium, thereby offsetting the additional weight from the battery (BMW Group 2010b).

However, what is most remarkable about the LifeDrive concept are not the product features it offers, but the broader business model options which it facilitates. First, it enables simplified, accelerated production processes (the modules can be built in parallel and components are easier to assemble), which in addition are supposed to be far more resource-efficient (BMW Group 2010a). Second, each module could be offered to customers using separate revenue models, i.e. one module can be leased while the other is sold in the traditional way. This also means that cars need not be replaced in their entirety – a customer may, for example, choose to upgrade the life module but keep the drive module.

The price of the i3 has not been revealed yet, but it is rumoured to come at a premium and cost just over €40.000 (Ewing 2010; Die Zeit 2011). The technology and materials used are certainly very expensive, particularly because there is a lack of scale. However, the improved production processes and revenue model options work in favour of an attractive purchasing price. BMW plans to invest € 400 million in new production facilities for the i3 and the i8. Martin Arlt, who manages *project i* at BMW, asserts that the i3 will aim to be attractive for all three customer groups mentioned above, including the economy type. The i3 – together with the plug-in hybrid i8 and more models yet to be announced – will be launched under a new sub-brand, *BMW i*, to distinguish them from BMW's existing product range (BMW Group 2011).

A daring initiative like *project i* is almost impossible to shoulder under economic pressure, when financial resources are already scarce. Without a burning platform, however, scepticism is very hard to overcome. BMW used company events with former American Secretary of State Madeleine Albright and German Foreign Minister Joschka Fischer to deliver the message that the megatrends of urbanisation and sustainability may make BMW's current business model obsolete in the future. BMW managed to win support from many of its horsepower-minded employees and project i has become a well thought-after area to work in within BMW (Reiter 2010).

BMW further assures that the i3 is just one aspect of its emerging ground-breaking mobility concept. It has founded a venture capital company to develop (digital) mobility services that "deliver extra comfort or smart advantages (intermodal travel, smart parking, recommendations, communication etc)" (BMW i 2011). It remains to be seen if and how BMW will live up to its promise to revolutionise individual mobility. If the i3 and the mobility services become a success, BMW will gain far more than a "green halo" (like Toyota with the Prius, Interbrand 2008). If not, its room to manoeuvre will be severely limited – even though BMW is in a relatively comfortable financial position now (Reiter 2010).

The changes in the automotive landscape may become even more radical than what has been revealed of BMW's vision so far. For example, a widespread adoption of car sharing business models would drastically reduce the number of cars produced. According to the world's largest car sharing company Zipcar, the number of cars needed per person is between 1:15 and 1:20 compared to owned cars (Zipcar 2010). Even if the ratio was only 1:5, the impact on demand would potentially be dramatic for OEMs.

However, car sharing business models have been around for a long time, and most industry experts that have been interviewed believe that it will remain just what it is today: niche. Yet at least for city dwellers who cannot charge a car at home or at work, car sharing complements the introduction of electric cars well because going to a nearby charging point would be no extra effort. Moreover, considering the relatively high purchasing price and low operating cost, car sharing may well turn out be the best way to afford a stylish EV. Maybe all it takes is an innovative company to design a compelling new car sharing business model – just like Apple did for online music and the smartphone market.

So far, established OEMs could rely on their strong brands and hard-to-imitate resources. However, the electric car game offers unprecedented opportunities for new challengers and ambitious incumbents alike. The variety of approaches in the industry demonstrates how unique the current situation is. While some can build on their first mover advantage (Toyota), most OEMs try to catch up through alliance building (and even Toyota is hedging its bets). Furthermore, while Daimler emphasises a rather gradual shift from current technology towards electrification (Mercedes-Benz 2010), archrival BMW discards many of its proven concepts and aims for nothing less than re-inventing the building of cars.

Moreover, successful Green Business Model Transformations of OEMs partly depend also on factors beyond the control of a single company, e.g. the public perception of the ecology and safety of EVs in general. Even developments in other industries (e.g. whether utilities can offer reliable charging at scale), and political goodwill will play a vital role (e.g., to set up a public charging infrastructure and create buying incentives) (IEA 2009c).

All these factors make it incredibly difficult for deciders to make robust choices now, but also highlight how valuable the proposed business model concept can be in this process: The companies that will succeed in the electric car game will be the ones that manage all business model components appropriately, and in such a way that they complement each other.

7.9 Industries Not Covered in Case Vignettes

The survey presented above covers a broad range of cases: consumer-facing vs. business-oriented industries, production-based vs. service-based business models, and base-of-the-pyramid vs. developed markets. Nevertheless, this industry survey is obviously far from exhaustive.

Adjacent to the described electric car industry, a whole ecosystem of new business models is about to emerge subsumed under the label *smart grid.* No universal definition of the term exists to-date, but it may be described as “an electricity network that can intelligently integrate the actions of all users connected to it – generators, consumers and those that do both – in order to efficiently deliver sustainable, economic and secure electricity supplies” (ETP 2010, 6). Smart grids are thereby the outcome of a convergence of the electric power, telecommunications infrastructure and information technology industry sectors (Leeds 2009, 10). In contrast to its less smart equivalents, it has the capability to allow data, as well as electricity, to flow in two directions between the supplier and the consumer. This way, it not only enables the large-scale introduction of plug-in electric vehicles as described above. It can also integrate a growing number of other distributed power sources effectively into the grid (DOE 2008, 12ff.). Industry observers expect the smart grid to revolutionise the energy industry and lead to numerous novel business models in the medium or long term (Klose et al. 2010). Traditional value chains will break down and be replaced by more complicated value networks (see Christensen 1997, 29ff.) involving many new players from diverse industries as mentioned above.

Smart grid business models alone offer more than enough content for a separate doctoral thesis. However, there are also other industries that might offer interesting cases to investigate. For example, raw material extraction industries are heavily reliant on their social license to operate – due to the significant environmental (and social) impact they generate in their areas of operation. Processing industries other than the covered food and chemical industries (e.g., metals, cement, and paper) have large environmental impacts that create opportunities for green business models.

This list could be continued, but, particularly two more industries should be highlighted as they have the potential to significantly improve the environmental performance of *other* industries: information and communications technologies (ICT) and financial services.

ICT plays a decisive role in many technical applications that reduce environmental impacts, including the smart grid (The Climate Group 2008). In addition, ICT can help companies to better manage their environmental performance. Industry giant SAP has discovered environmental sustainability as a new strategic growth area only a couple of years ago (SAP 2009, 13ff.). SAP's Vice President Sustainability, James Farrar, remarked during an interview that due to the widespread implementation of SAP software in corporations, SAP is in a unique position to help firms integrate their environmental management with other business processes. As mentioned earlier, Google has also started to become active in the sustainability domain (see 7.5). So has IBM with its "Smarter Planet" initiative (Palmisano 2010; IBM 2010d). Through the intelligent use of data IBM believes it can further positive change in numerous fields (IBM 2009).

The financial industry could play a critical role in sustainable development insofar as it has the power – at least theoretically – to deny financing to companies with unsustainable business practices. In this sense, the financial industry could become complementary to government regulation and lift the minimum standards of whole industries. Many challenges exist in this respect, including (but not limited to) the accurate measurement of sustainability and assuring adequate reactions on misconduct. However, first steps in this direction have already been taken by the industry. For example, the meanwhile widely adopted *Equator Principles* (see Equator Principles Association 2010) specify standards to assess and manage social and environmental risks in project finance. Being voluntary, the principles seem to work according to enlightened shareholder theory, i.e. banks meet stakeholder demands as long as a business case can be constructed (Macve & Chen 2010). HSBC, a leading bank with respect to sustainable finance, has additionally issued tailored risk policies for industry sectors including energy, chemicals, or mining and metals (HSBC 2010, 8ff.). As these policies get revised over time, the minimum standards rise and low performers will need to improve in order to remain compliant.

In conclusion, many spaces remain to be explored for exciting new green business model opportunities. Doing so in a systematic manner requires an abstraction of the specific examples that exist today. This industry survey already provides sufficient material to accomplish this task as presented in the next chapter. Before, the environmental performance of the described cases will be briefly discussed.

7.10 Environmental Performance Assessment of Case Study Examples

One of the biggest challenges that academics face is to match their conceptual work with reality. Green business models are no different in this respect. They tend to have a much higher level of complexity in practice than their abstract, idealised counterparts described in chapter 5. Giving a substantiated answer to the question of how green all the mentioned real or hypothetical business models currently are – or might be in the future – is a complex matter that goes far beyond the scope of this work. Moreover, an absolute, one-dimensional "rating of greenness" across all kinds of business models seems neither practical nor particularly useful. Nevertheless, the factual and hypothetical sustainability performance of the discussed business model examples needs to be evaluated at least broadly, in relative terms, and within its systemic context.

In accordance with the taxonomy developed in chapter 5, figure 7.4 shows a preliminary categorisation of the business models mentioned earlier (and some additional ones to provide a more complete picture).

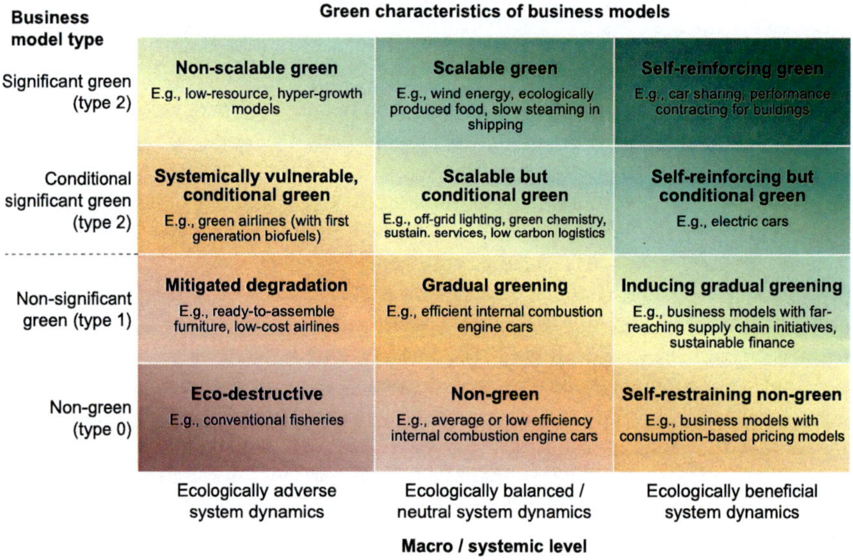

Fig. 7.4 Green characteristics of business models from the industry survey

In the following, each assessment will be explained structured along the four y-axis categories of business model types. Although the colouring represents a general trend of how "green" respective business models within each field will be, a comparison in absolute terms would not be meaningful. For example, a company with the size and influence of Wal-Mart has a much greater absolute positive effect on the environment even if it is only a little green, compared to a very small company applying a business model with an extremely high environmental performance. Moreover, no business models that meet the criteria for strong sustainability ("type 3") are provided.

7.10.1 Significant Green (Type 2)

Business models that fall in this category can safely be considered "green". Most examples from the industry survey will either fall under this or the next category (both type 2). A classic example that was not examined in depth because of its wider coverage elsewhere is renewable energy generation, like wind energy. The same category does apply, however, to business models that produce wind energy equipment, like that of GE Energy or Siemens Energy. Business models like these are dubbed *scalable green* in figure 7.4 as they can grow substantially without adverse feedback loops kicking in. This does not mean, of course, that there are no such feedback loops. Wind mills require non-renewable resources like steel to produce; hence production cannot be increased infinitely. Moreover, individual manufacturers may be more or less advanced in their environmental management practices. Relevant, however, is only that the overall environmental impact of these business models (i.e., their principal logic) is significantly better at scale than common alternatives, independent of factors of detail. Business models from the industry survey that would also fall in this category based on a preliminary, qualitative assessment are ecologically produced food (Frosta example), and slow steaming in shipping (Maersk Line example). However, a definite categorisation would require detailed impact quantifications of these and alternative business models that go beyond the scope of this work.

These business models are contrasted with those that are *non-scalable green*. These business models have significant green characteristics in principle, and thus are attributed to type 2. However, due to inherent systemic characteristics they may turn out to have an adverse overall effect on the environment. This includes business models that are very resource-efficient, but for the same reason grow strongly in volume. This can be the case if the resource efficiency allows for a much lower price point – which therefore attracts a much wider audience. Consequently, savings from unit-based energy or other resource usage are overcompensated by the success of the product; non-scalable thus refers to ecological, not economic aspects. A revolutionary car that is efficient and cheap would be a concrete example, as it may be bought by customers who could not afford one previously (the example of the *Tata Nano* only fulfils the condition of a very low price, though).

The opposite effect can be observed in business models with self-reinforcing green characteristics. Performance contracting for energy efficiency in buildings is reported to result in substantial savings of 20 to 30 percent of energy use (Siemens 2010a). This self-reported figure by Siemens is credible because energy savings are the business model's revenue source – the business would not be financially attractive were it highly overstated. This fact also explains the categorisation of *self-reinforcing green*: The company is reliant on achieving its promised energy savings out of pure self-interest; hence principal-agent dynamics (see chapter 6.1.3) actually work in favour of the environment, not against it.

It has been mentioned that electric cars may revitalise car sharing business models. As described earlier, car sharing drastically reduces the number of cars needed per person vis-à-vis the traditional ownership model. Of course, the net reduction of cars per person depends on the composition of the car sharing customers (people who had no car before, who give up their private car, and who have their shared car as a second car). But empirical data suggests that car sharing can realistically be assumed to significantly reduce the number of used cars overall (Katzev 2003), thereby greatly reducing the environmental impact from production. However, at least in terms of energy consumption and CO_2 emissions, the largest impact over the life cycle of a car lies in the use phase, not the production phase. Owning or leasing a car incentivises people to use their car often; consumption-based (e.g., per distance or time driven) revenue models have the opposite effect (Heiskanen & Jalas 2003, 190ff.). Again, customer composition plays a role for the overall effect, but even those who have had no personal car before have been observed to drive less after they joined car sharing (Meijkamp 1998). The world's largest car sharing company Zipcar claims substantial benefits from behavioural changes, too (Zipcar 2010), although some studies suggest this cannot be assumed universally for all car sharing schemes (Katzev 2003). From a technical point of view, the higher usage of shared cars leads to a faster fleet renewal which can be used to improve its efficiency (slightly reducing the gains from fewer cars per person, though). Overall, it seems fair to state that the vast majority of real-world cases will lead to a significant improvement of the environmental impact of individual mobility, and that positive systemic effects (consumption-based pricing) outweigh potential negative ones (e.g., cheap, convenient access to a car). Even this – at first glance seemingly clear – case of a green business model illustrates how tricky general assessments of a business model can be. The following group of business models are even more reliant on some boundary conditions that may prevent significant performance improvements from materialising.

7.10.2 Conditional Significant Green (Type 2)

Although the distinction to the previous category may be debatable in some cases, it is important to recognise the importance of boundary conditions for the assessment of the respective business models. Such boundary conditions can either lie outside of the scope of the business model itself (e.g., regional infrastructure), or be riddled with uncertainties that prohibit definite assessments (e.g. future technological developments, materialisation of systemic effects).

One example is solar-powered off-grid lighting (Osram and Philips examples). The pilot programmes clearly offer substantial environmental benefits, that is, no health hazard from burning kerosene, and substantial CO_2 emission reductions. Yet, if a reliable grid access was made available for all community members, the former benefit would disappear, at least for applications in homes. If the grid electricity was moreover generated from renewable sources, even CO_2 emissions would pose no advantage anymore. These are quite unrealistic 'if' conditions for most of the African continent and some rural regions in Asia, though. Nevertheless, the business model is *scalable but conditional green.*

In green chemistry, there are trade-offs between the different principles for green chemistry and engineering. For example, some products based on renewable feedstock may have disadvantages compared to petrol-based alternatives in other sustainability domains. Even the green chemistry product examples specifically selected for publication by Cognis are inferior to the baseline product in at least one category (Cognis 2010c). Results will likely be more ambiguous for other green chemistry products. While such trade-offs in environmental performance have been accounted for in principle in the methodology described in chapter 5.3, "green chemistry business models" seems to be too broad a category to be considered unconditional green.

According to a similar logic, sustainability services (BASF example) and low carbon logistics (Damco example) have the *potential* to be type 2 business models, but are highly dependent on their specific implementation type. If these services address sustainability only superficially, the environmental benefits will be rather limited.

A hypothetical green airline business model, as explained above, would need to make use of alternative fuels since other technical and non-technical measures have limited potential. However, the environmental effect of biofuels is controversial. In fact, their effect on the environment can be very negative depending on how they are produced (Fargione et al. 2008; Tilman et al. 2009; UNEP 2008). Second and third generation biofuels promise to solve this problem by not competing for arable land or leading to deforestation. However, scaling up a reliable supply for the required quantities without compromising on the environmental aspects is considered a significant challenge at this stage of development (ATAG 2009, 18f.). There is a risk that a boom will lead to adverse sustainability effects as it could already be observed with first generation biofuels (e.g., see BBC News 2007; Grunwald 2008b; Power 2008). Hence, the green airline business model – at this stage – is considered *systemically vulnerable, conditional green.*

In contrast, another green business model type in the transport sector, electric cars, is *self-reinforcing but conditional green.* That is, electric cars will substantially decrease CO_2 emissions from road transport, but only if the electricity is produced from low-carbon sources. It is self-reinforcing green because electric cars and renewable energy generation complement one another. If electric cars are integrated in a smart grid, they can help coping with fluctuations from renewable energy generation.

7.10.3 Non-significant Green (Type 1)

Many business models today can be subsumed under the label *gradual greening.* Their environmental performance is better than that of the average competitor, but is based only on incremental improvements (which will often become the industry standard shortly thereafter). One example would be an automotive OEM that produces particularly efficient ICE cars. The fleet emissions will be lower than industry average, but not represent a leap like in type 2 business models.

Rather ambiguous cases are business models that fall under the category *mitigated degradation*. These are business models that have already been optimised in their sustainability performance but show troublesome systemic effects in terms of their overall environmental impact. As mentioned earlier, low-cost airline business models also entail low emissions per passenger kilometre. However, low ticket prices entice people to fly who would not consider it otherwise, thereby resulting in significant additional traffic (Cairns & Newson 2006, 57ff.).

Another example is provided by ready-to-assemble furniture, with IKEA as its most prominent representative. This business model has the advantage that furniture packaging and transport can be organised very efficiently, and IKEA is likely seeing benefits that are similar to that observed in ecologies of scale in the food industry (see 7.2.2). However, it is not the size of the company, but the logic of the business model that should determine its categorisation. Furthermore, despite its many well-documented efforts to protect the environment (e.g., IKEA 2009, 49ff.; IKEA 2010), IKEA has still a long way to go and therefore is considered to apply a type 1 business model. For example, IKEA has been criticised for its supply chain practices with respect to wood. Deforestation is one of the single largest sources of man-made CO_2 emissions (larger than all transport combined), and IKEA sources significant amounts of wood from areas that are known for widespread problems with illegal logging (Goodman & Finn 2007). Yet, IKEA has missed its five-year target of sourcing 30 percent of its wood from certified suppliers by 2009 by a wide margin (IKEA 2009, 57).

With respect to the systemic level of its business model impact, IKEA puts a lot of effort into improving individual product units. But its furniture is stylish yet so cheap that everybody can afford it; and because it is not built to last, IKEA cultivates a throw-away culture for products that could principally last for decades (Shell 2009, 125ff.). Hence all of its efforts to reduce resource-usage, use recycled materials, and transport efficiently first have to compensate for the problematic consumption patterns that are inherent to its business model.

Other interesting cases are business models that cannot be considered type 2, but that have widespread positive effects on its industries or even beyond. One possibility is that such an effect is undeniable, but not attributable to the business model that induces the change. Wal-Mart's sustainability index (see 7.2.2) or sustainable finance (see 7.9) might become impressive examples for such a phenomenon. Wal-Mart or banks may not apply a green business model themselves, but through their activities they can oblige affected companies and their peers to improve their environmental performance. This, in sum, can result in effects potentially far greater than what the inducing companies could achieve within their own company doors. These improvements, however, are to be attributed to the companies that implement them within the logic of the green business model taxonomy.

7.10.4 Non-green (Type 0)

This last category obviously has not been the focus of the industry survey as it represents ordinary business models that make the current economic system unsustainable. In the best case, these business models are *self restraining non-green*, for example because their pricing model is consumption-based and therefore tends to contain its environmental impact this way. Most current business models will fall in the ordinary, *non-green* category. Automotive companies that sell ICE cars, but do not focus on efficiency more than their peers are an example. In the worst case, a business model features in addition dangerous feedback loops and thus have an *eco-destructive* effect. An obvious example are conventional fisheries. Before the introduction of sustainable fishing schemes, no competitor was incentivised to reduce its catch to sustainable levels, and the diminishing fish stocks provoked companies to use ever more destructive fishing methods (see 4.5.1).

The industry survey has revealed a broad range of factors that influence the environmental and economic performance of green business models. These factors differ in their level of relevance and have various managerial implications that need to be understood and managed in the course of successful Green Business Model Transformations.

Chapter 8
Managerial Implications of Survey Results

The survey in the last chapter shows that transforming firms operate under highly different boundary conditions regarding the characteristics of their organisations or industry. These boundary conditions can have a profound impact on the efforts required to succeed with new, green business models (Post & Altman 1994), and also moderate associated risks and opportunities (Aragón-Correa & Sharma 2003). In order to determine an expedient, low-risk approach a static view on boundary conditions is not sufficient – dynamic factors need to be considered as well. This does not only relate to the extrapolation of trends, but particularly to the competitive dynamics which the transformation itself may set in motion. The following will therefore investigate implications from observed practices and chains of events.

8.1 Overview of Managerial Implications

Static factors that influence transformation challenges are considered on three levels: industry, business model, and company. These factors are referred to as "static" not because they do not change over time (they do), but because a snapshot analysis can yield meaningful results. Factors are interrelated, e.g. if an industry is heavily affected by environmental issues, companies will mostly have strong competencies in environmental management. The three levels are also not strictly hierarchical: A company may be active in several industries or operate multiple business models. Conversely, a number of companies may operate the same business model. In some instances, business models may even stretch across traditional industry borders. Nevertheless, a multi-level analysis helps to prioritise and address transformation challenges effectively.

Dynamic aspects are harder to categorise and exhibit highly complex interactions. Hence, the assessment is limited to six exemplified dynamic patterns that may result from competitive action or collaboration: "first mover advantage", "creative destruction", "collective opportunity", "complementary competition", "green movement", and "green backlash".

A. Sommer: Managing Green Business Model Transformations, SPLCEM, pp. 221–257.
springerlink.com

Finally, the sum of external and internal influencing factors and managerial decisions result in a specific corporate journey. For example, companies under economic pressure differ in their choices from prospering firms. Firms that try to generate value through cannibalisation face different problems than those that pursue complementing or synergistic green business models. Understanding why companies have followed the path they have followed may therefore help other companies to increase their chances of success with their own Green Business Model Transformations.

8.2 Snap-Shot Analyses of Boundary Conditions

The following (non-exhaustive) table 8.1 lists important influencing factors, subdivided into general factors and those related to environmental sustainability:

Table 8.1 Influencing factors from static firm and industry conditions

Level	General factors	Factors related to environmental sustainability
Industry-specific	**Condition of industry structure:** stable vs. in flux or emerging	**Prevalence of environmental sustainability within the industry:** emerging and subliminal vs. established and influential
Business model-specific	**Familiarity of business model:** number and magnitude of changes to current business model elements • low familiarity increases the probability of failure • determines the likely source and strength of resistance	**Significance of environm. sustainability for business model:** effect of green aspects on economic viability related to … • … value capture aspects • … value creation aspects
Company-specific	**Change-related company conditions:** • Private vs. public company • Leadership: hierarchical vs. democratic, changes in top management • Organisational structure • Competitive position: market leader, challenger, underperformer, etc.	**Degree of eco-mindedness of the company** • history and attitude towards environmental sustainability • competencies in environmental management

8.2.1 Industry Level: Business Model Paradigms and Patterns of Firm Behaviour

8.2.1.1 Business Model Paradigms

I argue that companies act within *business model paradigms*, analogous to the concept of *technological paradigms* (Dosi 1982; informed by the Kuhnian conception of scientific paradigms, Kuhn 1996). That is, changes to business models usually happen along predefined trajectories determined by the paradigm (Teece 2008, 510f.). These changes can be technological in nature, but also be related to any other business model element. In the hitherto prevailing shipping business model paradigm, for example, "progress" includes faster delivery. Among other things this led to the ordering of super-fast ships. Yet, with the abandonment of this business model paradigm, Maersk Line seems to be following a new business model trajectory: Progress in the value proposition context has now shifted from speed to reliability and efficiency (and sustainability).

A business model paradigm, however, is not limited to the value proposition aspects that are derived as primary targets for improvement. It can also include the means, i.e. the value creation side. An obvious example is the combustion engine in the automotive industry, which is now challenged – once again – by the electric drivetrain. A lot of business model choices still exist for electric cars at this point in time (e.g., revenue models, value chain configurations, etc.; see 7.8). As the new business model paradigm becomes established, most of these principal choices will be selected and alternatives excluded by the paradigm (compare Dosi 1982, 152f.). Drawing upon institutional theory, a new paradigm results from the formation of certain regulative, normative and cultural-cognitive "rules" (see Scott 2001).

Consequently, a *business model paradigm* is understood here as

> *a set of prescribed characteristics that all business models within the respective paradigm have in common. These principles include basic design choices as well as the trajectory along which business models are to be improved.*

In industries that do not feature one, uniform market (e.g., owing to geographical separation or different sets of customer needs) several business model paradigms may coexist. The proposed concept of business model paradigms is related to that of *socio-technical regimes* (see Geels & Schot 2007), which have been described as "the mainstream, and highly institutionalised, way of currently realising societal functions" (Smith et al. 2010, 440). Socio-technical regimes are located on the level of *organisational fields* (see DiMaggio & Powell 1983, 148), and therefore include a particular industry plus other relevant actors (key suppliers, customers, regulators, etc.) as well as institutions and technological artefacts (Markard & Truffer 2008, 605). The business model concept incorporates an analogous unit of analysis for a micro perspective. Research in this field can thus principally be informed by findings on socio-technical regimes. However, the focus of the former obviously lies on firms that apply a certain business model, not on organisational fields from a broad societal angle (see Coenen & Díaz López 2010).

Socio-technical regimes interact with changes in the *socio-technical landscape* (exogenous context) and *niche-innovations* (Geels & Schot 2007). However, the exogenous context cannot always be as easily determined as the authors imply. It may be obvious in the case of a volcano eruption, but changes in customer perceptions are rather ambiguous. For example, consumer attitudes towards the environment are the result of broad cultural developments, but can also be actively shaped by corporations.

In any case, changes in customer needs can challenge existing business model paradigms. In an analogy to the preconceived selection of scientific questions and attempts to answer them, outdated business model paradigms eventually can no longer produce satisfactory solutions to the changed customer problems. Institutional factors including political support for an alternative paradigm and regulatory changes can also play a role, like the automotive sector illustrates (Nelson 1995, 77f.). Or, an invention creates the opportunity for a radically new business model (technology push). The advent of the digital era provides myriads of respective examples. Oftentimes, there will be combination of forces at work that overthrow a prevalent business model paradigm.

8.2.1.2 Industry Conditions as Catalysts or Obstacles to Green Business Model Transformations

Whether an industry is receptive to Green Business Model Transformations thus can depend on whether the current business model paradigm is stable or not. If the industry is unstable – that is, if it is just emerging or in a state of flux – new green business models outside the current paradigm will be easier to introduce than if the industry is very stable, with well-established practices and power structures. Bansal & Roth (2000, 730f.) use the term of *field cohesion* in a similar context. They argue that especially in environmental vice industries, the pressure to refrain from introducing innovative green initiatives that may challenge existing industry practices is strong. However, *specific shock* or *avalanche* types of external changes (Suarez & Oliva 2005) can suddenly destabilise a previously stable industry.

Maersk Line – albeit being part of an established industry – introduced slow steaming in the context of an industry-wide crisis of the dominant business model. In contrast, early attempts of launching electric cars failed, not least due to the strong inertia within the highly stable automotive industry at the time.

However, it is important to note that not all Green Business Model Transformations necessarily have to result in a business model paradigm shift in order to be successful: Frosta's green business model has been successful despite the fact that the industry has not widely imitated its approach – in fact, this was the prerequisite to derive competitive differentiation for the company. Moreover, Green Business Model

Transformations can operate within an existing paradigm, e.g. because a significant improvement could be achieved within the established paradigm.

In conclusion, the conception of business model paradigms can help to understand the particular transformation challenges that a company will face when introducing a new, green business model. However, in order to draw useful conclusions from the idea of business model paradigms, one needs to be able to delimit paradigms with sufficient precision and distinguish them from each other. If a paradigm shift is anticipated (or even intended), it is further necessary to determine the stability of paradigms, understand their trajectories as well as the systemic consequences of a shift (based on Dosi 1982, 161). The latter will be discussed in more detail in section 8.3.

A second factor, the prevalence of environmental sustainability within the industry, has a major impact specifically on whether a *green* business model (as opposed to any other type) can be established rather easily. If environmental sustainability is emerging or still latent within the industry, it becomes more difficult to anchor green characteristics in value propositions and find customers who are willing to appreciate it. Moreover, it will be more difficult to establish associated green business practices because players along the supply chain will not be eco-savvy. That said, the upside for a green business model may be extremely high under such conditions as it will likely be quite original and difficult to imitate for competitors.

Figure 8.1 summarises the discussion on the two industry factors described above:

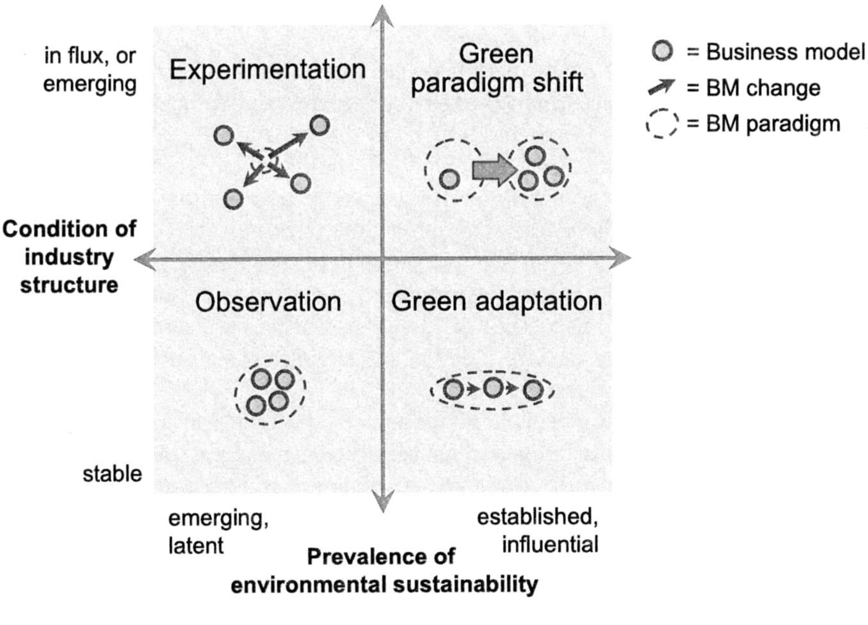

Fig. 8.1 Hypothesized behavioural patterns within industries

Companies that belong to a certain industry tend to follow common patterns of behaviour at a given point in time. Although individual companies from the same industry will always differ in how much they experiment and how prominently they deal with environmental sustainability, the majority will concentrate around a certain point in the matrix (or two, during transitional periods). The matrix further implies the existence of distinct industries. This is of course highly simplistic as industry boundaries are often blurry and shift over time (like the cases of electric cars or smart grids illustrate). Nevertheless, figure 8.1 can help decision-makers to understand the transformation challenges their company might face due to the condition of their industry (according to their definition).

The matrix consists of four quadrants, *observation*, *experimentation*, *green adaptation*, and *green paradigm shift*. It is important to note that the adjective "green" of the latter two labels is a result of the relevance of the force of sustainability within this work. In this context it represents a *disruptive*, i.e. gradually unfolding change of great amplitude (see Suarez & Oliva 2005, 1022f.). Sustainability may be interpreted as a selection force within the theory of evolutionary economics (see Winter & Nelson 1982). A generalised version of the matrix would not discriminate against other forces that might shape an industry.

The lower-left quadrant, *observation*, represents a stable industry which sees little progress in terms of sustainability. This does not necessarily imply an equilibrium in the sense of neoclassical economics (for a critical discussion of the neoclassical view see Nelson & Winter 1974). Nor does it mean that no adaptation is taking place. Yet, in the environmental domain, there is little search and experimentation activity. Most companies will only observe potential external changes (e.g. to legal requirements), but rarely make more than minor related business model changes.

Once environmental sustainability becomes more important, firms start to seek respective improvements and engage in environmental initiatives (*green adaptation* quadrant). However, resulting adaptations are usually only incremental and stay within the prevalent (but evolving) business model paradigm. This quadrant is a subset of the *transformation pathway* described by Geels & Schot (2007, 406ff.). The authors further propose three more pathways, *de-alignment and re-alignment, reconfiguration* and *technological substitution,* that all relate to transitions covered by the matrix. Unfortunately, the described pathway types are difficult to isolate even when applied to selected historical examples (Geels & Schot 2007, 415f.; Genus & Coles 2007). This diminishes their usefulness within the context of this publication.

In contrast to the lower quadrants, the upper ones represent a state of major disequilibrium. Either, the dominant business model paradigm did not allow for adaptations extensive enough to cope with increasingly pressing sustainability issues, or an unforeseen event has destabilised the industry. The latter may or may not be related to sustainability. The event may be endogenous to the industry (i.e., a spontaneous entrepreneurial innovation; see Schumpeter 1912) or exogenous (like the recent global financial crisis). If an industry enters a state of crisis, it shows extensive efforts of *experimentation* (upper left quadrant). Expressed in terms of evolutionary economics, firms not only search for more adequate organisational routines (see Witt 2008, 557), but they do so increasingly outside of the prevailing paradigm and explore rivalling ideas. That is, the search routines themselves get replaced, companies move from exploitation to exploration patterns (March 1991). However, as Gersick (1991, 22) remarks, empirical data suggests that a shock can also lead firms to rely even more persistently on old business practices, desperately seeking to optimise the current business model. Hence, firm behaviour may not be congruent at first.

If one of these experiments becomes the new dominant business model design (i.e. most competitors start to imitate it), a paradigm shift occurs. If this design happens to be based on – or leads to – a significant improvement in environmental performance, the industry has entered the quadrant of *green paradigm shift* (as explained above non-green paradigm shifts are not considered in figure 8.1). At this point many industry peers will still undertake changes that are radical in nature, but which now go in the same direction according to the new paradigm. Geels & Schot (2007, 405) propose that a niche-innovation is a candidate to break through when learning processes have stabilised, powerful actors support it, strong price/performance improvements have been made, and at least five percent of the market share are captured.

Certain business models that contradict the new paradigm may still survive in a niche, and business model innovation has sometimes in fact created a split of previously uniform markets (e.g., low-cost vs. full-service air transport). These circumstances may also explain the observation that, unlike in the past, many industries today support several fundamentally different business models at once (see Osterwalder & Pigneur 2010, 136).

8.2.1.3 Shifts in Patterns of Firm Behaviour

If slow steaming is indeed here to stay (Maersk 2010) it will represent a new green paradigm (see figure 8.2, left graph). That is, investments in accordance with the new paradigm and already achieved environmental performance improvements make it difficult to revert to previous operating practices – even if shipping capacity runs short due to growth in demand and if fuel prices decrease. Because the shipping industry seems to follow Maersk with slow steaming, it is currently considered somewhere in-between the *experimentation* and *green paradigm shift* quadrants. Future will tell whether the industry is really undergoing a paradigm shift or whether slow steaming turns out to be a short-lived experiment (dotted lines in figure 8.1). Maersk Line, for that matter, seems to be committed considering its recent shift in terms of vessels orders and equipment installations (Aston 2010; Marine Log 2010).

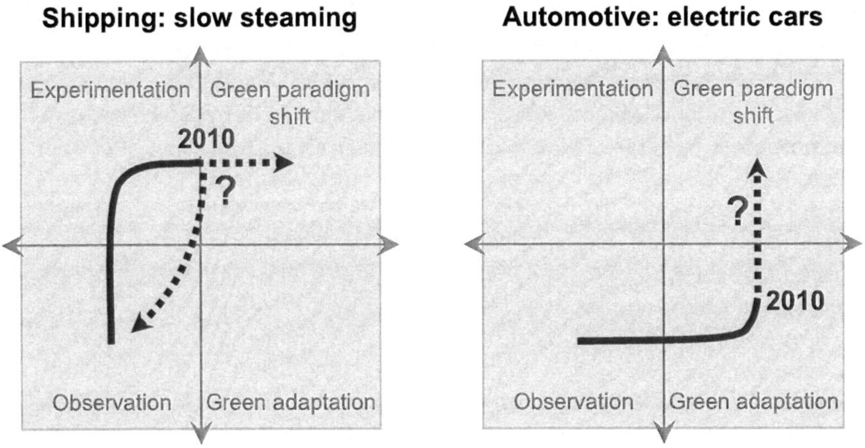

Fig. 8.2 Historic and potential future development paths of the shipping and the automotive industries

The automotive industry provides an example for a different path possibly leading to a green paradigm shift. While the shipping industry entered upheaval triggered by the financial crisis and was then steered by the competitive move of Maersk, the automotive industry has failed to respond decisively to the increasing environmental pressure related to climate change.

The industry has long countered criticisms regarding its CO_2 emissions with incremental improvements to production and vehicle efficiency. However, the pressure has grown so big that the eventual triumph of the electric car now seems inevitable (see figure 8.2, right graph).

Most industries will only occasionally enter into stages of instability and return to relative stability soon, like the theory of punctual equilibrium (e.g., Gersick 1991) suggests (see chapter 6.2.5). However, Eisenhardt & Martin (2000) argue that there are high-velocity industries in which successful business models are difficult to identify, and where change is characterised by simple decision rules and selection rather than planned adaptation. This suggests that these industries are in a constant state of either experimentation or paradigm change (also compare *hyperturbulence change*, Suarez & Oliva 2005, 1022f.). However, even in high-velocity markets, dynamic capabilities of firms are not considered to be unstructured. Hence, one may conclude that firms normally still operate within business model paradigms, although they are not as narrowly defined as in less dynamic markets.

Finally, most industry sectors can be expected to move towards the right on the x-axis in figure 8.1 over time – although new scientific findings, lasting solutions to environmental problems, or simply a shift in public perception could move a sector towards the left as well. An industry may even move from *green adaptation* to *experimentation* if an external shock occurs which is unrelated to environment issues, but dominates all other topics for the time being. If the solution turns out to be a Green Business Model Transformation, the industry would then follow an inverted 'Z'-shaped path.

Although the macro view offers valuable insights about why some transformations succeed while others do not, business model-specific factors play an evenly important role.

8.2.2 Business Model Level: Economic Effects of Environmental Sustainability

The greater the degree of change from the current to the new business model is, the higher the chance of failure tends to be (Linder & Cantrell 2000, 14). Not only are most organisations not experienced with major business model transformations, but the necessary mindset, resources and processes often need to be acquired first in order to make it work. If the business model interacts with a complex business environment, challenges are further amplified (Aragón-Correa & Sharma 2003, 79f.). However, a detailed analysis of the differences between the old and the new model will allow management to focus on the important transformation challenges and address them adequately (Johnson et al. 2008). This includes the anticipation of resistance from stakeholders (including, but not limited to employees) that are affected by the most significant modifications, or which have the necessary power and urgency to claim their interests, respectively (Mitchell et al. 1997). A detailed discussion of change management can be found in chapter 6 and is thus not repeated here.

8.2.2.1 Effects of Environmental Sustainability on Economic Viability

Of specific interest for Green Business Model Transformations is how the green characteristics of the business model influence its economic viability (although this question is of course also relevant for all other changes to the business model).

Firstly, green characteristics can work in favour of, or against the value proposition of the business model. In some cases, it may also be insignificant – or just create a shift in appeal between customer groups. For example, electricity from nuclear power is more attractive (i.e. eco-friendly) than electricity from coal to some consumers, but less to others – depending on how they weigh CO_2 and other emissions versus nuclear waste and the risk of an accident.

Secondly, green characteristics can affect the value creation side of the business model. That is, it may require rare competencies, increase supply chain complexity, increase cost, and so on. On the other hand, some green business models reduce cost, e.g. by being more resource-efficient. In some cases, the green aspects are even essential to the business model, i.e. it only exists in order to fulfil some purpose directly related to environmental sustainability.

Figure 8.3 categorises type 2 business models from the industry survey (see 7.10) according to the significance of their green characteristics on the value proposition and value creation aspects:

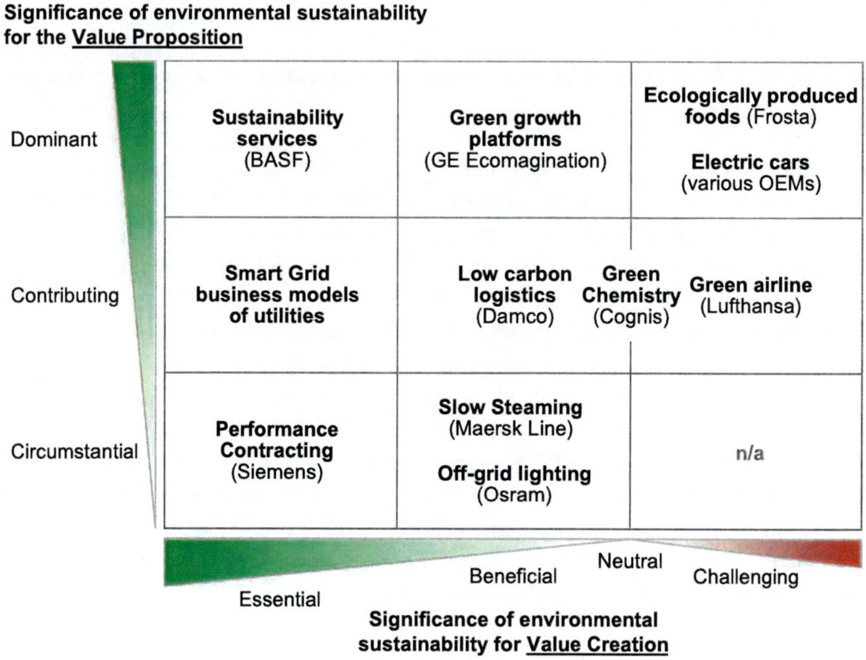

Fig. 8.3 Business model examples categorised by the significance of their environmental characteristics

To simplify matters, customer segments are not differentiated in the matrix. Moreover, it does not include business models whose green characteristics reduce the attractiveness of the value proposition overall; these business models are rarely successful, and none has been studied. It is important to note, though, that figure 8.3 makes no statement about how *green* a given business model is. Instead, it evaluates the economic consequences of green characteristics, including financial and non-financial aspects.

Starting with the top left of the matrix, sustainability services like that of BASF feature a dominantly green value proposition. Because the product that BASF sells is sustainability itself, environmental characteristics are also essential in value creation – without them, there would be no product.

Challenging is the categorisation of green growth platform like GE's Ecomagination: As the category is relatively generic, the content of such meta-models can differ widely. Even in the concrete example of Ecomagination the diverse products featured by the initiative are more or less directly related to sustainability issues (e.g. a wind turbine versus a water dispenser, both listed as Ecomagination products). The label Ecomagination has been created in order to give its environmental efforts a sticky name, to promote the marketing message. Although GE always emphasises that Ecomagination is about products that are green *and* offer a good operating performance or value proposition to the customer (General Electric 2006, 16), the latter claim has obviously existed before. Thus, Ecomagination as a meta-business model features – or adds – dominantly green value proposition aspects, although individual products under the initiative's umbrella may have other, more visible product features.

As Ecomagination was designed as a marketing programme, it was unclear a priori whether it would cost the company money on the value creation-side. However, it turned out that there were significant inefficiencies to be found, thus Ecomagination was also economically beneficial in this respect (General Electric 2006, 31). Nevertheless, this question posed a hurdle for getting support initially.

Manufacturers of ecologically produced foods and OEMs that plan to build electric cars usually both face a trade-off through sustainability. In either case, the new business models pose challenges on the cost-side for various reasons. Hence, advantages for the value proposition need to compensate disadvantages in value creation – otherwise the green aspects of the business model would make no economic sense (except for very limited niche offerings).

In Frosta's case, sourcing and production processes became more expensive as explained before. On the other hand, Frosta's Purity Command has fundamentally repositioned its products with consumers and retailers. Being dominant for the value proposition does not necessarily mean, however, that marketing should directly and exclusively focus on sustainability; sometimes, it is more serviceable to use proxy attributes instead. In the case of Frosta this was quality and taste. Frosta further realised that a success factor was to rely in part on traditional marketing messages and benefit from existing brand strength this way. GE's case is exemplary in this respect: "Ecomagination" relates to the previous slogan "imagination at work" and thereby combines the existing claim of innovativeness with environmental sustainability.

Electric cars would not be a topic today without climate change. The transition from ICE cars is so complex and expensive that it cannot be explained without the promise of zero local and low global emissions. Yet, additional cost for OEMs are high: extensive R&D, pricey batteries and new production processes makes the production of EVs significantly more expensive than ICE cars, even though the difference in life cycle cost is not as big. It is thus a great challenge for OEMs to create attractive business models for the mass market. It will be interesting to see whether radical concepts like BMW's i3 are better suited to achieve that compared to converted conventional cars (which most other OEMs are envisaging).

The trade-off described above is even more problematic in cases with challenging value creation aspects where green characteristics are only contributing to a broader set of important value proposition factors. In the case of green airlines, the trade-off may be unacceptable. According to industry experts, environmental performance will remain only a subordinate factor among others like safety, convenience, and – most importantly – price. If this assumption holds, expensive measures that are necessary to reduce the CO_2 emissions significantly will simply be rejected by the customer.

In the case of green chemistry, the positioning as a single point in the matrix is somewhat misleading. Indeed, green characteristics contribute to the value proposition of chemical products. However, whether or not these come at the expense of inferior functionality or lead to higher cost compared to conventional alternatives varies greatly between individual product categories. Hence, some green chemical products may in fact be disadvantageous in both dimensions while others are just the opposite. This relation is also dependent on macro-economic factors like the oil-price and the availability of renewable raw materials. On average, the value creation side is thus rated 'neutral', the value proposition effect "contributing".

Low-carbon logistics services like that of Damco work because they help customers save money by making logistics more efficient. Although the case seems to be similar to the sustainability services of BASF, the relevance of environmental vs. financial arguments tends to be reversed. Most customers appreciate the environmental improvements, but consider them a welcome side-effect. Hence, environmental aspects are classified as "contributing" / "beneficial" only. Some companies like Nike, however, would be willing to buy services like Damco's simply for the benefit to the environment as they need to meet corporate environmental targets (see Nike 2010, 113ff.). Should this target group become more commonplace, the business model would move upwards in figure 8.3.

Although only mentioned briefly as a potential area for further study, smart grid business models for utilities are a good example for environmental aspects that play an essential role in value creation. A critical driver of the smart grid is the – in many countries now obligatory – expansion of electricity generation from renewables. Utilities simply need a smarter grid to cope with the increasing decentralisation of power generation and resulting fluctuations. Smarter grids also encompass advantages on the value proposition side: For instance, with smart meters and differential pricing, customers can save money by avoiding peak load electricity (that needs to be provided by expensive back-up power plants). Although the smart grid infrastructure is a consequence of growing power generation from renewable power sources, it will enable various complementary business models with distinct features.

Business models on the lower row of the matrix heavily rely on benefits on the value creation side – usually through some sort of superior resource efficiency.

Hence, a green business model can hardly be challenging for value creation *and* encompass no or only circumstantial benefits for the value proposition.

Nevertheless, there are promising business models like slow steaming and solar-powered off-grid lighting which work despite the fact that customers do not (yet) appreciate environmental benefits for their buying decision: Off-grid lighting may succeed due to its lower price compared to kerosene lanterns, not because it is environmentally friendly. In slow steaming, fuel savings happen to be directly correlated with CO_2 emission reductions. However, customers would never accept it without enhanced reliability (and potentially lower prices) to compensate for the lower speed. Moreover, if Maersk Line succeeds with its attempt to reshape customer perception the trade-off will further shift in favour of sustainability.

The last business model to be discussed, performance contracting by Siemens Building Technologies, is positioned with environmental sustainability being essential for value creation, but only circumstantial for the value proposition. One may wonder why environmental performance is considered essential for value creation here while it is not for slow steaming (for which the environmental benefit also lies in direct energy savings). In contrast to slow steaming, performance contracting at Siemens directly relies on energy savings for the business model to be profitable. Maersk Line also optimises reliability, not only fuel savings (ships go faster if a delay is imminent). This also illustrates the difference to Damco's low-carbon logistics services: Here, the logistics chain is not only optimised for carbon efficiency, as Damco's Erling Johns Nielsen explained during an interview, but also for reliability and price. In contrast, the performance contracting revenue model is directly linked to energy savings, i.e. Siemens guarantees to realise energy savings that are used to finance the project. Siemens' customers mostly focus on the financial incentives – environmental concerns have only recently started to play a role with international companies who need to fulfil environmental targets (similar to Damco's experience). For now, the significance for the value proposition is still predominantly circumstantial, though.

8.2.2.2 Summary of Green Business Model Characteristics

Some categorisations are necessarily tentative. As the explanations above demonstrate, it is not always clear a priori whether green characteristics in sum turn out to be beneficial or challenging for value creation. Moreover, categorisations can shift over time, e.g. as macro-economic factors like the oil price shift or customer needs change (often in favour of sustainability). Thus, a static view will only give an incomplete picture. Nevertheless, some important lessons learned may be summarised as follows:

- If green characteristics are only circumstantial to the value proposition, companies should somehow ensure that customers noticeably benefit nevertheless from the advantages on the value creation side (like performance contracting does via its revenue model).

- If green characteristics contribute noticeably to the value proposition, companies do not need to limit their efforts in reducing potential trade-offs, but should also explore how they can shift the basis of competition towards sustainability (like green airlines may try to make passengers expect low carbon operations).
- If green characteristics are dominant to the value proposition, they must not be compromised – even if the value creation side is challenging as in Frosta's case.
- If green characteristics are essential for value creation, companies must nonetheless clearly demonstrate the value of the business model to the customer. Business models of this type tend to follow unfamiliar rationales that can lead to scepticism on the part of the customer. For instance, performance contracting may lead customers to believe that there is a catch; smart metering business models have already created strong pushbacks among U.S. consumers (Wald 2009).
- If green characteristics are beneficial for value creation, these characteristics can be used as show-cases to gain support – even if they are not critical for the economic success of the business model. For example, GE used "energy treasure hunts" to mobilise the organisation.
- If green characteristics are neutral or challenging for value creation, strong resistance is to be expected from representatives of the alternative, non-green business model (like has been observed with electric cars). Possibilities to deal with such issues include pilot programmes, separated implementation units, and more – a comprehensive review of measures can be found in chapter 6.3.4.

8.2.3 Company Level: Change Capabilities and Eco-Mindedness

Every organisation is unique, no matter which business model(s) it applies, and which industry it belongs to. Consequently, these specifics, too, are to be considered for a successful transformation.

8.2.3.1 General Change-Related Company Conditions

Contextual research of organisational change has examined many different factors in empirical studies (Armenakis & Bedeian 1999). Because this topic has been explored in detail in chapter 6, only exemplary cases from the industry survey are outlined in the following. General factors include company ownership, leadership, organisational structure, and the competitive position. Sustainability related factors concern the attitude and management competencies with respect to the environment.

One factor that has been mentioned several times during interviews is the difference between privately and publicly owned companies. Publicly-traded companies are widely believed to tend towards managerial short-termism (Laverty 1996; Marginson & McAulay 2008; empirical evidence suggests that the causation is more complex, though). That is, managers neglect some green business model opportunities to avoid short-term performance deterioration. In contrast, especially owner-led companies like Frosta AG are in the position to introduce venturesome green business models despite an uncertain, long-term payoff.

Large public corporations – contrary to popular belief – are *not* fated to be hardly adaptable (Kanter 2008) or non-innovative (Damanpour 1992). A large firm that uses the virtues of size while maintaining decentralised decision-making authority guided by shared values and goals (see Collins & Porras 1996) can show considerable innovative power. With Ecomagination, for instance, GE seems to have realised that notion pretty well. Yet Siemens in comparison, being more democratic, seems to have its strength in decentralised innovation rather than central platforms and initiatives.

Changes in leadership also have an influence on corporate transformations as research on punctuated equilibria shows – presumably, because new CEOs bring in fresh views, are less committed towards previous strategies, and are almost expected to implement some changes (Romanelli & Tushman 1994, 1145). This may provide an additional explanation why GE, with its then relatively new CEO Jeff Immelt, was able to launch Ecomagination.

Organisational culture is frequently considered a powerful barrier for transformational change (Mintzberg et al. 2009, 283ff.). But it has also been viewed as a source of competitive advantage (Barney 1986) that can be destroyed through defective change (Mintzberg et al. 2009, 298f.). The concept of culture certainly helps to explain the different approaches of the examined automotive OEMs. More importantly, it warns deciders against neglecting cultural change that may be a necessary constituent for successful business model transformations. This holds true for attempts to introduce a service model like performance contracting at Siemens or Osram's off-grid lighting business. It may also be a major challenge for utilities that want to compete in smart grid markets.

Organisational structure also affects transformation challenges. For example, Mintzberg et al. (2009, 323ff.) describe seven configurations of structure and power. For instance, *machine organisations* (e.g., automotive companies, airlines) tend to rely heavily on formal strategic planning. *Diversified organisations* (e.g., Siemens) have to deal with the autonomy of their divisions (i.e., firms within the firm). *The political organisation* is plagued by internal conflicts that make business model transformations hard to control. A company may (temporarily) turn political because of a transformation in the first place, but can also develop this configuration over time when it is protected from market forces for too long (e.g., in regulated energy markets). It is helpful to consider which of the seven configuration types are applicable. Among other things, such considerations help to determine an appropriate approach to incubate new business models (see Govindarajan & Trimble 2005). Siemens Building Technologies, for instance, first needed to learn at which point a new performance contracting unit is strong enough to be reintegrated into the core. This example also highlights the importance of geographical dispersion, which, in this case, was a barrier due to the strong regional character of the market.

Economic pressure has also been argued to induce major organisational change, but empirical results suggest a more complex relationship (Romanelli & Tushman 1994, 1144ff.). Industry leader Maersk did change in the wake of declining short-term performance, and so did Frosta, which saw its long-term survival threatened. On the other hand, BMW initiated a radical change project during good economic times. If the competitive position does not decisively influence the likelihood of a Green Business Model Transformation, it does at least have an effect on the corporate journey and how it is to be managed. This topic is explored in more detail in chapter 8.4 after competitive dynamics have been discussed.

8.2.3.2 Degree of Eco-Mindedness of the Company

The question how savvy a company is regarding environmental issues not only differs strongly between industry sectors, but also within (Sharma 2000). The variation of respective scores of SAM's annual corporate sustainability assessment (SAM & PwC 2010) also supports this notion. Furthermore, many companies cannot be considered advanced with respect to exploitation of opportunities from sustainability (Berns et al. 2009a).

Some companies claim to have a long history of sustainable business practice, that sustainability has become part of their corporate DNA. However, while this may express itself in strongly integrated operational environmental management, this does not automatically mean that the topic is addressed strategically, or vice versa. For example, one company that has implemented one of the most remarkable strategic environmental initiatives, General Electric, had a rather poor historic environmental track record (Fisher 2005). The toughest challenges that a company faces in the wake of a new green business model therefore also depend on the company's history: A company like GE has to make the operational part work first. Yet, a company with a "green DNA" may still find it incredibly hard to think of sustainability in strategic terms, despite – or maybe precisely because – of its long-established environmental management practices. Some companies like BASF in turn even manage to derive a business model from their long-standing sustainability experience.

Several authors have suggested models of environmental management approaches (e.g., Hunt & Auster 1990; Roome 1992; Post & Altman 1994; Winn & Angell 2000). These models usually represent a continuum of maturity, ranging from ignorant firms to environmental management champions. However, as argued above, this aggregate, outcome-based categorisation is not sufficient to derive specific transformation challenges. Hence, project managers need to consider the individual capabilities and shortcomings of the organisation. For Siemens Building Technologies this meant reconsidering hiring practices, Frosta had to redesign production processes and sourcing, and Osram needed to learn how base of the pyramid markets function. Many more examples have been mentioned, but the general lesson is that company-specific sustainability capabilities deserve due attention.

8.3 Dynamic Patterns of Business Model Transformations

Even if a company's business model has been highly successful for a long time, there may be a superior business model waiting in the wings. If that is the case, established business models will often prevail for some time even if the business environment has changed in a way that strongly discriminates the currently dominating model against alternative ones.

The reason for this is that either the alternative business model opportunity is still unknown to deciders, or that it is not recognised as feasible and competitive. The former usually results from unchallenged paradigms that prevent companies from looking for alternative models at all or makes them search too narrowly and exclude the periphery. The latter results from the problem that new business models will often be inferior in the beginning (like early cannons versus technically mature catapults). Reasons include the initial lack of scale (see Sterman 2000, 367ff.), immature elements of the business model (e.g., technology), or inadequate internal resources and processes to implement it (Johnson 2010, 164f.).

8.3.1 The Trade-Off between Economic and Environmental Performance within and between Alternative Types of Business Models

In the following, possible consequences of a Green Business Model Transformation are discussed in the context of competitive dynamics. Both the economic and the environmental performance dimensions are considered for this discussion (the social dimension is ignored, but could in principle be integrated in a joint sustainability dimension). Schaltegger & Synnestvedt (2002, 341) argue that the marginal return from environmental protection beyond compliance is decreasing, but initially positive (see also figure 3.2). This insight can be translated into the realm of business models, and economic versus environmental performance, respectively. Figure 8.4 (see below) compares established business models of the same type with an alternative, green one.

When considering actual implementations of business models most will not achieve the optimal, theoretically possible combination of economic and environmental performance (i.e. the maximum economic return at a given environmental performance). For each business model type, these hypothetical combinations constitute a function, from now on referred to as the *trade-off function* (depicted by the grey and the green curves in figure 8.4). The trade-off function is the boundary to the *business model option space*, which comprises all possible implementations of the respective type of business model (given current market conditions).

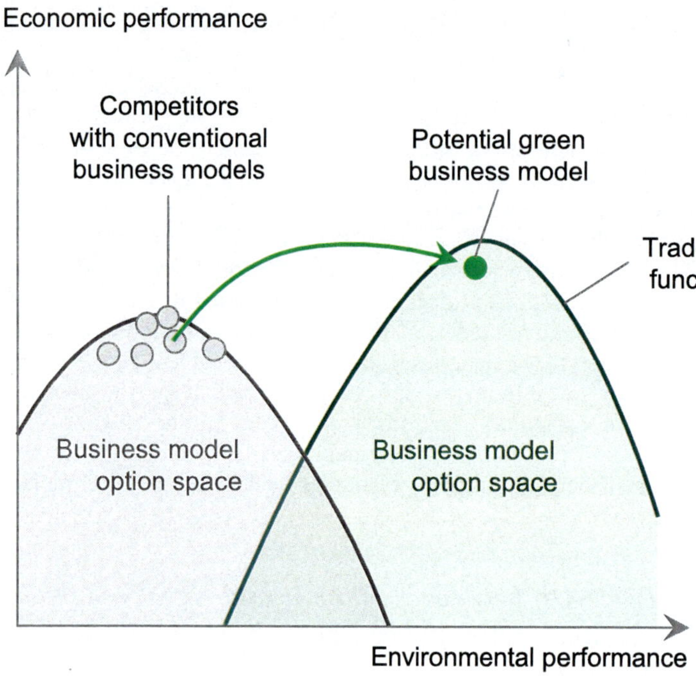

Fig. 8.4 Illustration of the performance of established business models and an alternative, green one

To discuss the matter of industry dynamics in intelligible terms, the following simplifying assumptions are used (as illustrated in figure 8.4):

1. A known alternative green business model opportunity exists.
2. Its potential environmental performance is considered superior in objective terms (setting aside all the difficulties of a single measure).
3. All competitors initially share the same trade-off functions.
4. Trade-off functions possess exactly one maximum economic performance.
5. The exact shapes of the curves are disregarded; only relative differences between curves are considered.

The first assumption is trivial if the considered firm is confronted with the choice to follow a pioneering competitor. However, whether a firm gets to choose to be the first to transform is "some combination of proficiency and luck" (Lieberman & Montgomery 1988, 49). A great deal of thought is devoted in chapter 9 to improve that proficiency and thus reduce the reliance on luck.

Assumption 2 excludes all the technical and conceptual problems of environmental performance assessment discussed in chapter 5.4. These would only obscure the strategic issues at hand. Thus, the abscissa (environmental performance) is only specified in abstract terms. The ordinate (economic performance) could be annual profit or another measure of economic success, although in this case, too, the details remain unspecified intentionally (including the considered time horizon).

Assumption 3 is necessary in order to keep complexity manageable; however, due to individual resources, companies will usually have differing trade-off functions. For example, one company may have a proprietary technology at its disposal or benefit from economies of scale that relax the trade-off in comparison to competitors.

Assumptions 4 and 5 serve the purpose to detach potential controversies about the exact relationship between the economic and environmental performance of business models from the discussion on competitive dynamics. This relationship has been argued to show decreasing marginal returns (Schaltegger & Synnestvedt 2002, 341; Schaltegger & Wagner 2006a, 10). This proposition is evident for pollution control or even for environmental protection in general, but does not necessarily hold true with respect to business models. For instance, branding effects may boost revenues only after a certain margin of environmental performance improvement is reached. Hence, the trade-off function of some business models may look like a bell curve (or an even more complex shape) rather than an inverted 'U'. The assumption of exactly one economic maximum seems reasonable for the majority of business models, although cases are conceivable in which a business model does not appeal to certain customer groups until a threshold of environmental performance is reached. This could theoretically lead to multiple local maxima. Eventually, however, benefits from further improvement will necessarily be dominated by increasing marginal cost.

Actual implementations of two different business models may exhibit the same economic and environmental performance in practice (the intersection in figure 8.4). However, a business model transformation (depicted by the arrow) will usually lead to a change in both. For the sake of simplicity, boundary conditions are considered steady during competitive moves (e.g., unrelated technological breakthroughs or regulatory changes that affect the trade-off functions are disregarded).

Figure 8.4 only shows differences in performance outcomes of the business models, not the underlying differences of business models that cause them. There are many potentially relevant performance dimensions in which business models differ, e.g., measures of quality or customer orientation (Johnson et al. 2008, 59). Industries can support various types of business models in parallel, for instance, to address different market segments. However, ultimately the complexity mentioned above can sensibly be ignored for the analyses that follow – as long as this is accounted for when interpreting their results to derive suitable courses of action.

The illustration in figure 8.4 may lead observers to assume that a business model transformation has a fixed impact on economic performance for all actors; however it does not. The whole business environment, including each competitor consists of open systems (see 4.5). Once one company transforms its business model, it changes not only itself, but also the larger system it belongs to. There are a number of different types of effects, many of which have been described at length by scholars of strategic management. Each effect has unique strategic implications. The subsequent analyses thus borrow from research of what is sometimes referred to as the positioning school (see Mintzberg et al. 2009, 85ff.), but also various other schools of thought. Six general *dynamic patterns* are explored in the following, each with two variants. Dynamic patterns may also occur in combination, although some are hardly compatible. They are relevant for any business model transformation, but occurrences involving green business models (observed and hypothetical ones) are especially highlighted.

8.3.2 "First Mover Advantage" Pattern

The idea of the first mover advantage is old. Ancient strategist Sun Tzu already described it in "The Art of War" (Sun Tzu 1963). The issue has also gained considerable attention in modern management literature. Lieberman & Montgomery (1988) name three principal sources for first mover advantages (FMAs): (1) *technological leadership*, (2) *pre-emption of scarce assets*, and (3) *switching costs and buyer choice under uncertainty*.

For instance, Toyota has been the first firm to sell hybrid cars in mass volumes (see figure 7.2), resulting in *technological leadership* (which may be more accurately labelled "innovation leadership", though, as it includes advances in products, processes as well as organisational forms). Toyota likely enjoys experience curve benefits (e.g., Ghemawat 1985) from its pioneering role. Its branding benefits from pioneering the hybrid market are undisputed (Interbrand 2008). Even though brand loyalty effects are considered to be stronger for convenience goods (Schmalensee 1982), these may lead future buyers of EVs to prefer a Toyota. Consequently, Frosta can hope to be rewarded for its pioneering role as well and become a preferred buyer choice in its product category.

To name another example from the automotive industry, BMW aims to gain technological leadership in carbon fibre technology with its i3. As a first mover, it may not only enjoy experience curve effects, but also secure critical assets at low prices, or even limit access to such resources for followers (*pre-emption*). Similar opportunities may arise from new battery technologies that require rare materials or patented manufacturing processes. What the examples from the automotive industry have in common is that competitors eventually *have to* follow: Once it becomes apparent that the new business model is attractive even for late entrants, it is likely to create a new business model paradigm. Back in 2007, some automotive managers might still have smiled at the Toyota executive according to whom all cars produced by Toyota in 2020 would be hybrids (see Economist 2007). This prediction will probably prove to be incorrect indeed, but possibly due to the introduction of BEVs rather than the continuation of ICE models.

Figure 8.5 (1a, left-hand side) illustrates the case generically: a Green Business Model Transformation offers significant economic gains for the first mover, but only limited ones for followers (the dashed line).

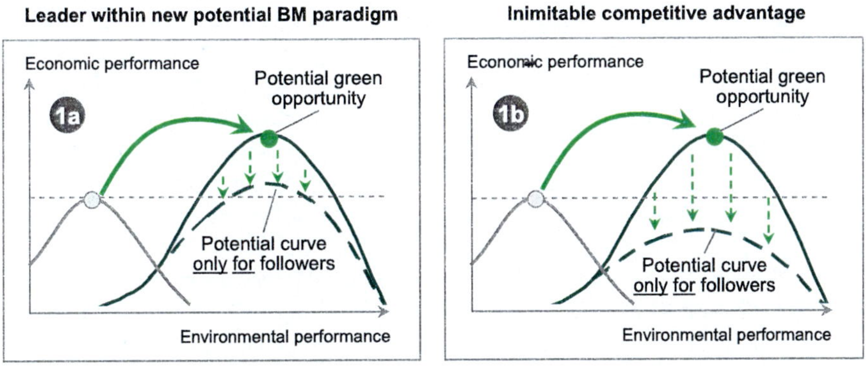

Fig. 8.5 First mover advantages from Green Business Model Transformations

BMW hopes that its LifeDrive concept with the carbon fibre chassis will revolutionise the industry. However, if BMW is indeed effective in building barriers for followers, the situation will be characterised more accurately by the right-hand graph of figure 8.5 (1b): it then becomes economically unattractive for competitors to follow. This way, a firm can sustain its differentiation for much longer, which can be an even bigger competitive advantage compared to (just) a better initial economic position. Of course, it is not always trivial to distinguish both cases in figure 8.5 under information asymmetry and uncertainty. Hence, a first mover may revert to signalling (see Porter 1980) in order to provoke desired reactions of competitors (e.g. to convince competitors to follow or not).

Frynas et al. (2006) propose a further mechanism to generate a FMA: political resources. They are defined as “firm attributes, assets, human resources, or any other resources that allow the firm to use the political process to improve its efficiency and profitability” (Frynas et al. 2006, 324). Political resources have been observed in automotive markets in the past, and may turn out to be even more influential in future electric car markets. Generally, green business models can be assumed to create a certain degree of political goodwill. This suggests that politics should be considered as a target group for some green business models (see 4.4) and be included in any analysis of potential FMAs.

“The early bird may get the worm, but the second mouse gets the cheese.”
American proverb

Empirical studies have affirmed that first movers are advantaged over their competitors (e.g., Robinson et al. 1994; Robinson & Min 2002). However, other research casts doubt on the general validity of FMAs (VanderWerf & Mahon 1997).

And indeed, there are good arguments why first movers may indeed be *disadvantaged* vis-à-vis followers: First of all, pioneers face greater uncertainty, and simply may get it wrong (Lieberman & Montgomery 1998, 1112f.): They may have the chance to shape customer preferences, and be able to acquire superior resources. Yet customers may reject the initial value proposition (e.g., see the cases of Osram and Frosta), and acquired resources may turn out to be less valuable than initially thought. Followers – having observed the developments in the market and the mistakes of the first mover – can then make the right choices without the burden of past commitments (financial or other) (Lieberman & Montgomery 1988, 47ff.). In the graphs of figure 8.5, this would mean that the result of the transformation (the green dot) is far from the optimum, leaving room for others to reap early mover advantages instead. In fact, in the U.S. Toyota's hybrid was launched seven months later than Honda's, the Insight, but Toyota still beat Honda in the marketplace with its more convincing Prius (Taylor 2006).

Even if most choices of the first mover are prudent, the firm has to make significant innovation investments (e.g., in R&D, buyer education, or infrastructure development). Imitation costs for followers ("free-riders") are usually much lower (Lieberman & Montgomery 1988).

As a result, scholars have argued for the strategy of *fast second* instead (Markides & Geroski 2005). Whether or not a firm can afford to wait also has to do with its own capabilities and that of its rivals: Firms with strong capabilities in marketing or manufacturing may be better off following (Lieberman & Montgomery 1998, 1113f.). Conversely, if the first mover is potent, followers have very limited options. For example, General Electric enjoys significant benefits from the pioneering Ecomagination initiative (see 7.5). Even powerful rival Siemens has not been able to challenge GE's subsequent image as a sustainability leader (SustainAbility 2008, 25), although a comparison of environmental credentials would suggest it to be at least on a par (SAM & PwC 2010, 54; Siemens 2010b). However, GE also took considerable risk, as a failure would have severely damaged its brand. Frosta on the other hand would principally be vulnerable to a marketing attack from giants like Nestlé or Unilever (see Urban et al. 1986). It can therefore not rely on first mover advantages alone. Nonetheless, Frosta can realistically hope that its competitive position remains unchallenged for now, not least due to incumbent inertia. Incumbent inertia, of course, can eventually turn against a former pioneer, too (Lieberman & Montgomery 1988, 48ff.). Toyota should therefore be wary not overestimate the advantages from its early entry in the hybrid market – they may not last for long.

8.3.3 "Creative Destruction" Pattern

Any novel business model can lead to Schumpeterian *creative destruction* (see Schumpeter 1912). The reason why the dynamic patterns shown in figure 8.6 specifically carry that label is that it most closely resembles the idea of an innovation destroying the economic performance of incumbents.

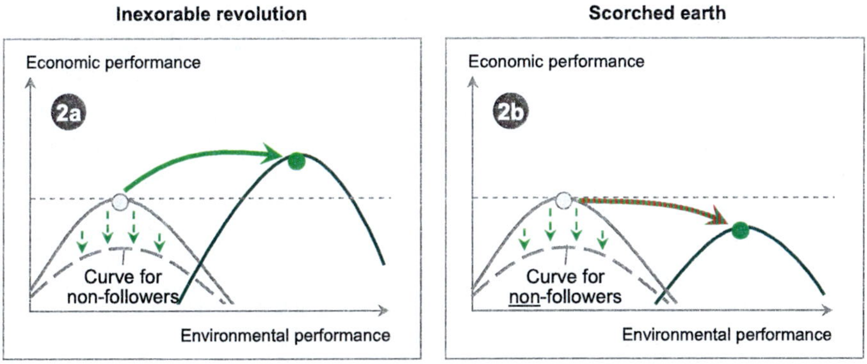

Fig. 8.6 Creative destruction through Green Business Model Transformations

FMAs in a sense are the other side of the same coin. In the FMA pattern, however, the economics of a new business model can remain obscure for non-followers for quite some time (especially profitability, but also market share); the deterioration of competitors' own numbers will be noticed rather quickly, though. The impact on the industry will thus tend to be perceived as being more drastic, and likely induce stronger competitive reactions.

Slow steaming may turn out to be a good example for what is called an *inexorable revolution* (2a, left-hand side of figure 8.6). Here, not only is the new business model superior, but it also damages the established ones. This effect can be expected to become visible very quickly even in dynamic and intransparent markets. One the one hand, slow steaming improves profitability on many routes under current conditions. Thus the new model offers a higher economic performance than the old one. On the other hand – and this point has yet to be confirmed in the coming years – slow steaming shippers like Maersk may be able to discredit others and hurt their sales. Two mechanisms can achieve this effect. First, the higher reliability made possible by time buffers through slow steaming may become expected by customers (who will streamline their own internal logistics processes accordingly). Shippers who cannot deliver that thus lose market share. Second, Maersk Line has started to make the CO_2 emissions of their shippings transparent to the customer. If environmental performance can be established as a factor in the buying decision with some customers, high emitters will lose further business. To amplify this point, Maersk Line has created a whole new sustainability strategy that includes a target for emission reductions of 25 percent until 2020. This requires additional measures, and could eventually even lead to a *greening war* (see 8.3.7). As soon as such fundamental changes occur within the customer base industry players will take notice. A rather quick industry-wide transition is the likely result.

The second graph of figure 8.6 (2b, right-hand side) may seem to be of little practical relevance at first glance. Which company would voluntarily damage its economic performance potential? The choice is rational under two possible scenarios: First, the pioneer has failed to achieve a favourable position under the established business model so far. Second, the pioneer wants to keep the profitability of the industry low in order to drive competitors out of the market or defend a leadership position.

An example for the first case provides the automotive industry. The rise of the electric car requires strong R&D expenses and threatens OEMs' market power (e.g., vis-à-vis battery suppliers). Hence, if expected value chain reconfigurations may well lead to lower profitability of the industry (see Porter 1980, 3ff.) – shouldn't that development contradict the interests of *any* OEM? Not necessarily, as the case of the Chinese automaker BYD Auto illustrates. Firstly, BYD originates from the battery industry; hence respective shifts towards higher supplier power would be beneficial for BYD as a whole. But even without this circumstance, BYD Auto could benefit from the development as established Western and Japanese OEMs would lose their previously advantaged position in ICE cars. Moreover, even if the profitability per car is hurt, the potential volume gains in the Chinese home market may overcompensate that effect.

With respect to strategies that put industry profitability under pressure the airline industry provides a (hypothetical) example: At the moment it seems unrealistic that the extra cost of CO_2 reductions through the extensive use of biofuels could be transferred to the customer. However, Lufthansa as a full-service carrier could try to lobby for respective regulations, or try to establish biofuel propulsion as a "hygiene factor" (analogous to Herzberg 1968) for a sizable customer segment that low-cost carriers cannot afford to lose. For companies with a cost leadership strategy like low-cost airlines this is indeed a troublesome scenario.

To provide a factual example from another domain, semiconductor chip maker Intel has designed its revenue model in a way that protects its advantage in product development: High margins of its latest chips are used to sell its older, low performance chips at very low prices. This way, mostly the profits of imitators are destroyed, and thus their ability to catch up technologically and compete in the premium segment remains limited (Sterman 2000, 374).

As the labels *inexorable revolution* and *scorched earth* suggest, these dynamic patterns are rather aggressive in nature. Although it is unlikely that aggressive business model transformations go unnoticed by competitors, they may still choose to ignore it, have other strategic priorities, or fail to respond due to organisational inertia (Coyne & Horn 2009, 92ff.). Nevertheless, the possibility of retaliation must not be ignored, especially in industries with high field cohesion (Bansal & Roth 2000, 730ff.). In extreme cases, retaliation can destroy the economic basis of the new business model for all competitors (and damage the profitability of the established business model). Such a development, of course, would not be in the interest of any firm. Companies unwilling to follow, however, may try to create a credible threat of retaliation and thereby discourage potential first movers.

Whether and how a competitor will react depends on various factors ranging from generic conditions of the industry to individual characteristics of each competitor (e.g., its history or the personalities of their leaders). How to assess all of these aspects, what measures to expect, and how to counter them is explained in detail in Michael E. Porter's classic book "Competitive Strategy" (Porter 1980).

8.3.4 "Collective Opportunity" Pattern

Collective opportunities arise when not only the early movers, but also followers could benefit jointly from transforming their business models. For instance, the new business model may provide access to new customer groups, reduce the dependence on rare (and expensive) resources, or allow companies to share infrastructure and cost. Figure 8.7 exemplifies two variations. The first, labelled *pathfinder for new BM paradigm*, illustrates a situation in which one company breaks new ground, thereby demonstrating the viability of the business model and reducing the risk for followers. If other companies decide to follow, not only can the followers expect an even higher potential economic performance, but that of the first mover automatically improves as well.

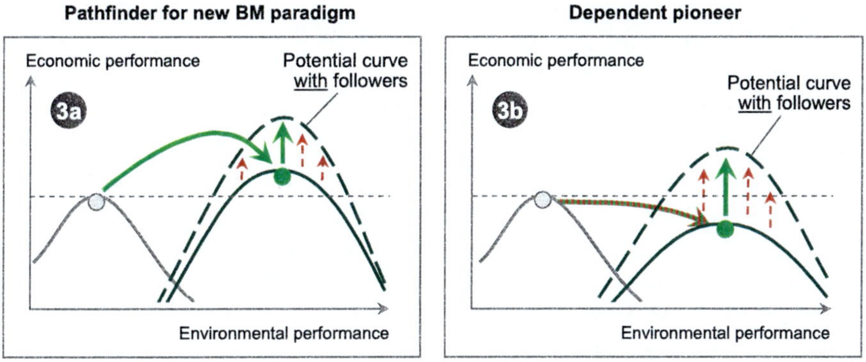

Fig. 8.7 Collective opportunities through Green Business Model Transformations

Again, slow steaming may serve as an example. As explained earlier, pioneer Maersk Line already enjoys cost benefits from slow steaming. Increased reliability is another advantage, but likely will be imitated by competitors if it turns out to be valued by the customers. However, this may not be a bad thing: if a large share of shipping companies can actually guarantee a certain arrival time, ports may be able to eliminate many inefficiencies and streamline their processes comparable to an airport. This saved time would in turn benefit the shipping companies, too. They could possibly even entice volume away from other transport modes by this means, thus expanding the total market size for all.

The introduction of a "grey" container fleet with shared assets (similar to airline alliances, see 7.7.2) would have similar effects. Yet, a group of companies would need to develop a *collective strategy* (e.g., Astley & Fombrun 1983; Astley 1984) to turn this vision into reality. In fact, the present-day standardisation of shipping containers was the result of such a past collective strategy (Mintzberg et al. 2009, 268).

The second variant, dubbed *dependent pioneer*, is different in that the first mover faces an economic-environmental performance decline. Only if others follow will synergies relief this disadvantage. A classic example is sustainable fishing: If a company adapts its catching methods and volume to sustainable levels, it loses revenue. Only if competitors follow the example will the (long-term) return of these measures be positive as fish stocks do not collapse. This dynamic pattern calls for explicit forms of collaboration (see Schneidewind 1998, 344ff.), e.g. in the form of initiatives like the Marine Stewardship Council (MSC).

Collective action can also be induced through unilateral action (Unruh & Ettenson 2010b). This requires the pioneer to have profound direct or indirect influence on the industry in order to change the rules of the game (Sterman 2000, 380).

One example for such a company is Wal-Mart. Not only is Wal-Mart able to set standards in its industry – like it currently does with its supplier sustainability initiatives – but it also has the power to internalise resulting cost savings (e.g., energy savings) *from the suppliers* (Aston 2009; Rosenbloom 2010). However, the whole retail industry can benefit from such activities. With its packaging initiative Wal-Mart envisions a 5 percent reduction in packaging over five years. Wal-Mart expects its supply chain to save US$ 3.4 billion (circa € 2.5 billion) as a result. If this practice spreads as Wal-Mart believes, the global retail supply chain will save nearly US$ 11 billion (circa € 8 billion) (Wal-Mart 2007, 58). Hence, Wal-Mart acts as a pathfinder not only in its own industry but also for its suppliers.

Another example is the automotive industry: Event studies show that high industry-wide R&D spending – which is in this case induced by the electric car challenge – can positively influence the general survival rate through spill-over-effects (Hage 1999, 611). Decision-makers are well advised, though, to actively exploit arising win-win situations as the benefits do not necessarily materialise automatically.

8.3.5 "Complementary Competition" Pattern

Complementary competition describes the effect when *non-followers* benefit from the business model transformation of a competitor (leading to a *soft revolution* or *niche migration* in figure 8.8). Assuming that the competitor does not leave the industry (i.e., its products remain substitutes for existing ones), main mechanisms behind such an effect will be a relaxation of price pressure or reduced overall demand for rare resources.

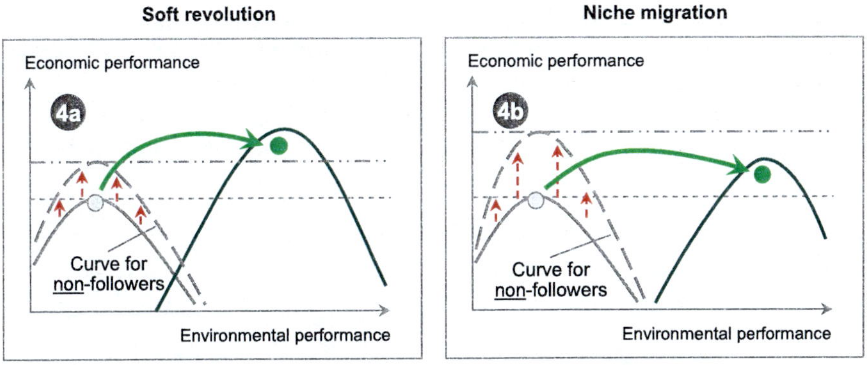

Fig. 8.8 Complementary competition through Green Business Model Transformations

In fragmented industries, the direct effect will usually be negligible when based on the transformation of just one competitor. In the case of Frosta it is thus doubtful whether the relieving effect on the rest of the food industry was very noticeable. Here, breaking a vicious downward spiral of eroding margins was at the centre of attention; however, this kind of deterioration of the trade-off function is not accounted for in figure 8.8 (as it is not related to the transformation). However, it may change the *dynamics* within the industry, e.g., away from price-based competition. If large players like Nestlé or Unilever managed to create sufficient awareness and acceptance for high-priced but quality green food, one of the patterns depicted in figure 8.8 would become imaginable: Not only would the pioneer create a high-margin segment, but it would ease the intensity of competition in the low-price-segment, too.

Maersk Line in part also tries to achieve this by directing customers' attention away from freight rates to overall logistics optimisation and environmental sustainability.

Whether complementary competition leads to a new business model paradigm or just a new niche not only depends on the relative potential economic performance of the alternative business model. Another precondition is that the market can sustain the new business model for the whole industry (i.e., it is mass market compatible, see also below). This question applies to all dynamic patterns above, although it becomes most obvious in the example of Frosta. Green chemistry also has yet to demonstrate that its principles can successfully be implemented broadly, and fundamentally transform the entire chemical industry accordingly (Anastas & Beach 2007; Anastas 2010).

8.3.6 "Green Backlash" Pattern

Until now, only dynamic effects on the *economic* performance of new green business models have been discussed. As the distinguishing feature, however, environmental performance is also subject to important dynamic effects. In extreme cases, these effects can threaten the very reason for their inception.

Figure 8.9 depicts two possible kinds of effects, both of which lead to deterioration of the average environmental performance of the new business model:

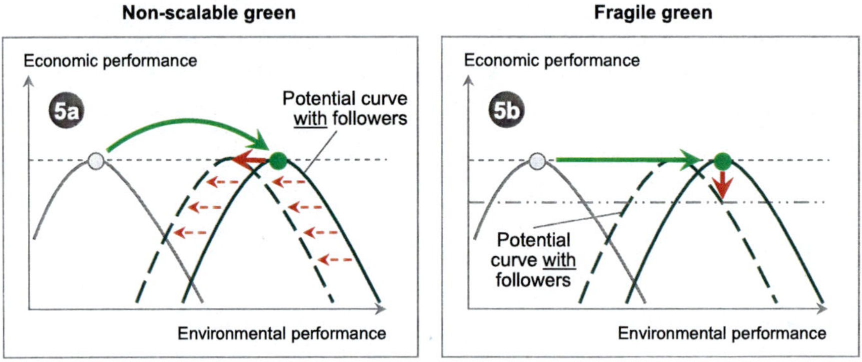

Fig. 8.9 Green backlash that can affect Green Business Model Transformations

The left-hand side, *non-scalable green*, describes the case when imitation worsens the environmental impact for all. Classic reasons are limited natural resources which the business model is dependent on. They can even be renewable, as in the case of first-generation biofuels. Here, strong market growth has led to highly adverse effects as human food supply and natural CO_2 sinks were affected. In this case, many argue that the trade-off function moved so far to the left that the supposedly green business model got actually *worse* than the conventional one (Grunwald 2008b).

Industries as a whole need to develop collective strategies in order to avoid such situations. Not only can the brand of all involved firms suffer severely, but regulatory measures (e.g., elimination of subsidies, bans) may follow. Even losing the social license to operate for an entire industry is not inconceivable (Sells 1994). Horizontal cooperation, e.g. in the form of industry associations, can be used as vehicles to implement (green) collective strategies that prevent patterns of environmental degradation (see Schneidewind 1998, 349ff.; Esty & Winston 2009, 82ff.).

An industry that has started to account for environmental risks is the airline industry. The International Air Transport Association (IATA) emphasises the necessity to use sustainable, i.e. later generation biofuels (IATA 2010, 47ff.). In Lufthansa's announcement of the first commercial flight route powered by a biofuel mix, the company thus highlighted that its biofuels suppliers are required to provide proof for the sustainability of their product (Lufthansa 2010).

The second pattern, *fragile green*, relates to the problem that followers may diminish the value of environmental performance for the business model. Either, the value proposition can be established convincingly with a lower environmental performance, or actual performance is not transparent. The latter opens up room for greenwashing. In both cases, if the greener competitors incur costs to sustain their high environmental performance, they lose competitiveness.

Because environmental sustainability is often laden with ambiguity and involves trade-offs between different dimensions of sustainability (see chapter 5), the question of environmental performance may become subject to fierce "lobbying wars" between competitors. If companies manage to discredit the environmental traits of a rivalling business model, this can have serious consequences for its economic viability. Green chemistry may face such problems at some point if volumes become large enough to seriously challenge conventional chemistry.

A maybe greater threat is if a green business model can be "downgraded" to cut cost, without the lower performance becoming evident. For example, should biofuels find widespread adoption in the airline industry, some companies might try to use cheap but less sustainable biofuels. If the other airlines do not manage to make regulators or customers understand and dismiss this practice, they may need to imitate the practice in order to survive.

8.3.7 "Green Movement" Pattern

Not all dynamics are problematic for the environment. Figure 8.10 describes dynamics that lead to an increase in environmental performance, a *green movement* within the industry:

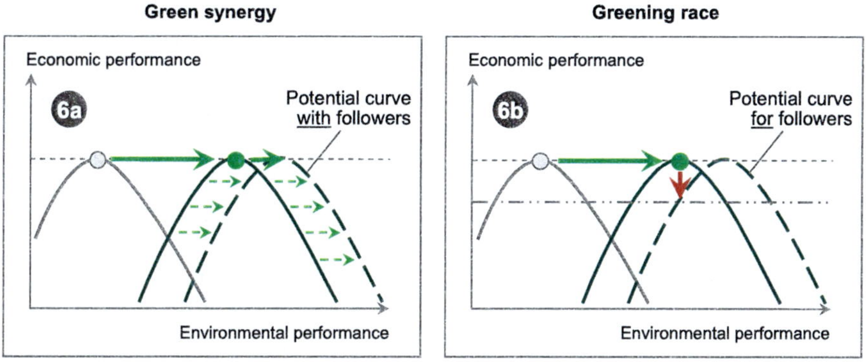

Fig. 8.10 Green movements induced by Green Business Model Transformations

Two principal reasons for green movements are discussed in the following: synergy and competition. *Green synergy* means that positive feedback loops make environmental performance increase with volume or the number of organisations that apply it. Such network effects are well-documented for market growth (e.g., Sterman 2000, 370f.), but also occur in the ecological domain. Electric cars are one example: Widespread adoption will enable applications that make energy generation more efficient (e.g., storage and virtual power plants). Cleaner electricity in turn improves the footprint of the EV that uses it.

The second dynamic pattern, *greening race*, is based on competition on environmental characteristics (as an analogy to "arms race"). This implies two things: first, environmental performance must be central to the value proposition, and second, a company that does not invest accordingly loses ground also in economic terms.

Changes in the external business environment (e.g. increasing transparency in the food business) may lay the foundation, and a decisive competitor move can then set the greening race in motion. This may be intentional or not. Wal-Mart, for example, certainly aims to set in motion respective dynamics with its supplier standards. Green corporate platforms for growth like Ecomagination do so within the company, although external organisations are increasingly invited to participate in the competition (e.g., General Electric 2010b).

All six dynamic patterns and its variations obviously represent ideal types that in practice will rarely occur in isolation. Moreover, the list of covered patterns is not exhaustive, and they consider only two rounds of action explicitly (that of the pioneers and potential followers). Advances in *game theory* (e.g., von Neumann & Morgenstern 1944) allow for formal modelling of competitive interactions like pioneers' marketing mix reactions to new entries (Shankar 1997). Most strategic issues, however, are too complex for game theory to provide definite answers; yet, it can inform strategic thinking (Mintzberg et al. 2009, 115). For example, it can be used to search for means to change the rules of destructive games to become more favourable in nature (Brandenburger & Nalebuff 1995).

From a practical perspective, formal modelling of competitor behaviour is not only complex, but may also be also unnecessary. Coyne & Horn (2009, 92) correspondingly argue that it is sufficient to ask merely three questions: "Will your competitors react at all? What options will they consider? Which options are they most likely to choose?"

8.4 Corporate Journeys towards Green Business Models

By and large, the industry survey demonstrated two things: First, Green Business Model Transformations are viable and desirable for a variety of types and business settings. Second, they are complex undertakings, and no simple recipe for success exists.

What all Green Business Model Transformations have in common – and what distinguishes them from other large-scale transformations – is that they establish environmental sustainability as a central aspect of value creation and/or capture. This link has to be robust and goal-oriented: robust in that the business model has to allow preempting or sustaining adverse competitive actions; goal-oriented in that it has to advance the long-term corporate strategy without compromising short-term needs (e.g., positive cash flows).

Hence, in this last section of chapter 8, the *process* of Green Business Model Transformations is recapitulated against the background of firms covered in the industry survey – with all their differences in history and behaviour patterns in the marketplace. This chapter is structured along three topics: motivation (or initial situation), strategic orientation, and transformation paths.

8.4.1 Motivation and Initial Situation

Why do established companies give up their current business models and try something new with an unknown outcome? The few companies that decide to do so have sometimes very different motivations and end goals indeed (e.g., Kotter 1995; Johnson et al. 2008).

Bansal & Roth (2000) argue that the general corporate ecological responsiveness is based on three firm motivations: competitiveness, legitimation, and ecological responsibility (ethics). These motivations are in turn influenced by the ecological, organisational and individual context, namely issue salience (certainty, transparency, emotivity), field cohesion (proximity, interconnectedness), and individual concern (ecological values, discretion). Although all of these motivations will likely play some role in Green Business Model Transformations, competitiveness stands out. At least four aspects in the characteristics of a company influence the way a company will approach a Green Business Model Transformation from an economic point of view:

1. the current economic situation (struggling vs. thriving)
2. the attitude towards frequent and radical change (antagonistic vs. welcoming)
3. the power and influence within the industry or broader business environment (low vs. industry leaders or national champions)
4. the behaviour patterns related to competitive and/or collaborative business practices (competitive vs. syncretic vs. collaborative)

As a result, the perceived urgency and room for manoeuvre greatly differs between the companies that have been investigated in the industry survey. For Frosta, the Green Business Model Transformation was a question of long-term survival. Maersk Line also saw itself in a difficult economic situation when it decided to act. General Electric or BMW on the other hand were not under immediate pressure, but took deliberate steps towards green business models. The advantage of the latter approach is obvious: More financial resources are at the firm's disposal, timing may not be as critical and a failure is less likely to have catastrophic impact. However, mobilising the organisation without a burning platform is more difficult.

Inertia will be less of a problem for a company that has established a corporate culture of actively addressing threats and embracing new opportunities. Notwithstanding common claims of an incumbent's curse, even large, incumbent firms can be receptive for radical change (Chandy & Tellis 2000). Dictated by the fast pace of their industry many new economy companies like Google can be considered open to radical change, trying new things frequently. Nestlé is an example for a company that is opposed to radical change (Wetlaufer 2001), representing the opposite end of the spectrum. Neither way is more successful per se; however, companies that are used to frequent, radical change will face less resistance, and be able to move faster and in bigger steps when attempting a Green Business Model Transformation.

The third aspect, the power and influence of companies has direct implications on the strategic choices they have at their disposal. A small company like Frosta cannot hope to exert substantial influence on supplier practices, customer perceptions or regulatory frameworks on its own. Nestlé is in a much better position to do so. The most extreme example in this respect is probably Wal-Mart. However, it should be noted that a Green Business Model Transformation itself can significantly increase influence of a company in the sustainability domain. GE's newly gained leadership position is a good example of this effect.

The question of competitive dynamic patterns has been discussed at length above. This leads to the question whether a company should view its transformation as a competitive move or rather take a collaborative approach and change *with* the industry. To a certain extent, both are possible at the same time. Lado et al. (1997) call this behaviour *syncretic rent-seeking*. Wal-Mart is one company that demonstrates this: for instance, on the one hand it initiates its sustainability index single-handedly. On the other hand, it invites competitors to join the effort (likely under Wal-Mart's terms and conditions, though). The automotive industry is another example for the coexistence of competition and collaboration, albeit the strategic alliances created today may only be a temporary phenomenon until the electric car industry has become more mature. Yet, in fact the interconnectedness of sustainability issues suggests that many Green Business Model Transformations will not be successful when applying the classic antagonistic view on competition.

The considerations above lead directly to the second topic, strategic orientation.

8.4.2 Strategic Orientation

The history of a company cannot be ignored. Although it does not fully determine the actions of a company in a given situation, certain patterns of behaviour will be more compatible than others depending on culture and competencies. This aspect is contrasted with the more or less deliberate choice of strategic long-term goals.

In fact, Green Business Model Transformations provide great opportunities to renew a company and throw some outdated corporate legacies over board. For example, GE has radically changed its stance on environmental sustainability, but did so based on its existing values.

Miles et al. (1978) proposed a well-known categorisation of firms according to their strategic orientation into *prospectors*, *defenders*, *analyzers*, and *reactors*. Moreover, numerous environmental management models have been developed (Kolk 2002). An analogous categorisation of strategic ideal types may help companies to frame their own efforts in the context of corporate journeys towards green business models. Based on the examples of the industry survey and the four characteristic aspects described above, five combinations of internal and external factors seem particularly conducive for Green Business Model Transformations (see table 8.2). The role of the environmental characteristics of the green business model differs accordingly and ranges from purely instrumental to idealistic.

Table 8.2 Firm types distinguished by their corporate journeys towards green business models

Corporate journey type	Economic situation	Attitude towards change	Power & influence	Market behaviour	Role of green characteristics
Drowning man	Struggling	Rather welcoming	Low	Any	Ensure survival
Tumbling giant	Rather struggling	Rather antagonistic	Leader	Competitive or syncretic	Help staying a leader
Game changer	Thriving	Welcoming	Leader	Any	Defining the environment
Opportunity seeker	Rather thriving	Welcoming	Any	Competitive or syncretic	Deliver economic returns
Do-gooder	Thriving	Rather welcoming	Any	Collaborative or syncretic	End in itself

The following provides a short description of each type with a representative example from the industry survey:

1. *Drowning man*: Companies facing a serious crisis that desperately seek a way out. They do not have a particularly influential position and thus have limited direct means of shaping their environment. These companies are characterised by a high sense of urgency and risk tolerance. Hence, changes tend to be radical as far as the financial situation allows. Example: Frosta AG.
2. *Tumbling giant*: Like companies from the *drowning man* category, these companies face economic pressure that represents the main motivation to take decisive action. However, these companies in contrast are central players in their respective industries or countries. As a result, they often suffer from incumbent's inertia, but also have a number of additional levers at their disposal to influence the environment. Example: A.P. Moller - Maersk.
3. *Game changer*: Companies in the powerful position to shape future industry standards and influence the political agenda in their favour. They appreciate the opportunities arising from sustainability trends and intend to leverage them accordingly. Although their business model transformations may not always be ground-breaking on their own, they have large societal impact by re-shaping whole industries. Example: Wal-Mart.

4. *Opportunity seeker*: Companies that consider themselves to be in a unique position to create value through a Green Business Model Transformation. Usually, this position will stem from internal capabilities or a specific market position that distinguishes them from their competitors. However, due to the lack of immediate urgency, business model transformations will often have a pilot project character until the concept is proven. A company may be both an opportunity seeker and a game changer. Examples: GE, BMW.
5. *Do-gooder*: These companies are comparable to opportunity seekers in that they are in a special position to realise a given green business model. However, their motivation originates from societal needs rather than economic opportunities, and the initiative tends to be bottom-up rather than top-down. Main challenges compared to opportunity seekers differ accordingly: getting top management attention vs. mobilising the organisation, securing resources vs. decision autonomy, and establishing a commercial mindset vs. resisting greenwashing. Example: Osram.

Internal and external factors like the ones listed in table 8.2 can facilitate or hinder the adoption of proactive environmental strategies (Aragón-Correa & Sharma 2003). However, they are certainly not sufficient to exhaustively capture the full complexity of observed corporate behaviour (see Lee 2008, 64f.). Nor is the research sample large enough to rule out the existence of other important types. Even assigning the company examples to types is not trivial, a common problem of respective classifications and typologies (Kolk 2002, 24f.). Yet the purpose of the typology above is educational rather than theoretical; it is merely supposed to help managers framing the challenges associated with their Green Business Model Transformation.

In conclusion, the industry survey refutes the argument that only a very limited, special set of companies should consider a Green Business Model Transformation. Whether a company is struggling for survival or exploring new opportunities, whether it solely relies on its own strengths or seeks a more collaborative model – in each case it is critical to understand how environmental sustainability can affect competition and be used to the company's advantage. For instance, management may think of sustainability in terms of proxies: Can sustainability stand for quality, lead to resource efficiency, improve the corporate image, or help the business model in other respects to become successful? Obviously, the company also faces the question at this point whether a *green* business model serves the company best to achieve its goals, or whether an alternative course of action (e.g., rigorous cost cutting) would serve them better.

Assuming that a Green Business Model Transformation is indeed an expedient measure, management also needs to reflect on how to "vaccinate" its business model against attacks from competitors or stakeholders: Is the transformation low-profile or will competitors feel threatened? Can they somehow be made allies rather than adversaries? Can shifting the basis of competition or strategic manoeuvring be used to make competitors act according to the company's interests? Is the business model adequately protected against criticisms from NGOs? Can support from governments be facilitated?

All these questions emerged as critical for the success of new green business models and will therefore receive due attention in the management framework presented in chapter 9.

8.4.3 Transformation Paths

Green Business Model Transformations can not only serve different purposes, the kind of *progress* and *mechanisms* of transformation differ as well. As shown in figure 8.11, the transformation process can be *continuous*, *staged* (e.g., by geography or product category), or take place with a single *"big bang"*.

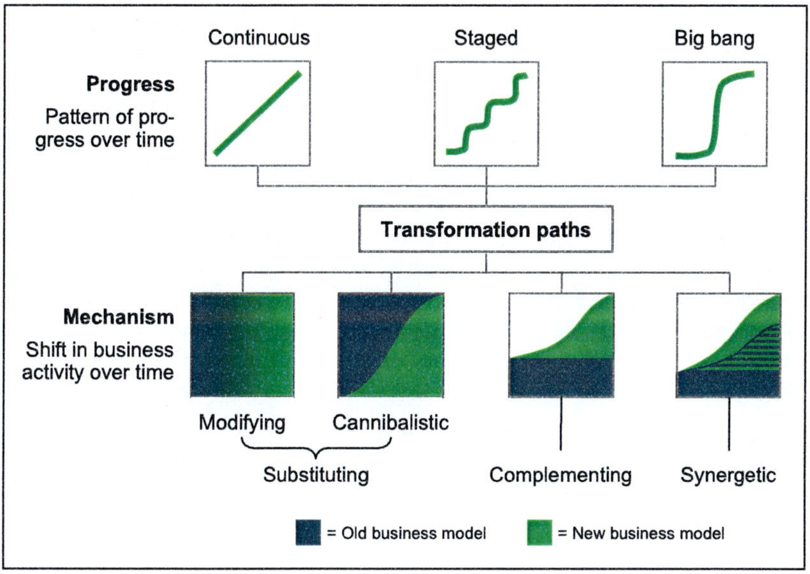

Fig. 8.11 Transformation paths differing in kinds of transformation progress and mechanism

At the same time, business activity during the transformation may shift via four mechanisms: The current business model may be directly *modified* (e.g., like Frosta by changing production processes and products), or the new business model may be introduced as a competing offering (like electric cars). In the latter case, adoption by customers leads to *cannibalisation*. The new green business model may also *complement* existing offerings (e.g., like Osram's, by introducing a completely new offering for a previously unserved customer group). Lastly, the new business model may be *synergetic* with existing business models beyond simple resource sharing (like the sustainability services of BASF benefit its other businesses). Although many business models are relatively easy to allocate to one of the mechanisms, they are not as clear-cut in all cases (e.g. a low-cost subsidiary of a full-service airline will be partly cannibalistic but also address complementing customer segments). In addition, management can steer the degree of cannibalisation to a certain extent (i.e., targeting slightly different customer segments) or derive (cost) synergies from business models that do not fall under the “synergetic” category. Furthermore, the substituting mechanisms do not need to be a zero-sum game: the company may gain market share, or increase the overall market size with the new business model (see the example of Maersk).

A company usually has some degree of freedom to select timing and progress of its transformation path, depending on the design of the business model.

Business cases of new synergetic and complementing business models will usually be rather straight forward in the sense that simple methods like discounted cash flow can be applied. Because neither modifying an existing business model, nor product cannibalisation are easily reversible, the case is different here. Frosta knew from the outset that it would lose more than the invested money if it tried to reverse its transformation. Similarly, GE knew that a failure of Ecomagination would not only hurt its brand, but also confidence in management’s ability to define and carry out strategic initiatives.

Strategies for product cannibalisation (e.g., Conner 1988) additionally need to consider potential actions of competitors. Generally it is unfavourable to cannibalise profitable products, often also due to existing assets or sunk cost (R&D, marketing, etc.) (see Lieberman & Montgomery 1988, 48f.). Yet even if the new business model has financial disadvantages over the current one, it may be better to lead rather than to follow. In fact, Connor (1988) implies that the ideal strategy for an incumbent may be to be the first to develop a superseding business model, but then wait with its introduction for the first competitor to do so. However, other criteria than protecting existing rent streams also need to be considered, e.g. the opportunity to set new standards or be recognised as a pioneer. All these considerations are highly relevant for the emerging electric car market, for example (Avadikyan & Llerena 2010).

The question of whether the new business model should be separated from the organisation has already been discussed in chapter 6.5.2.4. Especially for cannibalistic transformations which lead to serious conflicts, a separate unit will usually have the greatest chances of success. The same may apply to complementing business models if they require very different mindsets and steering mechanisms to be successful. In order to manage arising trade-offs and conflicts effectively O’Reilly III & Tushman (2004) suggest to create an *ambidextrous organisation*. That is, the new business model project is kept structurally independent from the core business, but senior management actively coordinates both business models. This way, synergies can still be realised while harmful interferences are avoided. New incentive systems that reward overall success rather than that of individual units may need to be introduced for this purpose.

If products are directly replaced, an integrated functional project design is considered equally promising (O'Reilly III & Tushman 2004, 76). Transformations of synergetic new business models lend themselves to an integrated approach. Because synergies are not limited to sharing resources but also promise to generate additional revenue, these business models will usually be considered an opportunity within the core business so that the risk of harmful interference is much lower.

The illustrative graph presented earlier in figure 8.4 and the following assumes that progressing incrementally along the trade-off functions is afflicted with poor economic performance for most of the time. This implies that firms should "jump" instead of seeking a gradual transition to the new business model. This conclusion will be correct for most business models as they cannot perform until all changes to business model elements are completed and they start complementing each other. However, such a "jump" also implies that the business model design needs to be fairly sophisticated from the outset. When design parameters are difficult to determine theoretically (as in Osram's case), companies may thus choose to introduce the business model in stages and test and refine the design at limited scale first. This possibility does not exist for all business models, of course. For instance, it would not have made sense to implement only parts of Ecomagination first, or limit it to a certain region.

Ultimately, firms face the difficult task to choose a transformation path that is compatible with its strategic goals (e.g., of being a leader), maximises expected returns, and balances these with the associated risk. The last chapter intends to help firms master this and all other challenges related to managing Green Business Model Transformations within a comprehensive management framework.

Chapter 9
Managing Green Business Model Transformations – A Framework for Management Practice

The challenges associated with green business transformations are huge: They require bold decision-making under conditions of great uncertainty, consideration of hidden systemic effects, coordination of complex organisations, and adaptation to unforeseeable events.

A suitable management framework must therefore combine sound analytics with creativity techniques and deliver a systematic process to cope with the complexity of the transformation in large, established corporations while maintaining enough leeway to address the inevitable small and big unexpected challenges that will materialise.

9.1 Framework Overview

Despite the non-linear nature of change, it is still useful to define phases and steps to structure the tasks that need to be completed when managing Green Business Model Transformations as if they were in fact linear. Directly representing change in its non-linear form may be interesting from an intellectual point of view, but impractical if the goal is to help busy managers understand and follow a proposed framework.

A. Sommer: Managing Green Business Model Transformations, SPLCEM, pp. 259–346.
springerlink.com

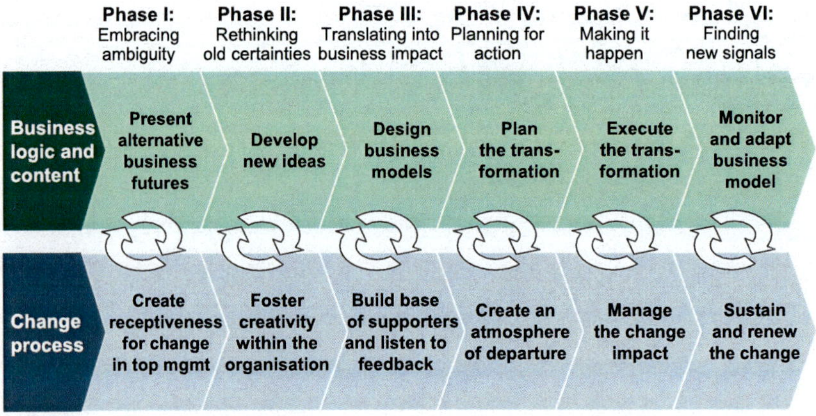

Fig. 9.1 The Green Business Model Transformation framework: business logic and change process layers

Two different but interwoven layers of change each deserve due attention: One refers to the business logic and content of what is to be changed. The other is the more hidden aspect of the change process that includes people's mental models, emotions, and actual behaviour. While the former is easier to capture in PowerPoint format and often receives the most attention, the latter is equally important for the success of any business model transformation. Figure 9.1 illustrates this dualism over the course of the proposed six phases of the Green Business Model Transformation framework:

As empirical studies show, it is initially important to conceive of ambiguity as something to be embraced rather than feared and avoided (see chapter 6.3). Moreover, major but deliberate change is known to require the full buy-in of top management. That is why *Phase I* is concerned with creating receptiveness for change among members of the top management team. They are confronted with alternative futures for the business environment and directly involved in their creation.

Once the soil has been made fertile for new ideas to improve the business, *Phase II* is used to generate them. Both the creativity of small, high-calibre teams as well as larger segments of the organisation should be utilised to think of new or better ways to meet the customers' needs by adopting environmentally friendly practices.

During *Phase III*, the project team translates ideas into business models. At this point it is crucial to sound out important influencers within the organisation, leverage their expertise, and start building a base of powerful supporters for the change.

At the start of *Phase IV*, the silhouette of the upcoming change is already visible to high-level circles. The project team develops detailed plans on how the change is to be realised. In order to reduce later resistance, the wider organisation now has to be prepared for the transformation.

Phase V is not only concerned with implementing the actual changes, but also includes dealing with the impact of the changes. As with every major change initiative, there are winners and losers – the latter of which must not be neglected. The why and the how of the change need to be explained over and over again. New behaviours need to be reinforced while old ones should be actively discontinued.

During *Phase VI*, the success (or failure) of the change is monitored. Since no business model design or implementation is perfect, adaptation will likely be necessary. Furthermore, newly introduced change is known to be frequently reversed if regressive tendencies are not actively prevented (see section 6.3.4). Eventually, new signals may be found that can trigger new transformation activities.

The six phases are laid out in the following section. Explanations will be especially elaborate for Phases I through III because these phases contain new management techniques not described elsewhere. Large, established firms can generally be expected to possess substantial experience related to phases IV through VI from previous major transformations. Hence, these respective sections are relatively brief, but include references to practical examples from chapters 7 and 8 for illustration. Figure 9.2 provides an overview of the content-related aspect of the Green Business Model Transformation framework:

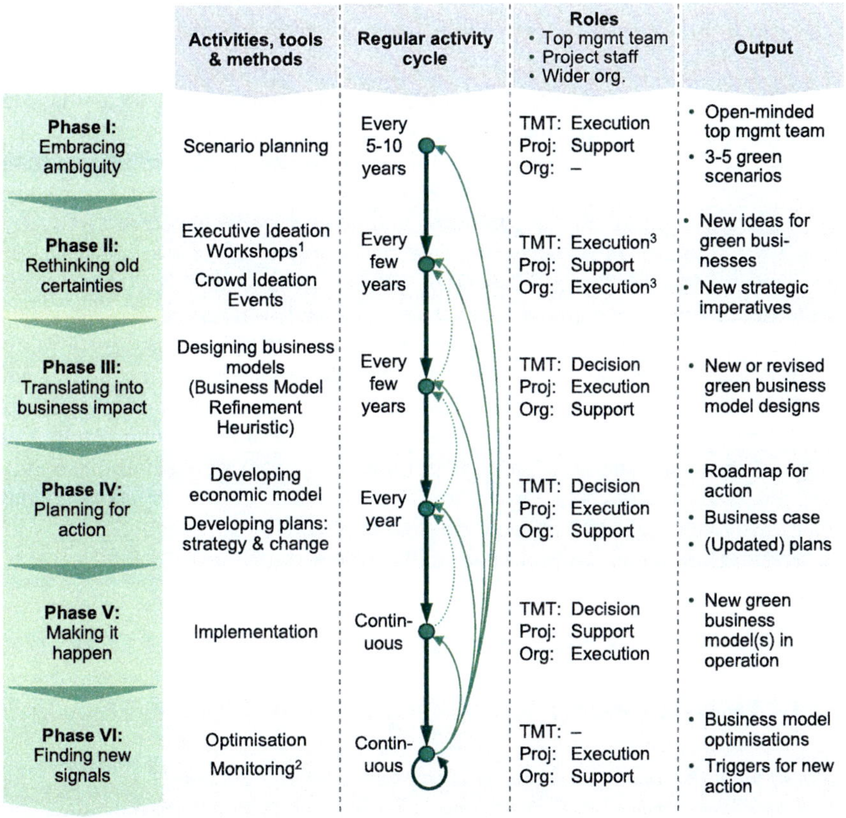

1. Using techniques including business model decomposition, industry analogies, stakeholder perspectives, or green business model prototypes.
2. Items to be monitored include business case assumptions, the external market, stakeholders, the general business environment.
3. Idea generation workshops are conducted with (top) management teams; collecting the wisdom of the crowds can involve parts or most of the organisation (like IBM's "Innovation Jams") by means of electronic collaboration software like wikis, forums, etc.

Fig. 9.2 Overview of the content-related aspect of managing Green Business Model Transformations

As illustrated in figure 9.2, the overall process is not linear but involves a regular activity cycle with periods of repetition ranging from up to a decade or more for the first phase to a year or even less for the later phases. Moreover, there are multiple exception-based iteration cycles at different stages of the process. It is vital to respond promptly to unexpected developments or deviations from plans.

While the first two phases require deep involvement of top management teams, the execution of later phases is predominantly in the hands of dedicated project staff or the organisation at large. The different management tools and methods of the six phases are designed to result in specific outputs that help firms to progress in their particular context. Some firms may already be advanced with regard to certain topics while others venture into completely uncharted terrain. Therefore, *problem context*, *management tools* and *change management challenges* are investigated for each phase in the following analysis. Although the content and change-related aspects of the framework are separate for the sake of conceptual clarity, it is important to remember that they are inseparable in practice. In fact, some key individuals, often including CEOs, will need to play important roles with regard to both aspects.

Before the six phases are discussed in detail, a last cautionary remark needs to be made: The proposed management framework cannot guarantee success for all conceivable types of Green Business Model Transformations. Some successful change will necessarily be emergent in nature and show strong characteristics of random experimentation that cannot be reproduced in the same way as a machine produces a good it was designed to produce. There is some art involved in creating successful new business models (Teece 2010, 190). However, the framework has been designed to incorporate creative and iterative elements so that it has *the capacity* to produce a broad range of successful Green Business Model Transformations and very few unsuccessful ones compared to an unmanaged process. Practitioners who intend to use the management framework should also consider the fact that different realisations of the framework can be more or less successful depending on the particular company that uses it. Hence, the proposed framework should not be applied like a recipe from a cookbook, but instead be understood as a well-organised kitchen with effective tools and valuable cooking ingredients. However, a good meal is still not produced without gifted cooks and skilful kitchen staff.

9.2 Phase I: Embracing Ambiguity

9.2.1 Problem Context

"Stocks have reached what looks like a permanently high plateau."

Irving Fisher, Professor of Economics, Yale University, 1929 (Wikipedia 2011), shortly before the start of the Great Depression

"[The] central problem of depression-prevention has been solved."

Robert Lucas, Nobel Prize Laureate in Economics, 2003 (Krugman 2009), a few years before the start of the global financial crisis

In his best-selling book "The Black Swan", Nassim Nicholas Taleb (2008) argues that managers are trained to deal with the normal and continuous, but unable to cope with unpredictable, large-impact events ("black swans"). These events are considered highly unlikely and seem to contradict common knowledge – until they happen. Managers often find it very difficult to give up their old mental models that have served them well for many years.

Fig. 9.3 Dog picture to demonstrate emergence in perception (attributed to R. C. James; as stored in the Wikimedia Commons 2007)

Pictures can be a powerful way to demonstrate how the brain processes information and recognises patterns (or not) (e.g., see de Brabandere 2005). Consider figure 9.3: Many people will see only random black and white spots. Some will recognise a pattern in the form of a Dalmatian dog (hint: the dog walks from right to left with the head at the centre of the picture). The dog cannot be found by independently identifying its body parts, but is suddenly captured as a whole instead (see Gestalt theory, e.g. Wertheimer 1925; Köhler 1971).

Analogously, many conventional business tools fail to identify "the dog" – in the form of an opportunity or risk – because they search for specific signals and do not consider the picture as a whole. Moreover, if managers have never seen a Dalmatian dog before – or consider it impossible (or very unlikely) to occur under the given circumstances – they will not recognise it even if the signals are strong.

Sometimes, a pattern is not hard to see but ambiguous instead. In figure 9.4, some see a duck while others recognise a rabbit. In fact, it may be both – or neither. The interpretation depends on knowledge about the world, as well as expectations stemming from specific circumstances. The owner of a rabbit is more likely to see one in the picture, just like someone who sees the picture on Easter as opposed to October (Brugger & Brugger 1993). Moreover, naïve observers (e.g., young children) are not capable of spontaneously switching interpretations. If informed that there is an ambiguity, however, many of them can recognise the alternative interpretation (Mitroff et al. 2006).

Fig. 9.4 Duck-rabbit picture to demonstrate ambiguity in perception (Jastrow 1899; as stored in the Wikimedia Commons 2006)

In the world of business, ambiguity is frequently present (albeit not always recognised as such). Instead of preparing the business for various eventualities, managers usually seek certainty – even if it is unachievable. Companies rely on more and more sophisticated forecasting techniques, naively ignoring the fact that the future sometimes unfolds in a way that is disruptive and not a continuation of the past (Cornelius et al. 2005, 94). At the same time, managers are reluctant to invest in new technologies before they can be sure which alternative will prevail. Instead, the usual reaction is to further invest in the firm's existing technology, thus worsening the lock-in problem.

With respect to sustainability, the duck-rabbit metaphor frequently represents a risk-versus-opportunity perspective: Managers initially tend to consider sustainability issues a threat to the approaches that have been used successfully in the past. Gersick (1991) argues that mindset breakers are necessary to overcome such views. Sustainability should not be considered a continuation of earlier problems, but a "new" phenomenon that can provide opportunities as well.

During the first phase top management is thus supposed to learn to notice the relevant patterns, recognise ambiguity, and deal with it adequately. The answer to increasing ambiguity in the business world, however, is not trying to eliminate it; that's simply not possible. It is also not a good choice to fear it and become defensive; that only delays the inevitable. Instead, managers should try to embrace ambiguity – not by making irresponsible big bets, but by preparing the firm for different futures.

9.2.2 Scenario Planning

One technique to overcome the problem of narrow in-the-box thinking as described above is scenario planning. Accurately predicting the future is usually impossible, as notoriously inaccurate past predictions of oil prices illustrate (Shell 2008). In contrast, scenario planning can enable managers to investigate whether the current business model logic still holds in radically different business environments that *may* become reality. Unlike contingency planning or sensitivity analysis, scenario planning considers various uncertainties and changes in multiple variables simultaneously – instead of only one at a time. Hence, scenarios are stories that describe possible (but necessarily unlikely) future worlds, thereby also shedding new light on the current reality (Schoemaker 1995, 26f.).

This way, scenario planning can help to overcome problems (like overconfidence or biases in predicting future developments) that are associated with different classes of knowledge: "Things we know we know", "[t]hings we know we don't know", and most problematic, "[t]hings we don't know we don't know" (Schoemaker 1995, 38). Therefore, scenario planning can be used to challenge managers' mental models and make them consider changes they would otherwise ignore – changes like a Green Business Model Transformation.

9.2.2.1 When to Use Scenario Planning

Scenario planning has become a well-known management tool since Shell prominently used it during the 1970s (Wack 1985b). As a result, experiences with the tool are numerous – despite the fact that only some firms use it, and these only once or twice per decade. Schoemaker (1995, 27) summarises circumstances under which the use of scenario planning is especially rewarding:

- When a company faces high uncertainty, especially if combined with insufficient strategic thinking and costly surprises in the past
- When (substantiated) opinions strongly diverge within the company and a common language and framework is desired
- When the industry is or will be subject to significant change or competitors are using scenario planning

This description is strikingly consistent with the concomitants of Green Business Model Transformations.

Schoemaker (1993, 195) discusses scenario planning in comparison to other strategic thinking techniques: lateral thinking and brainstorming (e.g., De Bono 1970), synectics (e.g., Gordon 1961) and morphological analysis (e.g., Zwicky 1969), the Delphi method (e.g., Linstone & Turoff 1975), dialectic reasoning (e.g., Mitroff & Emshoff 1979; Schwenk & Cosier 1980), requisite decision modelling (e.g., Phillips 1982), and dynamic systems analysis (e.g., Forrester 1958; Senge 1990; Sterman 2001). Some of these techniques are used in later phases. They differ in terms of problem scope and degree of systematisation, as well as in their suitability to identify strategic issues, to bound uncertainty, and as an internal communication device. Variations of lateral thinking techniques are used in Phase II as they are broad in problem scope and a good complement to scenario planning. The required systematisation is brought in during Phase III with dynamic systems analysis and a newly developed technique that shares some similarities with requisite decision modelling.

However, Phase I is dedicated to scenario planning as it is the only one of the considered strategic thinking techniques that is fully capable of identifying strategic issues and bounding uncertainty, and in addition can serve as an internal communication device (Schoemaker 1993, 195).

9.2.2.2 The Conventional Scenario Planning Process

Veteran scenario planning firms spend millions to extensively prepare and support the scenario building process. Scenarios are thereby not only narratives, but substantiated by sophisticated economic models that can be used to explore the range of uncertainties, for example by using Monte Carlo simulation (Schoemaker 1993). Schoemaker thus distinguishes between *learning* and *decision scenarios.* The former allow focusing further research on specific issues of interest while the latter enable informed decisions.

However, not all companies are willing to devote as many financial and human resources to a scenario planning exercise as Shell does. In fact, this may not even be necessary if the goal is to support a Green Business Model Transformation.

De Brabandere & Iny (2010) assert that scenario planning exercises can even be successful if they are conducted in a timeframe as short as one day. The prerequisite for success is the preparation of the relevant material on key trends and that top management personally participates in the scenario building workshop. Otherwise, there is a significant risk that the developed scenarios are not embraced by top management later on.

Moreover, it is possible to define green scenario prototypes that significantly simplify the scenario building process in practice. Companies merely have to adapt them to their specific needs, which makes it significantly easier to build scenarios that are of high quality both individually and as a set. Before the specific scenario planning process to support Green Business Model Transformations is discussed, the conventional scenario planning process is laid out: first in general terms, then by using the example of Shell's 2001 Global Scenarios project.

The suggested approach of Schoemaker (1993; 1995) is representative for typical scenario planning processes and briefly outlined below. It consists of ten steps (shortened from Schoemaker 1993, 197):

1. Define issues of interest: time frame, scope, and decision variables
2. Identify major stakeholders: roles, interests and power positions
3. List trends and predetermined elements with an impact on issues of interest

4. Identify key uncertainties: influence on variables and interrelations
5. Construct two extreme scenarios, one with all positive and one with all negative outcomes of the uncertainties[20]
6. Assess the internal consistency and plausibility of these scenarios
7. Develop new, plausible *learning scenarios* that cover a wide range of outcomes
8. Assess stakeholder behaviour for these scenarios and identify topics for further study
9. Develop quantitative models if possible and simulate uncertainties
10. Reassess variables and retrace steps 1 through 9 to arrive at *decision scenarios*; they can then be distributed to aid in decision-making under uncertainty

According to Shell, the critical transition from conventional research to scenarios (starting with step 5 in Schoemaker's approach) involves the development of a scenario structure. This structure is determined by one or more focal questions. Answers to these questions correspond to branches along which key uncertainties play out. Scenarios then represent certain paths in the scenario structure. There is no single best way to develop a scenario structure (deductive, inductive, or normative), but any approach will include iterations (Shell 2008, 44ff.).

Schoemaker clarifies that the ten steps represent a roadmap – rather than a recipe – that may take "well over a year" (Schoemaker 1993, 196). The fact that scenario planning exercises can take such a long time is not a matter of experience. On the contrary – Shell's 2001 Global Scenario project took almost two years – and even more than that if the application and dissemination of the scenarios are counted as well (Shell 2008, 25). However, experienced firms may produce better scenarios and have learned how to make them more effective as a tool to communicate and to support decision-making (Wack 1985a; 1985b).

Shell's four-phase process started in late 1999 and was led by a *scenario director*. This person's responsibility was to orchestrate the project team along with supplementing experts so that a balance between unconventional and plausible thinking would eventually result in relevant, high-calibre material. A core team of 15 that included specialists of complementing disciplines and "process" staff (manager, editor, and support staff) was responsible for building and disseminating the scenarios. The full team consisted of 15 additional members who helped build and apply the scenarios at the level of the business units. During the research phase, the full team was selectively supported by 200 specialist contributors (Shell 2008, 25ff.).

Prioritising the right areas of enquiry – those that are relevant yet thought-provoking – was among the greatest challenges. Luckily, the project could benefit from Shell's ongoing research activities. Some selected topics were directly related to environmental sustainability: "long-term energy scenarios to 2050;

---

[20] There are other techniques to create first scenario themes, e.g. by selecting the most important uncertainties and crossing them; or by contrasting high versus low continuity, degree of preparedness, turmoil, etc. (Schoemaker 1995, 29).

development paths for renewables and unconventional liquids; fuels and drivetrain evolution; [...] and sustainable cities" (Shell 2008, 33).

The scenario building process itself included interviews with leaders and three major workshops: *orientation* (five days), *building* (five days), and *affirmation* (one day) (Shell 2008, 25ff.). The first workshop aimed to provide focus and deeper understanding for further research. After the following research period, the second workshop included the actual scenario building: Separate teams worked on the scenario structure in parallel, later combining and refining initial scenarios which were eventually presented and scrutinised by four executives who had not participated in the previous parts of the workshop. During the last main workshop, the full team scrutinised the final drafts of the scenarios once more. Furthermore, implications for business units and Shell as a whole were examined. Lastly, the team discussed scenario names in terms of possible interpretations depending on industry, language, or culture (Shell 2008, 57).

The project had 1,000 primary and 2,000 other recipients, not including external recipients who were given access to the published materials. As also intended within the framework in this monograph, Shell emphasises the importance of systematically applying the scenarios as part of strategic review processes, leadership workshops, and even individual decision-making (Shell 2008, 29ff.).

Shell's scenario planning process described above is certainly one of the most extensive examples of such a project. Within the context of a Green Business Model Transformation framework, one or two years may seem too long a duration for Phase I. Even with a narrower scope, a smaller team and a compressed time frame, Shell's process will take too long to be useful in many practical contexts as envisioned in this publication. Therefore, I have developed an accelerated process that is tightly integrated and customised for Green Business Model Transformations (for a brief description of a tested, general approach to developing scenarios more rapidly, see de Brabandere & Iny 2010). Before introducing this process in detail, general quality criteria for scenarios are presented, as they are vital for the success of any kind of scenario planning exercise.

9.2.2.3 Quality Criteria for Scenarios

Quality criteria can be distinguished into those that make individual scenarios powerful, and those that are necessary for a set of high-quality scenarios to work in conjunction.

Scenario experts (e.g., Wack 1985a; 1985b; Schoemaker 1995) emphasise that a scenario of high individual quality must first of all address the main challenges that the organisation faces. Furthermore, it has to be plausible to a large enough portion of the audience. This requires internal consistency of underlying assumptions and that these assumptions are not counterfactual (i.e. scientifically disproven). Also, the scenario should relate to current mental models of the audience instead of becoming an abstract intellectual exercise. At the same time, the scenario has to challenge current beliefs and incite people to take a fresh, long-term view. It should shed light on critical uncertainties and distinguish them from *predetermined events*. Predetermined events are events that have already happened (or are known to happen), but whose consequences are not fully apparent yet (Wack 1985b, 77). In conclusion, a scenario must be possible but unlikely – it is *not* an attempt to predict the future.

Some formal characteristics help to improve the effectiveness of individual scenarios. First of all, a good scenario requires an interesting, cohesive storyline. Like a motion picture, it should convincingly demonstrate how certain events unfold and lead to the described end state (Cornelius et al. 2005, 94f.). This end state should possess a certain stability, i.e. exist for some time, which makes it worthwhile for a company to prepare for it (Schoemaker 1995, 30). Moreover, the story has to be told in a way that appeals to people and fires their imagination (Shell 2008, 54). Some form of conflict will usually be involved. A very concrete, detailed description (e.g. of one day in the future life of a certain individual) is much better suited to create a deep, intuitive understanding of a scenario than a comprehensive but abstract description. In addition, people also tend to find such scenarios more credible[21] (Schoemaker 1993, 201). Cartoons or other visualisations can be used as valuable complements to written text (Shell 2008, 60f.). And last but not least, a catchy but 'unproblematic' name (low ambiguity, no cultural issues) has to be found for each scenario (Schoemaker 1995, 29; Shell 2008, 57).

Scenarios do not function in isolation; they need other scenarios to contrast them. Usually, three scenarios are the minimum to capture a wide enough range of future possibilities, and four (some argue five) are the maximum that can be handled by the audience (Wack 1985a, 146). Even if scenarios are of high individual quality, they do not necessarily work effectively as a set. For that, the set has to meet the following criteria:

In combination, the scenarios should incorporate great diversity in how the world might develop and encompass the major contrasting views on important issues. Moreover, the scenarios have to be orthogonal to each other. This means that they need to differ in most relevant dimensions, without affecting the plausibility and cohesion of the individual scenarios (Schoemaker 1995, 30). Even if the scenario structure includes only two focal questions, there are many implications on other dimensions that should be used to differentiate them. For example, if two of four scenarios can be viewed as variants of the other two, the set is not sufficiently diverse.

[21] The so-called *conjunction fallacy* makes people erroneously infer that a conjunction is more probable than one of its constituents. For example, if a woman is described as emancipated and concerned with social justice, most people consider it less likely that she works as a bank teller than that she works as a bank teller *and* is active in the feminist movement (which is logically inconsistent) (Tversky & Kahneman 1983).

9.2.3 Management Tool: Accelerated Green Scenario Planning

Accelerated Green Scenario Planning is a rapid scenario planning approach customised for Green Business Model Transformations. The use of *Green Scenario Prototypes* is the main innovation of this new management tool. It greatly speeds up the process while reducing the risk of inferior outcomes. Green Scenario Prototypes are a set of predefined scenarios that represent generic possible future worlds – seen from the angle of environmental sustainability. The term *prototype* refers to the fact that the scenarios still need to be customised and detailed. However, they already provide a rich framework of four complementing futures that should sufficiently cover most relevant aspects. Before explaining the Green Scenario Prototypes in more detail, the process of Accelerated Green Scenario Planning is presented; it consists of five steps (see figure 9.5):

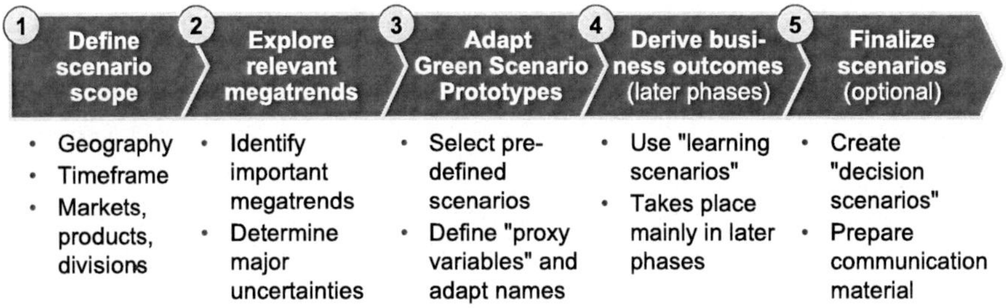

Fig. 9.5 Overview of Accelerated Green Scenario Planning

9.2.3.1 Step 1: Define Scenario Scope

The first task is to define the timeframe of the scenarios. Factors like the industry clockspeed (see 6.3.1) and important political or symbolic dates can form the basis for this decision. A timeframe of about 20 years is assumed for the Green Scenario Prototypes, i.e. they correspond to the year 2030. Deviations from this timeframe – which are advisable for short-lived industries – need to be considered when using the Green Scenario Prototypes.

Environmental sustainability is already predefined as the overarching theme. However, a company may wish to further limit the scope of the scenarios to make them more focused. For example, a company may decide to focus on a certain country or region instead of using a global scope (however, Green Scenario Prototypes are based on the perspective of the (Western) industrialised world). Furthermore, the scope may be restricted to certain markets, products, or divisions. This way, conglomerates may focus on the areas that are most affected by environmental sustainability. However, focusing too early can also have disadvantages: Firstly, divisions with hidden implications can be much more valuable to examine. For decisions with obvious uncertainties, the relevance of sustainability will already be much better understood. Moreover, a conglomerate can adapt scenarios to the needs of certain divisions at a later stage, based on its company-wide scenarios (e.g., see Shell 2008, 72f.).

A decision memo concerning the scenario scope will usually be the first deliverable of the newly-formed project team[22].

9.2.3.2 Step 2: Explore Relevant Megatrends

If the company has not already done so on a different occasion, the project team needs to conduct a systematic analysis of relevant megatrends in order to provide content for the scenarios.

Up to a certain level of detail, megatrends are already incorporated into the Green Scenario Prototypes. However, without a good internal understanding of relevant megatrends, a company will find it difficult to determine implications and provide substantiated decision support.

The megatrends that should be researched are not limited to those that have an immediate impact on the company; sometimes, megatrends that affect the business environment or society at large can have a non-obvious, but significant impact on businesses in the long run.

Based on the identified megatrends, predetermined events and key uncertainties should be made explicit. For this task, the scenario project team may need to consult a diverse range of internal and external experts.

9.2.3.3 Step 3: Adapt Green Scenario Prototypes

Step 3 features an essential scenario workshop with the top management team. During this workshop, participants fill the Green Scenario Prototypes with life. They assign possible outcomes of key uncertainties and predetermined events to scenarios, thus making them tangible. In case the various outcomes are sufficiently covered in three of the four prototypes, their number should be reduced. Conversely, a fifth scenario may turn out to be necessary.

One of the most critical tasks in Phase I is to select what I call suitable *proxy variables*. Proxy variables are representations of the different scenario worlds. For example, one proxy variable could be the "Person of the Year 2030" of Time Magazine (also see de Brabandere & Iny 2010, 2). Depending on the scenario, a different (fictional) character is suited to represent the scenario. The associations that are linked to these persons help develop an intuitive understanding of the future worlds. Assuming the workshop participants made up the "Judges of the International Court for Crimes Against Nature" as one representation of the "Person of the Year 2030", the scenario would imply a strong sense of environmental

[22] Below, I will use the term "project team" to refer to the staff that are assigned full-time or who dedicate most of their capacity to support the Green Business Model Transformation. Different people will be assigned to the project team during each phase, and the team will vary in size. Hence, the term project team does not refer to certain individuals but to the organisational unit that manages the transformation.

protection and major geo-political influence on sustainability issues. In conjunction with other proxy variables, participants can thus create vivid images of alternative futures.

BCG's scenario planning expert Luc de Brabandere asserts that proxy variables need to address three levels in order to cover a wide enough range of issues: the business environment and the society at large (e.g. the Person of the Year 2030); industry, market and competition (e.g., a typical customer); and internal factors specific to the company (e.g., a future workplace at the company). Section 9.2.4.3 provides sample proxy variables for the four Green Scenario Prototypes that the project team can use for inspiration. Workshop participants may also adapt scenario names if alternatives can be found that work better in the context specific to the company.

Finally, top management can have first discussions about implications of the scenarios on strategy and business models.

Step 3 is the core task of the scenario planning exercise and as such needs excellent preparation as well as top management participation. The megatrend material should be provided in an easy-to-digest format in advance so that participants can prepare themselves for the adaptation workshop. Furthermore, the project team has to prepare a set of suitable proxy variables for the workshop. However, participants from top management should take the lead in choosing the actual variable "values" for each scenario.

The workshop itself should be moderated by experienced scenario planning practitioners. If the scenario project team has done its homework properly, experience shows that as little as one full workshop day may be enough to create high-quality learning scenarios (de Brabandere & Iny 2010). Even within such a short timeframe, a workshop can produce some eye-opening and vital insights for top management.

9.2.3.4 Step 4: Derive Tangible Business Outcomes

Step 4 is mainly conducted during later phases of the Green Business Model Transformation framework. However, in order to provide a consistent view on the overall method, it is included here as well (detailed instruction will follow). Tangible business outcomes are derived from the developed learning scenarios in various ways:

- Reviewing of strategic imperatives – either as part of a regular annual strategic planning process or in a separate workshop (still part of Phase I)
- Using insights from scenarios to concentrate focus and enrich ideation workshops and events (Phase II)
- Designing business models: identifying research needs not yet sufficiently covered by scenario research and validating assumptions (Phase III)

- Planning, implementing, and monitoring new business models; the green scenarios can be used to support respective decisions under uncertainty (Phases IV through VI)

Partly in parallel to step 4, the project team can execute the fifth and last step of Accelerated Green Scenario Planning. Though optional, this step is highly recommended in most instances.

9.2.3.5 Step 5: Finalise Scenarios

This step has two aspects: Transforming learning scenarios into decision scenarios, and upgrading the scenarios to become a primary internal and external communication vehicle for the company.

Creating decision scenarios means that the project team re-uses (quantitative) results from business model design efforts and potentially conducts additional analyses to enhance them. The goal behind these activities is to provide decision-makers with a substantiated "fact base" to deal with ambiguity and uncertainty in their domain. Often, such decision scenarios prove to be a valuable complement to conventional forecasts. Depending on the scope of the scenarios, they can help to anticipate important trends, to develop strategies (at the global, country-wide or business unit level), or to make project-related investment decisions (Cornelius et al. 2005).

If a company decides to use its green scenarios to communicate with a wider audience, they may need some face-lifting. If not done already, producing graphic illustrations is highly advisable. Text and content may need to be refined to suit the new target audiences (e.g., the public at large) and formats (e.g., brochures, books).

While all of the above-mentioned activities have to be coordinated by the scenario project team, the communications department should hold primary responsibility for generating and placing communications material.

9.2.4 Green Scenario Prototypes

For the purpose of this work, I have developed four Green Scenario Prototypes. A project team can select a subset that is especially relevant to its company and potentially complement the set with a new scenario. In any case, it is necessary to translate the scenario prototypes to conform to the specific context of the company so that recipients can relate to scenarios personally. To do this, suitable proxy variables need to be selected that describe the scenario in a powerful way. Moreover, some aspects of the scenario prototypes may become reality much faster than anticipated while others may turn out to be counterfactual in the future. In these cases, the scenario prototypes need to be updated in order to remain plausible.

The idea to define scenarios for sustainable development is not new. Based on the initiative of the United Nations, the Millennium Ecosystem Assessment (MA) has produced four scenarios for the year 2050 ("Global Orchestration", "Order from Strength", "Adapting Mosaic", and "TechnoGarden") (Millennium Ecosystem Assessment 2005, 71ff.). The scenarios focus on the duality of industrial and developing countries and differ as to how sustainable development is addressed (i.e., regional vs. global, reactive vs. proactive approaches). This results in varying population sizes and economic growth and has implications for the enhancement or degradation of ecosystem services, respectively (also see 3.3.3).

The World Business Council for Sustainable Development (WBCSD) has developed three scenarios (also for 2050) that directly address businesses, albeit in a general way (WBCSD 1997, 14ff.): "FROG!" describes a future in which the world continues to follow the primacy of economic interests at the expense of sustainability. "GEOpolity" represents a scenario in which rethinking takes place and, through globally orchestrated policies, the economy is redirected towards a path of sustainable development. Finally, "Jazz" describes a world in which free market powers successfully cope with social and environmental challenges through innovation, experimentation, rapid adaptation, and voluntary cooperation between stakeholders.

The scenarios mentioned above are instructive for developing the Green Scenario Prototypes but cannot simply be adopted as they stand. The four MA scenarios have been developed against a natural science rather than managerial background and the WBCSD scenarios are already over a decade old. Moreover, all scenarios seem to assume that the world is indeed at risk and merely focus on whether mankind finds an adequate response. Although this basic assumption might find broad consensus in most scientific circles (e.g., IPCC 2007; Stern 2007) dissenters do exist (e.g., Lomborg 2001). More importantly, environmental sceptics can still be found in boardrooms, at least when it comes to tough economic decisions (e.g., Hickman 2010). An alternative scenario structure has thus been developed.

9.2.4.1 The Scenario Structure

For the scenario structure, three focal questions have been identified (see figure 9.6):

1. How serious is the planet's condition with respect to environmental sustainability today (current path of development)?
2. Will catastrophic crises be prevented by 2030?
3. Which shaping forces – political institutions, business, or the civil society – and corresponding mechanisms will be dominant in 2030?

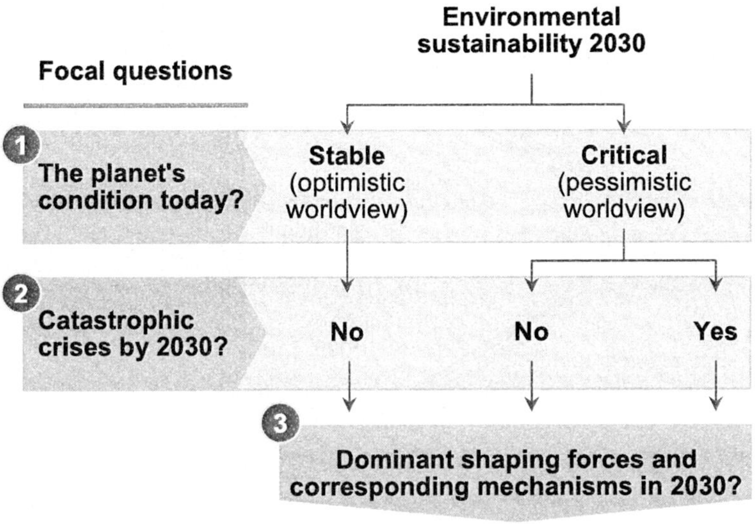

Fig. 9.6 Focal questions and scenario structure of the Green Scenario Prototypes

As indicated above, the first focal question is important as not all decision-makers agree on whether issues like climate change, water scarcity, pollution, etc. are really as threatening as many scientists and activists claim. Hence, in order to include environmental sceptics, it is necessary to have at least one optimistic scenario in which many of the planet's problems will self-regulate and a business-as-usual approach and conventional market mechanisms are sufficient to prevent any severely damaging events. Hence, the optimistic view implies that the answer to question two will be that no catastrophic crises will occur.

In case the warnings of environmentalists turn out to be substantiated, the question is whether or not humanity will be successful in changing the current path of development and can prevent catastrophic conditions by 2030.

Depending on the outcome of these two questions, it is possible to hypothesize which shaping forces will dominate sustainability affairs in 2030. Three main forces are distinguished: Political institutions like governments or international organisations (e.g., the United Nations), "business" in terms of single companies or representative bodies (e.g. industry associations), and civil society, consisting of individual citizens or NGOs. Depending on each group's interests and the power balance between them, certain mechanisms are likely to have major influence regarding other parties' actions (see figure 9.7):

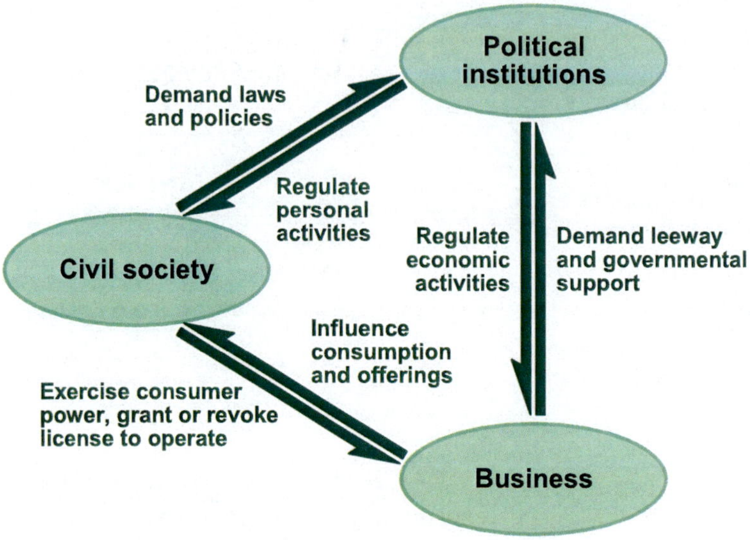

Fig. 9.7 Shaping forces and corresponding mechanisms of influence

Civil society may demand certain laws and policies from its political representatives; these in turn may regulate activities of both citizens and corporations. Businesses can try to thwart any restrictions and instead ask governments for support (tax breaks, government aid, etc.). Corporations also have influence on consumption patterns and can build (possibly informal) coalitions to influence product offerings or to enforce certain industry practices. Civil society can exercise its power towards the corporate world in various forms ranging from an uncoordinated use of buying power to organised boycotts to revoking a company's or industry's license to operate.

9.2.4.2 Four Green Scenario Prototypes

I have derived four antithetic prototypes from the developed scenario structure. Each represents a certain class of scenarios that share the same basic assumptions about sustainability and the world. Each needs to be customised later to the areas of relevance that are specific to the company. Nevertheless, each prototype can be described up to a certain level of detail. Figure 9.8 presents an overview of the four Green Scenario Prototypes and their characteristics.

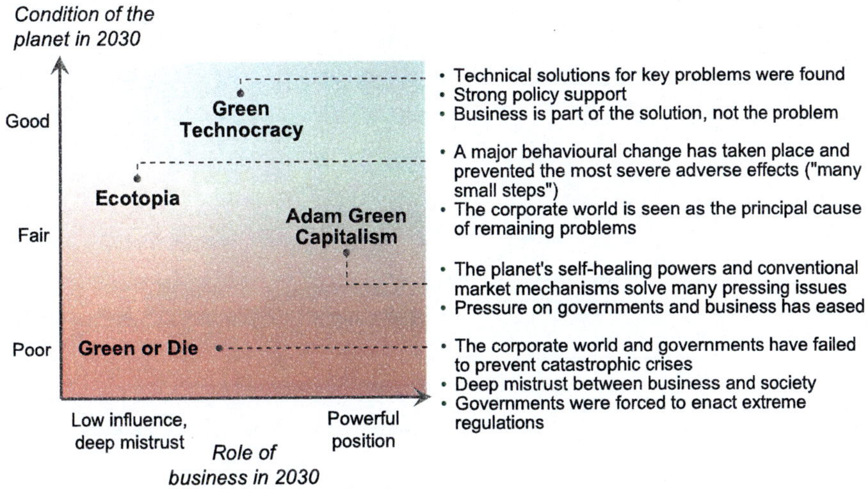

Fig. 9.8 Overview of the four Green Scenario Prototypes and their characteristics

Green Scenario Prototype: "Adam Green Capitalism"

The first Green Scenario Prototype is based on an optimistic view regarding the planet's environmental issues and is called "Adam Green Capitalism". The chosen scenario name is deliberately provocative in order to create vivid associations, referring to Adam Smith, whose legacy[23] is usually considered a symbol for free-market (or laissez-faire) capitalism (e.g., Smith 1863). The "green" component of the name indicates a "new flavour" of capitalism in which green is a visible economic factor as it dominates many advertisements and product designs. Yet, because many "Cassandras" that environmentalists had warned against turned out to be comparatively harmless, the pressure on governments and the corporate world to act decisively never reached a critical level. On the contrary, indications of real, but non-catastrophic environmental problems are often considered to be fearmongering nowadays.

[23] It has been argued that Smith's works are misinterpreted to a certain degree. He in fact explicitly advocates some forms of governmental intervention in his writings (e.g., Buchholz 2007). However, for the sake of naming a scenario it is the common rather than the correct interpretation that is relevant.

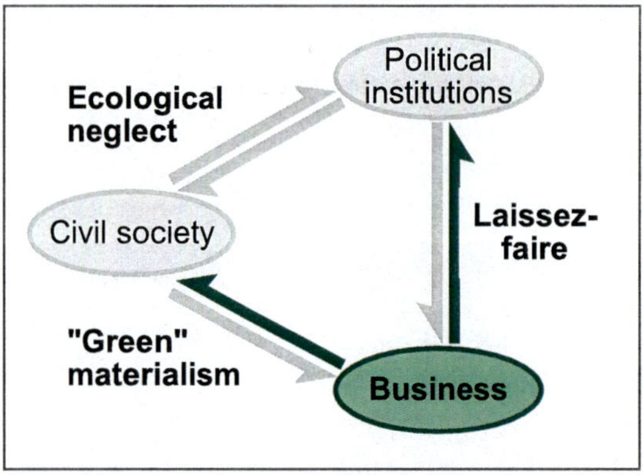

Fig. 9.9 Dominant shaping forces and their relationships in "Adam Green Capitalism"

As a result, businesses enjoy maximum freedom to use environmental sustainability in opportunistic ways. Companies widely use green claims to boost their sales or polish their corporate image, but usually refrain from addressing environmental problems fundamentally. Hence, "green" becomes just another product feature (like quality or cost) and sustainability issues are largely in the hands of marketing and communications departments.

Consumers on the other hand are confused by the large number of green claims and product labels. Yet, governments do not feel pressured to regulate business practices against the wishes of corporate lobbying groups. NGOs frequently reveal corporate greenwashing, but civil society does not exercise enough power to effectively contain such corporate practices. Instead, governments largely rely on voluntary agreements to limit environmental impacts of economic activity (see figure 9.9).

Green Scenario Prototype: "Green Technocracy"

The second scenario also assumes that humanity can prevent environmental conditions from becoming too serious. In fact, under the leadership of the corporate world and in conjunction with political institutions, environmental challenges are overcome and previous damage can even be largely restored (see figure 9.10).

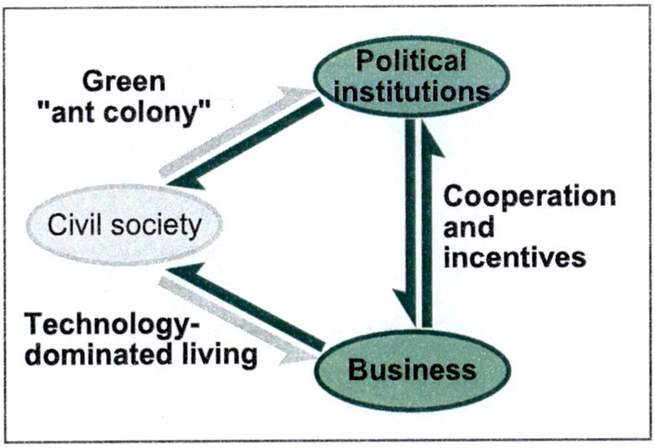

Fig. 9.10 Dominant shaping forces and their relationships in "Green Technocracy"

This scenario, called "Green Technocracy", is based on a pronounced belief in technical progress. Large-scale governmental funding of research in key areas shows tremendous success: Clean energy has been made cost effective and scaled up globally (e.g., compare CCS, DESERTEC (2009)), breakthroughs in agriculture, water and pollution reduction technologies avoid further degradation and allow mother earth to recover from the most serious environmental damages. The development of new technologies is flanked by strict laws that regiment the way of living for citizens. Technology thus dominates many aspects of daily life: Individual activities related to transport or consumption are coordinated by means of (digital) technology, particularly in urban environments (for illustration see IBM's current vision of smarter cities, IBM 2011b). For example, "energy and waste management software" controls people's homes and workplaces. Resource-intensive activities are tightly regulated. However, due to the great success of the implemented measures, resistance within civil society against the coerced changes remains low and people quickly get used to their new lifestyle.

Governments provide businesses with generous incentives to develop green tech and invest heavily in green infrastructure. Other political interventions are largely constrained to ensuring equal rules of competition.

Green Scenario Prototype: "Ecotopia"

The other two remaining prototypes are based on a pessimistic world view and assume that the world today indeed teeters on the brink of disaster. However, in the "Ecotopia" scenario, most great environmental problems are successfully defanged by a deliberate, far-reaching change in attitudes and behaviour. The scenario is called "Ecotopia" in reference to ecological utopias like the one described by Ernest Callenbach in his novel "Ecotopia: The Notebooks and Reports of William Weston" (Callenbach 1975).

In the "Ecotopia" scenario, civil society is the driving force of change (see figure 9.11). In stark contrast to "Adam Green Capitalism", humans overcome many materialistic habits and instead seek a life in balance with nature. Technology is not rejected generally, but is subordinated to desired social forms of living.

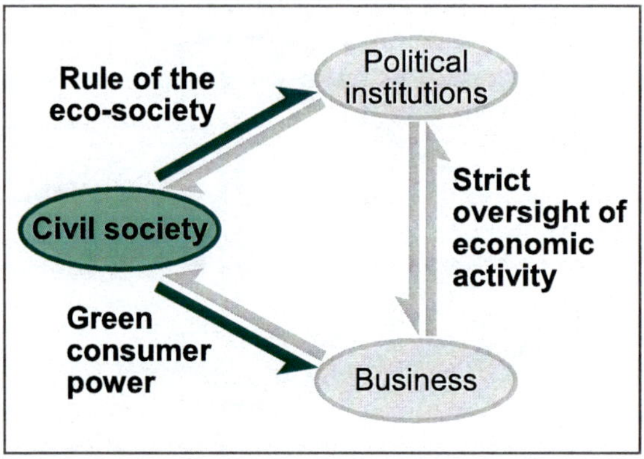

Fig. 9.11 Dominant shaping forces and their relationships in "Ecotopia"

Civil society has reclaimed the condition in which governments and the economy serve the interests of the people – something which was perceived to be lost. As such, governments were pressed to provide far-reaching legal protection against any environmental hazards. Large lawsuits are pursued in cases of violation of respective laws. On a similar note, consumers exercise their buying power to shape an ecological, demand-based economy. Many products are bought locally to support communal structures, and imported goods need to fulfil strict fair trade standards in order to avoid the threat of a boycott.

Green Scenario Prototype: "Green or Die"

The fourth and last scenario is labelled "Green or Die". It is a rather gloomy scenario in which the worst case has actually materialised: Severe environmental crises have claimed many lives, made some cities uninhabitable, caused huge flows of environmental refugees around the world, and even led to escalating armed conflicts.[24]

The global economy is also heavily affected: Some resources have become extremely expensive and most governments have enacted extreme laws in order to cope with the dire situation. Products and even entire industries which were considered main contributors to the crisis have been banned; all others have to meet strict environmental regulations.

[24] When I first developed the "Green or Die" scenario in late 2009, the described catastrophic environmental conditions and their societal consequences seemed highly unlikely within the given timeframe – even for a scenario. Since then, the Deepwater Horizon oil spill in the gulf of Mexico and the tsunami and subsequent nuclear crisis in Japan demonstrated that even within the industrialised world, environmental mega-disasters can occur at any time.

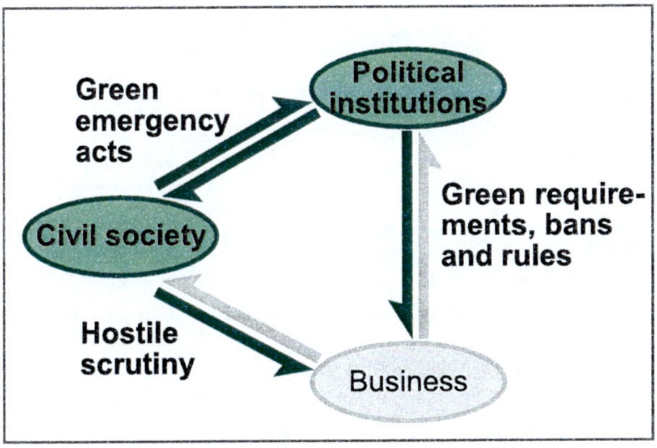

Fig. 9.12 Dominant shaping forces and their relationships in "Green or Die"

The civil society has developed a deep mistrust of the corporate world – which sometimes erupts violently against identified villains. Restrictions to the familiar lifestyle are accepted perforce (see figure 9.12).

9.2.4.3 Applying the Green Scenario Prototypes in Practice

The general method of Accelerated Green Scenario Planning has already been laid out. After the scenario scope has been defined in step 1, the relevant megatrends need to be explored (step 2). For each scenario, these megatrends may play out in different ways (if that is not the case, the scenarios are likely too similar).

There are numerous trends that either contribute to the problems of environmental sustainability or to their solution. Moreover, what is a challenge for some firms is usually an opportunity for others. Trends can be attributed to three broad domains: the economy, society, and the environment. In contrast to the traditional view that regards society and the environment as mere constraints, it is useful to view economic trends as embedded in societal trends, which in turn are embedded in environmental trends (Senge et al. 2008, 102ff.). Figure 9.13 lists examples of such (mega)trends based on Esty & Winston (2009, 34), Z_punkt (2008), and observed trends from the industry survey (chapter 7). For example, consider a company which has realised that the Next Billion consumers (*economic trend*) in India provide a huge opportunity for selling its products. However, the increasing problem of pollution and waste (*environmental trend*) in this country would be worsened due to a lack of appropriate recycling systems. Depending on different stakeholder reactions, this could have various consequences in the future: For example, local governments could tax or prohibit products. Or – potentially even more problematic – global NGOs may pillory the company via social media (*societal trend*) – eventually leading to consumer boycotts and brand damage in home markets. However, in seeing these issues early, management can identify related risks and prepare for eventualities.

Scenario planning is specifically designed to help companies anticipate such concealed chains of events and find countermeasures. Yet the scenarios may not work or may even be misleading if the available fact base is poor (not all trends in figure 9.13 are relevant to every company and some important ones may be missing).

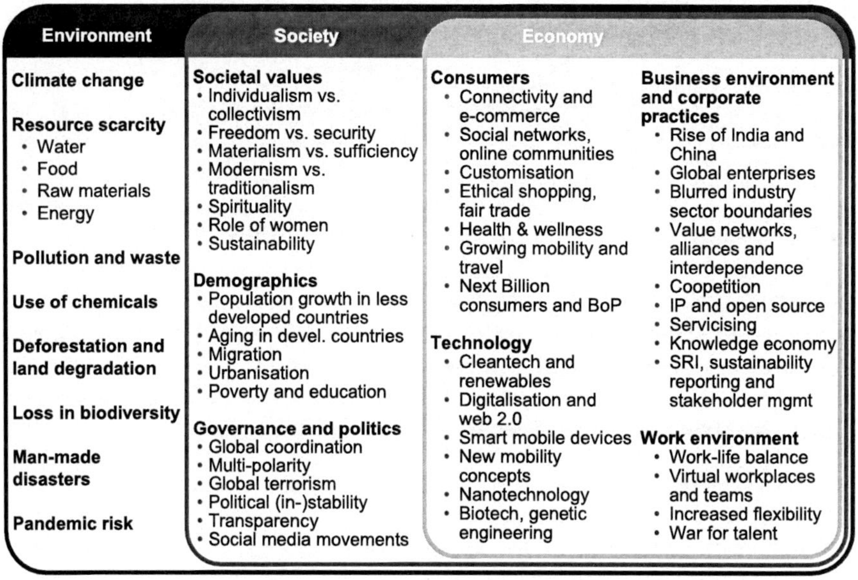

Fig. 9.13 Examples of environmental, societal and economic trends (partly based on Z_punkt 2008; Esty & Winston 2009, 34)

Step 3 (see 9.2.3.3) is about making scenarios tangible: The Green Scenario Prototypes need to be selected and then modified to appropriately represent the specific circumstances that are relevant for the company. The audience can relate to and benefit from the scenarios much better if the proxy variables are carefully selected and vividly described. Table 9.1 shows examples of potential proxy variables – three each from the domains of business environment and society, industry, market and competition, and internal company specifics. The selected proxy variables then have to be woven into a vivid, convincing story that also indicates how this picture of the future has developed as it has.

Table 9.1 Examples for proxy variables and implications of Green Scenario Prototypes

		Green Scenario Prototype			
Domain	**Proxy variables**	"Adam Green Capitalism"	"Green Technocracy"	"Ecotopia"	"Green or Die"
Business environment and society	Environmental regulation (general)	Voluntary commitments instead of regulation	Mandatory use of software that controls the environmental impact of daily life	Subsidies for local production and consumption (energy, agriculture, etc.)	Environmental GAAP; bans of certain industries and products; strict regulation of pollution and resource usage
	Capital markets	Environmental concerns are generally ignored	Dominance of clean tech investments; global infrastructure projects	Local "green micro-financing"; large projects difficult to finance	Labour & financial capital is abundant, natural capital extremely scarce
	Time Magazine "Person of the Year 2030"	The founder and CEO of "GreenAds Inc.", an advertising agency	The inventor of cold fusion technology	The 1st president of the "United Green Nations"	Judges of the "International Court for Crimes against Nature"
Industry, market and competition	Consumer behaviour	Fashion / lifestyle trend: "Green is the new black"; consumer confusion & greenwashing	Humans embedded in a system of smart infrastructure; networked life	Shopping local; fair trade is the standard for imports	Expensive energy and resources; eco-educated consumers
	Model production plant 2030	Mega-plant that realises unprecedented economies of scale; large, glossy information centre attached	Extreme miniaturisation; fully automated; digital integration of processes with partner network	Plant integrated into the natural environment; autonomous renewable energy supply	Closed-loop systems in all relevant processes; fully automated eco reporting to authorities
	Supply Chain Management	Opportunistic sourcing: cost, quality, or even green	Focus on super-efficient megacities; self-optimising supply chain systems	Fragmented supply chain; public private partnerships	Extensive environmental auditing; "Eco-shoring"
Internal company specifics	Positioning of sustainability within the company	Dominance of marketing and communications	Part of the R&D department; most board members with scientific background	Decentralised; strong emphasis on community relations	"Chief Sustainability Officer" replaces CFO position
	Job application process	Company buys certificates to offset CO_2 emissions of travels to interviews	Environmental track record to be submitted (as recorded by mandatory software)	Completion of an "Environmental Volunteering Year" prerequisite for a job	Job interviews per videoconference only to avoid any travel
	Corporate car policy	Only for executives; "green" luxury models	Hierarchised driving allowances: emissions per km x distance travelled	No company cars, every employee gets free public transport	No company cars; no parking available at the company's facilities

The practical implementation of step 4 is explained in detail in the later phases of the framework. A practical example for step 5 is well documented by Shell in the report "Scenarios: An Explorer's Guide" (Shell 2008) and thus not repeated here.

9.2.5 Change Management Challenges

There are two major change management challenges in every scenario planning exercise: The first is to find and motivate the right people to participate in the project. The second is to ensure that the resulting scenarios are actually used in a meaningful way.

Scenario planning requires a rare mixture of domain experts and senior managers: Contributors should be able to let go of their current assumptions about the world and their business. These people usually are among the most powerful and brightest members of an organisation and thus indispensable in their regular jobs. And yet, significant time commitments are inevitable for such a project.

During the project, participants need to deal with the unfamiliar condition that all ideas count – independent of the rank of the person proposing it – and that the process (at least initially) requires fostering divergence instead of consensus.

According to Shell (2008), scenario project teams usually go through five phases: They start with elation and let go of assumptions. Then, the team is hit by information overload, followed by disorientation about which scenario ideas to use and how. Departing from this motivational trough, "some get swept away" (Shell 2008, 51) before elation is finally reached again when the power of the created scenarios becomes apparent. It is the responsibility of experienced scenario practitioners to guide the team through this journey. Hence, if the organisation has no experienced scenario practitioners within its ranks, it is advisable to seek external support.

The second challenge has already been touched upon briefly: Along with a systematic use and integration of the scenario planning in regular strategic management activities of the company (see section 9.2.2.2), it is essential to involve executives in the scenario building process. Even if a set of scenarios is of the highest quality, it is hard for executives to extract deep insights if they see it for the first time after the final scenarios have been developed. There is a personal learning process involved that cannot be skipped without losing some of the benefits.

Often, a leadership team may then find the scenarios intellectually interesting, but ignore them to a large extent when it comes to action. The first Shell scenario teams also had to witness this happening. Instead, managers' mental models need to be dismantled and reconstructed in order to change their actual behaviour (Wack 1985b).

Apart from early involvement, it is necessary to provide bridges within the scenarios to decision-makers' daily jobs and their deepest concerns. If the audience is a regional or divisional management team, it may thus be necessary to augment the scenarios with some details of specific relevance for this group (Wack 1985b, 87f.; Shell 2008, 62ff.).

9.3 Phase II: Rethinking Old Certainties

9.3.1 Problem Context

"Who the hell wants to hear actors talk? The music — that's the big plus about this."

Harry Warner, co-founder of Warner Bros., 1927 (Maney 2005)

Now that Phase I has (hopefully) been completed successfully, top management and other addressees of the scenario planning project can see the Dalmatian dogs and realise that the duck may in fact be a rabbit – or neither or both (compare p. 288f.). It is now that the leadership team needs to rethink old certainties. Business practices and strategies that were known to be effective in the past may not be so anymore – already today or in the foreseeable future.

The quote of Harry Warner on sound in films may seem foolish today, 80 years later. It is not. In fact, what will become technically feasible in the (near) future, what customers will like or buy, and what will be profitable or not, has repeatedly been badly misjudged by acknowledged experts. Prominent examples of such products include the telephone, computers, television (Wikiquote 2011b), and – more recently – compact fluorescent light bulbs (Kanellos 2007) and hybrid cars (Taylor 2006).

Harry Warner's misjudgement teaches us two lessons: First, even – or especially – experienced experts do not always see the "next big thing" or disruptive change coming. Second, one does not have to get the details right in order to move in the right direction: Warner Bros. nevertheless decided to invest in sound technology – albeit erroneously believing that the opportunity lay in music, not voice (Sperling et al. 1998, 91ff.). This is just one more example of why embracing ambiguity is so important.

If senior managers (who tend to be experts in their industry and function) manage to take a fresh look and combine it with their experience, powerful new ideas and reorientations will often be the result. There are two main approaches to successfully rethinking old certainties: top down or bottom up.

The first uses the combined intelligence, creativity, and experience of the top management team and relies on certain workshop techniques. Such types of workshops will be called *Executive Ideation Workshops* below.

The second approach relies on collective knowledge and intelligence: Great ideas are buried in the heads or desks of the many members of large organisations. In some cases, these ideas simply don't make it to the top (see Uzzi & Dunlap 2005); some ideas still need to be combined with others to become useful; and sometimes people are not even aware that they have had a great idea. During *Crowd Ideation Events*, the wisdom of the crowds is collected systematically and then sorted, judged, and rearranged by senior managers. They thereby transform them into viable business ideas (for an overview of the process of innovation see 6.5.2).

9.3.2 Management Tool: Executive Ideation Workshop

As mentioned above, *Executive Ideation Workshops* require a limited number of high ranking participants, one or more moderators, and potentially additional experts to widen the horizon of the group. Osterwalder & Pigneur (2010, 142f.) strongly emphasise the importance of diversity of ideation teams, including the level of seniority (for a summary of empirical research on team effectiveness see 6.3.2.3). However, excluding key executives from the workshop is problematic as they will not participate in the learning process and may later oppose unfamiliar green business model designs that challenge the existing one. Apart from creating top management buy-in, high seniority of the participating managers facilitates the creation of strategic insights. Business unit managers should also be involved if the created insights are to be used outside the headquarters (see Mintzberg 1994, 112). The experts could be internal or external; it is also a possibility to invite company representatives from other industries, customers or other stakeholders to bring in new perspectives.

Independent of the specific techniques used, these workshops function by confronting participants with unfamiliar questions or perspectives. One famous example is General Electric's "destroy your business": CEO Jack Welch gave senior managers the task to develop business plans that were capable of destroying GE's brick-and-mortar-based business models (Kachaner et al. 2008, 14f.).

Various workshop techniques are suitable to foster creativity and strategy development. Three of them have been selected and integrated with Phase I. Furthermore, a fourth technique specifically adapted for Green Business Model Transformations is presented in table 9.2:

Table 9.2 Overview of Executive Ideation Workshop techniques

Technique	Leading questions	Scenario cross-check
A Business model decomposition	• Which elements provide competitive advantages? • What options for change exist (in theory/practice)?	What are the implications of the green scenarios on the analysis?
B Industry analogies	• What would leading firms from other industries do? • What business models do they use?	
C Stakeholder perspectives	• Which are the most relevant stakeholders? • What would they like the company to do?	
D Green business model prototypes	• What green business model prototypes could work for this company? Why / why not?	

All four techniques can be combined or applied independently. For example, a big workshop involving a large part of the leadership team may be designed to feature parallel working sessions, each using a different technique. The results are presented to the other groups and discussed. In a smaller workshop design with fewer participants, a single group may apply two techniques sequentially. A couple of years later, a similar workshop using two other techniques can be conducted.

Although all techniques may in principle be combined with each other, some work together more effectively.

Business model decomposition leads to a deeper understanding of the company's business logic and is thus to be considered the most fundamental technique of the four. Hence, it can be well combined with *industry analogies*, *stakeholder perspectives* or *green business model prototypes*.

9.3.2.1 Technique A: Business Model Decomposition

This technique has an almost compulsory character since a deep understanding of the current business model is a prerequisite for designing new green business models in Phase III. For this purpose, business models are decomposed into components and elements (see 4.4): Each of them is then examined to identify opportunities and threats. Probing questions are used to rethink old certainties (also see Osterwalder & Pigneur 2010, 140f.). This is particularly powerful if used in conjunction with the green scenarios developed in Phase I.

For analytic purposes, the business model definition presented in chapter 4.4 should be used. In management practice, decomposition exercises often consider the value chain only (e.g., Kachaner et al. 2008, 15). Respective analyses usually take other business model components into account only indirectly, although it is sensible to do so explicitly. In order to reduce the task complexity for the participants, the business model decomposition exercise may be split into several parts. For example, one group could examine the value chain and another the remaining business model components. However, it is crucial to revert to the holistic view on the business model at some point. Therefore, this approach is only recommended for business models with rather loosely coupled elements.

Business model decompositions can be carried out as follows: First, a fact base of the business model(s) under examination should be prepared for the workshop. This can be done by conducting interviews with relevant individuals within the organisation. During the workshop, business model components and elements are scrutinised systematically, one by one. Asking the right questions is the biggest challenge. Fortunately, many good probing questions are of a general nature and not specific to a company or industry (Coyne et al. 2007, 73ff.). The following probing questions apply to any business model element and can be used as a guideline (although they may need to be rephrased slightly in some instances):

- Which business model elements provide a competitive advantage? Which are under competitive pressure in terms of cost, quality, speed, responsiveness, etc.?
- Which elements have seen the most / the least changes in recent years?
- What other configuration options are used by competitors? Are they successful this way? Why or why not?
- Are completely different options conceivable? If not, what could be done / needs to change in order to make them technically and financially viable? Could partnerships or alliances help?
- Could competition be forced to follow a business practice to realise industry-wide economies of scale?

- Could competitors be prevented from imitating a new, innovative configuration of business model elements?
- How are business model elements interrelated? (thoughts on this question can be considered preliminary work as it will be central to the approach described in phase III)

Questions can also be tailored to specific business model components like the examples in table 9.3 (partly taken or adapted from Coyne et al. 2007, 74):

Table 9.3 Probing questions for business model decompositions (partly taken or adapted from Coyne et al. 2007, 74)

Value propositions	▪ What is the biggest hassle of purchasing or using our product? ▪ Do some customers spend a large portion of what our product costs them to adapt it to their specific needs? What modifications are they making? ▪ How would we customise our products to every customer if cost played no role? ▪ Could we still meet the needs of a significant subset of customers if we stripped 25 percent of the hard or soft costs out of our product?
Target groups	▪ For which current customers is our product the least suited? Could this situation be rectified by removing one specific barrier? ▪ Which customers use or purchase our product in the most unusual way? ▪ Do any customers need vastly more / less sales attention or cause much higher / lower service and support costs than others?
Key resources	▪ Which of our key resources would also be valuable in other industries? ▪ More specifically, do we create any type of information as a by-product that would be valuable to firms from other ndustries?
Key processes	▪ Which value chain segments create value? Which don't? ▪ What if they were in- or outsourced? ▪ For which value chain segments are new entrants to be expected? ▪ What would we do differently if we had perfect information about our suppliers, partners, customers, competitors, etc.?
Financial logic	▪ If we could price every customer individually, how would we do it? ▪ What would happen if a competitor started a price war tomorrow? ▪ If our current main source of revenue disappeared, what would we do? ▪ How would we cut our cost by 50 percent if we had to?

In order to free workshop participants from remaining mental straightjackets, participants should be asked to pose these questions against the background of each of the green scenarios (split in different groups or sequentially). This also ensures that the sustainability aspect gets due attention, although unrelated insights should of course be taken into account as well. Because of the large number of potential questions, each combined with three to five green scenarios, it is crucial to select only a few questions that are believed to work the best (additional ideas for questions may be found among the SWOT analysis questions of Osterwalder & Pigneur 2010, 217ff.). The project team may want to run an internal test to identify the most promising candidates. Discussing more than two or three questions per group in an hour is likely to lead to shallow results.

During the decomposition analysis, it is important to clarify which business model layer is being examined. For conglomerates, there is a "meta layer" on which individual businesses share resources, processes or customer segments. Such meta layers can represent a competitive advantage or disadvantage vis-à-vis competitors. For example, a company may be able to cross-sell to the same customers by using global key account management. In adverse cases, a company may suffer from uncoordinated customer acquisition – in the worst case, different divisions or regions directly compete with each other (possibly unknowingly). With respect to environmental sustainability, problems may be more subtle. For example, one division may address customers based on a green value proposition, while another undercuts that approach with unsustainable business practices affecting the same customers.

Analyses of individual business models will often be easier and more tangible. When a company applies different business models in parallel, e.g. varying from country to country, it is important to focus on one or a few main variations. However, the full range of models in operation obviously has to be considered in later phases.

The decomposition analysis will likely result in strong confusion among participants: In one scenario a business model configuration works well, while it fails miserably in another. It is unlikely that many dominant solutions will be found for any business model element. If that is in fact the case (and solutions are also determined to be those already deployed), the company either belongs to an extremely stable industry, or, much more likely, the participants have not given up their old mental models yet. Without a change in mental models, consequences of the different scenarios do not get fully thought out. At this point, the moderator or an external expert can try to provide mind breakers that contradict conventional wisdom about the current business model. These mind breakers may be examples from other industries or thought experiments. If this attempt fails, the team can repeat the exercise with a different workshop technique. If this does not yield any results either (and Crowd Ideation Events also seem unlikely to be very fruitful), there is no point in investing more energy and resources. The Green Business Model Transformation should then be aborted.

However, if the expected outcome materialises in the form of shattered old certainties and many seemingly contradictory conclusions, the challenge is to systematically exploit the newly generated insights. These insights can serve three purposes: (1) to create a pool of ideas for new or improved business models, (2) to develop new business strategies, and (3) to develop contingency actions in case certain scenario events eventuate.

Ideas should be judged based on the criterion of possible competitive advantage, not financial impact – this aspect should deliberately be ignored at this stage. Ideally, individual business model ideas are combined into one (or several) coherent draft. This draft will still be a rough outline rather than a sophisticated concept. However, the underlying rationale and main improvements should already be formulated.

Often, new business models also require a shift in strategy in order to be successful. The strategic imperatives, i.e. which business model draft fits to which strategy, have to be clarified. At this point, the exercise may already reveal that a new strategy is required anyway – independent of any changes to business models.

The last of the three purposes is to develop contingency actions. This task reflects the problem of ambiguity; different scenarios often imply different (re-)actions. Several outcomes are conceivable:

First, the circumstances may justify immediate actions to rectify current business model shortcomings. Either the identified risks are intolerable, or risk mitigation measures would require relatively low effort. As scenario veteran Pierre Wack from Shell writes (quote by C. W. MacMalion of the Bank of England): "No time is as usefully wasted as that spent guarding against disasters that do not in the event occur." (Wack 1985a, 146).

Second, two or more alternative paths are possible, but impractical to pursue in parallel. In this case, it may be reasonable to keep alternative options open and delay the final decision on which alternative to pursue (see real options, p. 179). For example, after years of rumours about Macintosh computers with chips from Intel instead of PowerPC CPUs, Apple made a surprising revelation: Apple had developed its operating system OS X for both processors over the course of up to five years. When CEO Steve Jobs officially announced the switch to Intel in 2005, he explained that Apple had waited until it finally became apparent that future PowerPC computers would not be competitive with Intel-based alternatives (Apple 2005; CNET News 2005).

Third, if even a just-in-case option (like the one Apple used) is not feasible, the company should at least prepare contingency plans. If the contingency eventuates, this preparation not only allows for faster reactions but also avoids longer periods of confusion within the organisation.

While the first part of the business model decomposition workshop consists of specific analyses, the conclusion is very similar for the remaining three techniques. Ideally, one or more promising business model drafts have been generated until the end of the Executive Ideation Workshop. Visualisation techniques (e.g., Osterwalder & Pigneur 2010, 146ff.) may be used to discuss these drafts and make them more tangible. However, time will not always suffice to produce tidy and concise business model drafts already during the workshop. For this reason, Phase III includes a step to work up new business model drafts for further analysis.

Figure 9.14 is a flow diagram which shows the typical process of Executive Ideation Workshops for all four techniques:

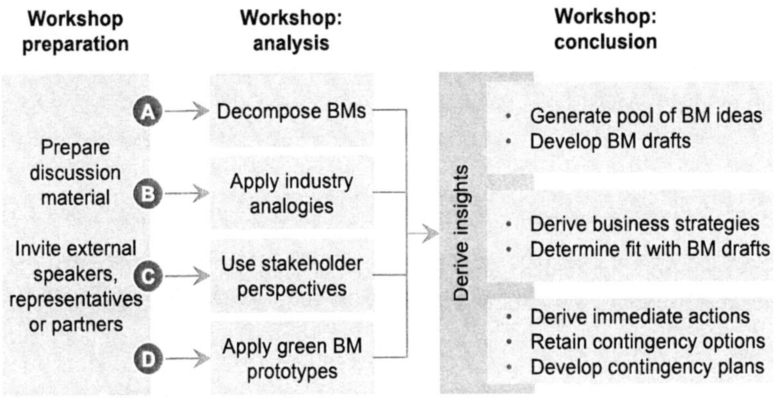

Fig. 9.14 Typical process for Executive Ideation Workshops

9.3.2.2 Technique B: Industry Analogies

Rumour has it that Bill Gates once stated at a computer exposition: "If GM had kept up with technology like the computer industry has, we would all be driving $25 cars that get 1,000 miles per gallon."

GM allegedly later replied: "Yes, but would you want your car to crash twice a day?" (Lowe 1998, 173)

This joke illustrates how some historic, sometimes arbitrary circumstances become unwritten laws within an industry. But why do we accept that operating systems crash so often? Why do cars tend to be larger and heavier than the previous models? For example, today only around five percent of the mass that is moved by precious fuel belongs to the actual passenger.[25]

The industry analogies technique requires more sophistication from organisers than the other techniques. Selecting good analogies is more of an art than a science. For example, industries as diverse as Hollywood and pharmaceuticals both use a blockbuster model and can hence be contrasted with each other in a meaningful way (Kachaner et al. 2008, 15). The differences that can be found in the analogies are often more enlightening than the similarities: Is there an equivalent to a star actor in the pharmaceutical industry? Can the pharmaceutical industry also develop alternative revenue sources like DVD and computer games that complement box office sales? The aim is not to create the perfect analogy, but to think about the company and industry in novel ways.

The selection of appropriate industry analogies is highly specific to the company; hence only limited advice can be given in general terms. Interesting analogies may

[25] This calculation is based on one passenger. For example, a Volkswagen Golf I from 1974 weighed around 800 kg (Wikipedia 2010c). The kerb weight of the Golf VI from 2008 is between 1.2 and 1.5 tons, depending on the configuration (Volkswagen 2010).

be found in industries that once used the same type of business model, but evolved since then. Inspirations may also stem from companies that serve similar customer segments, but with a different type of product, channel or service strategy. Individual companies that have reshaped or created new industries can also be of interest. Examples like Google or Apple have in many ways showed how the hidden laws of an industry can be broken to generate breakthrough success. Johnson (2010, 131) presents a list of 19 types of business models that can be used as a starting point when searching for appropriate analogies.

Once a suitable industry or company has been identified, the organising team should seek to invite some industry representatives to the workshop. Because they do not work for a direct competitor, they may be willing to share interesting insights about the inner workings of their respective industry. A comparison of different value chain segments will be much more stimulating with discussion partners that have first hand experience instead of only desk research type of knowledge.

One special type of industry analogy is the *Model T analysis* (Gottfredson & Aspinall 2005): It refers to Ford's famous one-size-fits-all Model T car from the 1920s and aims to reduce unnecessary (i.e. costly) complexity. The first step is to identify the analysing firm's own (theoretical) Model T product. Then complexity is added carefully to the business model baseline until the cost grows more than the additional benefit from extra variety.

Similarly to industry analogies, but even more detached from familiar business problems, *metaphors* offer insights stemming from different disciplines like sports or biology (e.g., von Ghyczy 2003; Kachaner et al. 2008, 15). Once again, the benefit does not so much lie in seeking a perfect match of the metaphor, but in investigating the realms where it breaks down. However, the risk of overusing metaphors is even greater than with industry analogies.

9.3.2.3 Technique C: Stakeholder Perspectives

Jack Welch's "destroy your business" initiative is nothing else than a variation of the stakeholder perspective technique. In this case, the stakeholder is a competitor or rival. Other potential perspectives include that of customers, suppliers, government representatives, NGOs, shareholders and owners, or any other relevant stakeholder (Kachaner et al. 2008, 16).

The owner perspective may be especially interesting for public companies with a strong investor relations focus. Taking the viewpoint of various types of shareholders can probably help to improve the share price by better meeting their expectations (Kachaner et al. 2008, 16). However, most interesting in the context of a Green Business Model Transformation is taking the perspective of an (imaginary) business magnate who completely identifies him- or herself with the company. What risks would that person take for the long-term success of the firm? Ignoring all the usual obstacles to taking bold and fast decisions within large, political organisations – what would be possible? Such an entrepreneurial view may conflict with the often fast-paced management reality that is dominated by quarterly results. However, exactly this contrast may spur creativity.

Considering the viewpoints of stakeholders who – in contrast to owners – are not necessarily sympathetic to the company can also yield surprising results. For example, the practice of working together with formerly hostile NGOs has seen a strong increase in popularity in recent years. Listening to former adversaries can bring about great ideas on how to make business models more sustainable – both environmentally and economically (Esty & Winston 2009, 183ff.).

The most obvious example for the stakeholder perspective technique relates to customers: Many companies have successfully innovated by putting themselves in the place of their customers or by engaging in intensive dialogues with them (compare GE's "dreaming sessions" in chapter 7.5). Osterwalder & Pigneur (2010, 126ff.) propose to create "empathy maps" of customers in order to understand them better. The activities mentioned above can complement quantitative data from market research and potentially reveal important latent needs and preferences. In addition, firms may animate customers to directly function as innovators (see section 6.4.4). Crowd Ideation Events (see below) are especially suited to utilise this possibility.

9.3.2.4 Technique D: Green Business Model Prototypes

Green business model prototypes are taken from real-world examples of companies from various industries. At first glance, this technique therefore has some similarity to industry analogies. However, using green business model prototypes focuses on possible compatibilities, not deviations from otherwise similar business models. Examples of green business model prototypes are described in chapter 5.5. They include low pollution models, low waste models, dematerialisation models, smart models, servicising models (product-to-service), performance contracting models, renewable models, eco-consulting models, base-of-the-pyramid models, do-good models, health models, and green meta models.

During the workshop, various models are reviewed for their applicability. They can be presented on the basis of real-world examples to make them more tangible. None of the business models will be pure in reality. Often, models are mixed. Moreover, very few companies manage to realise uncompromisingly sustainable business models. Instead, business models will be sustainable in relative terms, i.e. compared to the industry benchmark. Relative sustainability also relates to the question of what impact the business model has on operations of suppliers and customers, not just internally (compare 5.4). For example, Siemens reported emissions of about four million tons of CO_2 equivalents for 2008, but claims to have saved 34 million tons for its customers during the same time period through the environmentally-friendly products it sells (Siemens 2009).

9.3.3 Management Tool: Crowd Ideation Event

Linus Pauling coined the phrase "The best way to have a good idea is to have a lot of ideas." (Uzzi & Dunlap 2005, 2). Many firms try to respond to this circumstance with employee suggestion systems. Results vary from company to company, but all have one problem in common: Problems, solutions, decision opportunities and deciders do not all coincide by chance very often – just as the garbage can model of organisational change suggests (see chapter 6.2.6).

Crowd Ideation Events overcome this problem: For a few days, deciders are made available, decision opportunities are provided, and problems (or problem domains) are presented systematically. The solutions (and further problems) that are buried within the organisation just need to be collected. Depending on the topic, external audiences like suppliers, customers or even the general public may also be invited to participate. As mentioned before, IBM's massive "Innovation Jam" resulted in a one-billion-dollar-venture in green business, among other things (see 6.5.2.1 and 6.5.2.2).

9.3.3.1 Event Preparation

From a project management perspective, there are three major workstreams involved in the preparation of Crowd Ideation Events.

The first is the *content workstream.* It deals with conceptual tasks: defining the goals of the event and which domains are in focus (e.g. a certain division, customer needs, or functional topics like green procurement, etc.). Several topics may be selected, but it is important to structure them adequately: A too narrow focus can prevent the truly innovative ideas from emerging while too little structure makes it hard for participants to contribute constructively and for organisers to evaluate the contributions subsequently (for a practical example see IBM's topic structure from 2006, Bjelland & Wood 2008, 34).

The *event management workstream* is responsible for scheduling, availability of key people and all other conventional event management tasks.

The third workstream, the *technical infrastructure workstream*, may be considered to be a part of the second. But due to its importance and very different required skill set, a separate section is dedicated to technical infrastructure options. One of the biggest risks of failure for such an event is that technology hinders the flow of ideas instead of fostering it. IBM's experiences show that even one of the most technology-savvy companies struggles to master the challenges associated with groups of thousands of collaborating people. Moreover, without smart tools to structure and evaluate the vast amounts of generated information, it becomes unmanageable and remains of limited use (see Bjelland & Wood 2008).

9.3.3.2 Technical Support Infrastructure

Several IT tools can be used, individually and in combination, to cope with the described challenges of mass collaboration:

Ordinary *web pages* usually function as entry points to explain categories and subjects. There is no limit to creativity in this respect. Apart from static text and graphics, Flash animations, recorded lectures or videos can be used. The design of these entry point web pages will influence the materials later produced by the audience: A more conservative presentation of the topics will tend to result in substantiated, but incremental suggestions rather than unprecedented, groundbreaking ideas. The latter can be facilitated by a more provocative design, paired with brain-teasing questions. For example, some of the questions proposed for business model decomposition could be put up for discussion. The risk, of course, is that this could lead to a large amount of irrelevant contributions.

Forums are the classic way to organise discussions online. Forums are structured websites on which users can make posts which other users can reply to. The main problems associated with forums are the many repetitions and comments of low relevance. That is why a thought out forum structure is essential. Capable moderators who steer discussions and provide relevant cross references to related discussions or sources of information also help. However, IBM's experience shows that moderating a continuous forum discussion with many, many participants is very difficult (Bjelland & Wood 2008, 37).

The length of a forum necessarily increases with every contribution, be it valuable or not. This issue is largely resolved by so-called *wikis*.

Wikis are special websites which can be altered by all (or a large number) of its users. This way, users can delete unnecessary information and correct mistakes. Controversies are resolved in separate forums without bothering the reader. The online encyclopaedia Wikipedia is the most prominent example of a wiki to date. IBM used wikis to collect ideas, but also to jointly develop rough business plans of selected, especially promising ideas. However, results were mixed as people did not consistently build on previous ideas of others instead of their own. Moreover, large amounts of parallel access attempts caused technical problems. In order to rectify these problems, the next Jam included real-time chats with limited participation (Bjelland & Wood 2008, 35).

Chats (or *videoconferencing*, once facilities and bandwidths are widely available) can accelerate conflict resolution and thus make discussions more productive. Once key issues have been resolved, wikis are ideal to document and refine ideas and concepts as necessary.

One major shortcoming of IBM's Innovation Jams is the all-ideas-are-equal approach. Participants have no means to quickly identify the most promising ideas before the organising team has reviewed them – after the session is over. A way to overcome this issue is the use of software that features a *voting* system. Dell has implemented a web site for its customers and other interested persons where ideas are posted and users can decide how high they are ranked. Users can both vote for or against a particular idea; older ideas lose points over time as votes on them get less frequent. From the time of its launch in February 2007 until December 2009, more than 13,000 ideas have been posted and received over 700,000 votes. Dell has so far implemented 390 of theses ideas. Since December 2009, Dell also organises so-called "Storm" events that follow a concept similar to the one proposed above: Dell posts certain topics relevant to its business and the community has a limited time to discuss it (Dell 2010).

Dell's approach has the advantage that less time is wasted on discussing obscure ideas – they disappear at the bottom of the ranking. On the other hand, some ideas may be ignored by chance, e.g. because two other good ideas are posted at the same time. Also, some really good ideas may not make sense to many people – the judgement of the crowd will likely be inferior to that of senior managers in some critical cases. IBM tries not to lose these ideas by using *text mining* software instead. It automatically categorises forum entries based on recurring keywords. Human brain power is still needed, however, to make sense of the pre-structured information (Bjelland & Wood 2008, 37).

9.3.3.3 Event Execution and Information Processing

IBM abandoned its two-session design of the 2006 Innovation Jam and instead organised one 90-hour session for the 2008 event (IBM 2008a). A shorter duration is conceivable for smaller organisations – as long as the target audience has sufficient time to attend.

As already explained, the use of moderators to guide discussions is advisable. Furthermore, technical staff must be available to keep the technical infrastructure up and running for the duration of the event. However, as Edward Bevan, the vice president for technology and innovation programs at IBM puts it: "The online portion [...] is the rather the large part of the tip of an iceberg." (Bjelland & Wood 2008, 40). Much more critical is the later processing of the information and translating it into viable business opportunities and business model drafts. As explained, technology can support this process. However, senior managers must become involved at some point – otherwise much of the idea collection effort will be wasted.

Judging from IBM's experience, Crowd Ideation Events – by design – are less suited than Executive Ideation Workshops to create breakthrough ideas. The value lies more in augmentations to existing ideas or businesses (Bjelland & Wood 2008, 39). Moreover, Crowd Ideation Events obviously require considerable amounts of time and money (commercial technical infrastructure solutions are available, e.g. see IBM 2010b; Salesforce.com 2010). However, they are a great method of unearthing the ideas of employees, customers and other community members within a short period of time. Moreover, it may be the only practical way to get such a large pool of knowledge to contribute to challenges as complex and diverse as the ones related to environmental sustainability. Therefore, Crowd Ideation Events are an ideal complement to Executive Ideation Workshops: Green business model drafts and ideas that have been produced can be scrutinised, refined and extended before the design of the actual business models begins in Phase III.

9.3.4 Change Management Challenges

It comes as no surprise that small group workshops face different change management challenges than massive online collaboration events.

Executive Ideation Workshops entail two barriers to success in particular: The necessary commitment of busy, high-ranking managers and an atmosphere of unrestricted creativity during the workshop sessions.

The former can be ensured if the CEO and other board members insist on the importance of the exercise. They should lead by example and ideally participate themselves for the whole time. During the workshops, little psychological tricks can have astonishing effects on motivation as Coyne et al. (2007, 75) illustrate: "We once had the top six executives of a $100 billion company working full tilt because each had bet $20 that his team could come up with the best idea." Any symbolic reward may work just as well.

Creating an atmosphere of unrestricted creativity may be even more of a challenge. One important pre-condition is that hierarchies stay outside of the workshop room. Ideas of the CEO should not count more than any other idea during the session. Simply announcing "no hierarchies" as a rule for the workshop may not be sufficient at a company with a strong hierarchical culture. However, the use of small groups of only four or five certainly helps, especially if combined with efforts of the moderator. Coyne et al. (2007, 78) further suggest putting people who are known to be dominant in one group so that they do not neutralise other, more reserved characters. At the end of the workshop, the most promising ideas should already be selected. Thereby, participants whose ideas are not selected feel less passed over than if it were done later outside of their presence. This already contains one potential source of resistance at the time of the workshop. Nevertheless, later additions to workshop suggestions should always be considered. The workshop or other occasions may inspire some great ideas that surface at a later point in time. It is naïve – and counterproductive – to assume that idea creation and strategy making can be fully captured with any workshop or event format (also see Mintzberg 1994, 108).

If mobilising a group of executives is difficult already, mobilising a crowd of thousands of employees or other individuals can require truly creative approaches. A Crowd Ideation Event should preferably be promoted almost like a product launch. It should be announced repeatedly until the "big day". Incentives for participation like a competition for the best ideas can be a great motivator. After the event, it is important to communicate that the ideas were acknowledged and what is going to happen to them afterwards. This could be done in a similar fashion as employed by delivery firms that allow customers to follow their orders online. If this approach causes problems in terms of confidentiality, the progress of ideas and the winners of the contest may be announced at a later, but predetermined date.

9.4 Phase III: Translating into Business Impact

9.4.1 Problem Context

"Invention is not the product of logical thought, even though the final product is tied to a logical structure."

Albert Einstein (Pais 2005, 131)

Transforming a great idea into a lucrative one is one of the toughest challenges in the business world, even for the most respected companies (see chapter 6.5.2.2). Often, this is due to systemic effects of business model changes that are neglected. For example, customers do not appreciate the seemingly obvious advantages of the new business model; cannibalisation effects harm existing businesses; necessary resources or processes are not available or are inadequate; costs explode due to unforeseen new requirements; intended prices are not achieved; competitors react in unexpected ways. Often, several of these effects occur at once, reinforcing each other (Sterman 2000).

There are at least four reasons why business model design flaws are often only detected once the business model is in operation (the business model may of course also fail due to poor implementation, which is covered in Phase V):

1. Respective analyses are sometimes selective, i.e. certain domains that are not familiar to the analyser are not considered. For example, a marketing expert may be aware of a problem source, but only consider its consequences on branding and pricing. An involved engineer would have understood the effects of that issue on production processes, but is not aware of the problem. This lack of coordination is often structural and not easy to resolve.
2. Another risk is oversimplification. Decision-makers usually do not see much more than a slide show of the proposed new business model (and maybe a prototype if a physical product is involved). Important assumptions are often listed only as a footnote or omitted entirely. Complex but critical issues are thus oversimplified to fit into board meeting format. Even if problems are discussed by specialists, these issues may not make it beyond the middle management layer. Decisions are thus deliberately based on incomplete information (which is, of course, inevitable to a certain degree). Once top management has reached a decision, flawed conclusions are rarely reversed on the basis of previously known issues that have not been adequately considered. However, cognitive simplification processes in strategic decision-making obviously do not only occur at the board level. For a comprehensive discussion of the topic see Schwenk (1984).
3. The desire to conform to the expectations of top management may also play a role in poor decision-making. For example, if the CEO had the initial idea, its advantages may be overemphasised at the expense of well-justified objections.
4. The reasons for failure may also be genuinely unpredictable, for example, when a competitor invents a superior substitute product, thus making the business model obsolete from one day to the next. However, these true

strokes of fate are the exception rather than the rule – for all other cases the proposed management tool for designing business models can significantly reduce the risk of introducing a severely flawed business model.

9.4.2 *Designing Business Models*

Phase III consists of five steps in total. The first two steps will usually have been completed (at least in part) during Phase II. After the last step, the first formal go / no go decision will be made to go ahead with the business model or to abandon the effort. Figure 9.15 provides an overview of Phase III:

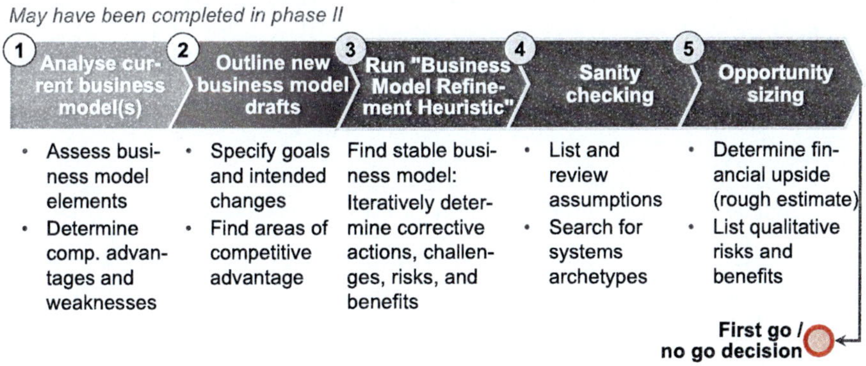

Fig. 9.15 Overview: steps for business model design

As mentioned above, the first step, a thorough analysis of current business models, should normally have already been completed during Phase II. Such an analysis includes the assessment of all business model elements as well as areas of competitive advantages and weaknesses vis-à-vis competitors.

Ideally, Phase II has produced one or more solid business model drafts. If the ideas are still rather vague, the next step is to outline new business model draft(s) based on the previously generated ideas. More precisely, goals and intended changes to business model elements are specified. The thereby envisioned competitive advantages need to be made explicit so that they can later be examined in terms of significance and durability.

The newly developed *Business Model Refinement Heuristic* is executed in step 3. It starts with an initial business model draft and iteratively aims to identify corrective actions that are necessary to implement the business model and overcome practical challenges. It also determines emerging positive and negative side effects in the process. The heuristic stops once no more consequences of previous changes are found. Cumulative changes then represent the final business model. A detailed description of the management tool is presented in the next section.

Step 4 serves two purposes: It provides another quality gate and in addition forms the basis for reliable monitoring of the business model once implemented.

The project team should document all assumptions corresponding to each business model element in a spreadsheet. This list should also include important implicit assumptions. For example, if no specific reactions from competitors are expected, this is also worth noting.

Each business model element of the final business model design should then be categorised according to the following criteria:

- Has the element been changed (yes / no)?
- What is the proposed change (list of changes)?
- What are the underlying assumptions (list of assumptions)?
- What is the confidence level of the underlying assumptions (low / medium / high)?
- How difficult is it to quantify the effects of the changes on financial and non-financial indicators (simple / challenging / very hard)?
- Does the element contain, or is it part of an important feedback loop (yes / no)?
- Can the feedback loop be simulated (yes / no)?
- Can it be tested, e.g. through experiments or with real customers (yes / no)?

The systematisation above will help the project team to provide decision-makers with a balanced view on the new business model design in an easy-to-digest format, e.g. with a traffic light rating.

Step 4 also includes the search for system archetypes (see 4.5.3). Respective feedback loop structures can elude the type of reasoning used in the Business Model Refinement Heuristic and may require quantitative modelling in addition (part of Phase IV).

The final step in Phase III is dedicated to opportunity sizing. The aim is not to develop a detailed economic model at this stage; this exercise is reserved for Phase IV. Instead, the project team calculates the potential upside in rough terms (and communicates it as such). Back-of-the-envelope calculations in a simple spreadsheet should suffice. Additionally, the team lists qualitative benefits and risks for consideration. These risks and qualitative benefits should already have become apparent during steps 3 and 4.

Throughout Phase III, additional research may need to be conducted in order to explore and substantiate different business model design options. This may include desk research, but also first field tests if possible (also see Osterwalder & Pigneur 2010, 160ff.).

The material produced at this point serves as a basis for the first decision-making point. If the potential financial upside and qualitative benefits do not justify the associated risks and effort, the steering committee or responsible decision-making body should opt to discontinue the project. If there is more than one alternative business model design, only the promising ones are to be pursued in Phase IV.

9.4.3 *Management Tool: Business Model Refinement Heuristic*

9.4.3.1 Basic Idea

The proposed methodology is inspired by optimisation algorithms used in computer science. Many conceptually simple problems cannot be solved mathematically precisely because the calculation effort increases exponentially (or worse) with the "size" of the problem. One famous example is the problem of the travelling salesman: A salesman plans to visit a number of cities to do business. To be efficient, he wants to take the shortest route. The number of possible routes increases extremely fast with the number of cities (for example, to connect just 10 cities there are already 181,440 possible routes[26]). Therefore, an algorithm that just calculates all possibilities and then selects the optimal one will quickly become impractical.

Analogously, humans can hardly figure out the simultaneous effects of changes in business model elements on each other (especially if feedback loops are involved). Computer scientists often use fast heuristic algorithms to find a "good" (but usually non-optimal) solution for such problems. I propose to do the same for designing business models.

In fact, managers use heuristics (or "rules of thumb") all the time. Similar looking problems are treated with the same type of solutions, based on experience. However, overly simple heuristics in this sense often break down when the problem is of a certain complexity (Schwenk 1984) – as the design of business models usually is. Hence, the proposed heuristic simplifies the complexity of designing business models differently: It transforms multidimensional interdependencies into a number of single dependencies that are analysed sequentially instead.

A critical observer may argue that such a heuristic contradicts the basic principle of systems thinking to look at problems holistically instead of breaking them apart. There is some truth to this; however, management practice shows that the alternative to analysis is usually not holistic thinking, but no or faulty thinking (be it due to neglect or limited human apprehension). Step 4 (sanity checking) is specifically designed to ensure that the holistic perspective on business models gets due attention.

9.4.3.2 Procedure

The heuristic is based on a number of *rounds*. In each round, the changes or effects on business model elements from the previous round are examined with respect to their further implications on all other elements. Because a business model does not exist independently of its external environment, interrelationships with other important stakeholders – especially competitors – need to be considered as well.

[26] For a symmetrical travelling salesman problem with n cities, the number of possible routes is (n-1)!/2 (if the direction of travel between cities does not matter).

Figure 9.16 shows a simplified, schematic example for the procedure of the Business Model Refinement Heuristic:

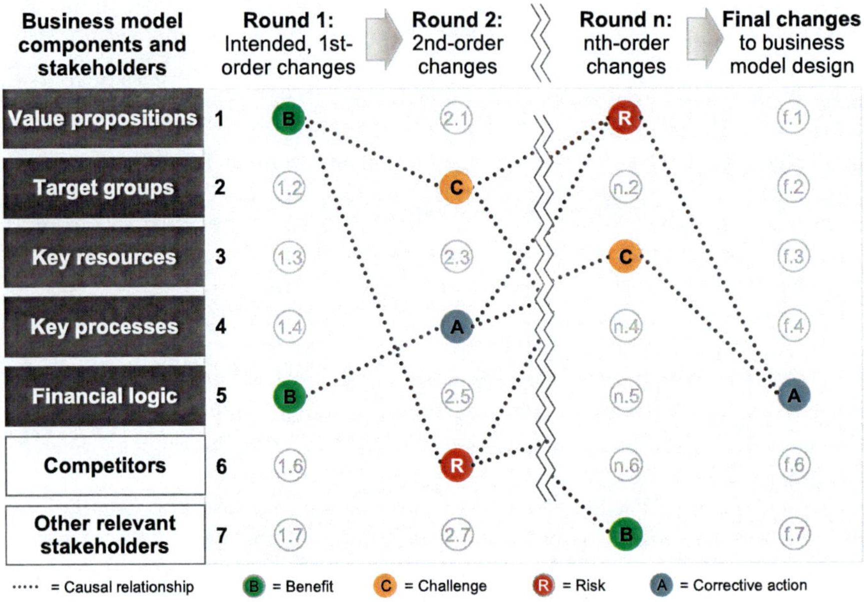

Fig. 9.16 Schematic representation of the Business Model Refinement Heuristic

Benefits, *challenges*, *risks*, and *corrective actions* are distinguished. The difference between challenges, risks, and corrective actions may seem especially blurry at times. A *challenge* is a problem that needs to be solved one way or another if the business model is supposed to work. In contrast, a *risk* represents a problem that may or may not materialise. Risks require contingency actions rather than solutions. *Corrective actions* provide solutions or reduce risks. If a challenge can be resolved with a single corrective action, it may be better for the sake of clarity to omit the challenge as it has no further implications[27].

In figure 9.16, the analysis starts with 1.1 (notation: "round.component"): The business model draft features a new value proposition. Since it is the first round, there are no implications from earlier rounds. The procedure now continues with the analysis of 1.2. However, there are no intended changes to 1.2 through 1.4; hence these components can be skipped.

There is a change to the financial logic (1.5) which needs to be recorded. For example, the business model may include a new source of revenue. Since there are no changes planned for 1.6 and 1.7, the first round ends now.

[27] This decision involves some judgement, of course. A challenge may have no further implications initially, but such implications may arise in combination with other challenges or risks that surface at a later point in time. In case of doubt, challenges should rather be kept until subtle interrelations of any kind can be ruled out.

The second round starts with 2.1. However, there are no implications from either 1.1 or 1.5 on value propositions, hence the procedure moves on to 2.2. And indeed, the new value proposition (1.1) has an implication on target groups. For example, it may require a new, more intense type of relationship with the customer (a challenge). Corrective actions that fulfil this requirement are figured out in later rounds.

There are no implications on 2.3, 2.5 or 2.7, but the key processes component (2.4) needs to be investigated. For example, the new source of revenue may require a delivery process that is more tightly integrated with the customer and thus needs to be adapted. Hence, 2.4 is considered a corrective action that already contains the respective challenge (which is therefore not shown explicitly).

Moreover, there is a risk associated with competitors (2.6). For example, the new value proposition may carry the risk that competitors copy it quickly. Several corrective actions can be determined (in later rounds) that make it harder for competitors to imitate it.

Benefits, challenges, risks, and corrective actions are analysed analogously in the following rounds. This primarily includes implications of the previous round, i.e. 2.2, 2.4 and 2.6 for round 3. However, earlier implications (1.1 and 1.5) should not be ignored completely. Although they have already been checked in previous rounds, these implications may still be relevant in conjunction with other implications that were only found at a later analysis step.

The procedure continues until no more implications from the last changes are identified (this last round is not shown in figure 9.16 since f.5 is the last implication). At this point, all cumulated changes and implications are combined to specify the final business model design.

After the heuristic is finished, initially conceived competitive advantages may have been replaced with others. Some business model elements may still be identical to the initial draft while others turn out quite differently after a thorough analysis.

In practice, the analysis will need to be done at the level of business model elements rather than components (figure 9.16 is simplified for the purpose of graphical representation).

9.4.3.3 Methodical Limitations of the Heuristic

Although the Business Model Refinement Heuristic resolves or mitigates many flaws that are common in business model design, it has its weaknesses, too.

One problem is that the algorithm is path dependent. This means that the order in which business model elements are analysed can influence the result of the procedure. For example, if there are two alternative corrective actions to deal with a challenge (e.g., changing the configuration of a key process internally or entering a partnership instead), the business model element that happens to be first is more likely to be changed[28]. This choice in turn influences later business model design choices.

---

[28] One might argue that the initial choice could be reversed once the second alternative is considered. However, this would again require non-linear thinking, the very difficulty this heuristic is designed to defang.

To mitigate this problem I recommend to assign the task to at least two teams working in parallel. In order to reduce path dependency as much as possible, the order of components and elements should be different for all teams. Once all teams are finished, they can discuss diverging choices to achieve the most optimal solution.

Teams working in parallel also help with another problem: In contrast to computer algorithms, the executing "hardware" is not a machine but a human being; and the analysis steps do not consist of mathematical calculations, but of complex judgements. In principle, the heuristic can be executed by a single individual. However, due to the diverse knowledge domains that business model design touches upon, it is advisable to create teams that can cover most of them. Moreover, the reasoning of one individual is often more error-prone than that of a group. For the same reason, software firms use *pair programming*, which has been shown to improve correctness for complex systems in certain settings (Arisholmet al. 2007).

The heuristic also does not necessarily terminate within a reasonable number of rounds. This may seem like a rather theoretical problem, but unfortunately it is not. New business models may cause price wars with competitors, rely on self-reinforcing loops like word of mouth, etc. Not only do system structures like these require many iterations until their final outcome can be estimated, they also usually elude qualitative methods like systems thinking (compare chapter 4.5.2). In these cases, the use of formal models and computer simulation may become necessary.

Furthermore, some implications are not self-referential, but require in-depth analysis instead, e.g. through market research. All these problematical points need to be recorded and coordinated with the parallel work teams.

In both cases, uncertainties should already be investigated in this phase if doable with minor effort. Otherwise, a rough estimate has to suffice for the moment. The project team must present such issues as risks during opportunity sizing (step 5) and, should the go / no go decision be affirmative, further investigate them during Phase IV for the economic model.

The discussion above shows that the heuristic is not a simple one-day exercise. Some iteration will be necessary when new insights make previous estimates and conclusions obsolete.

Finally, the size of the gap between the found business model design and the optimal one is unknown. In practice, this means that one can never be sure that there is no superior business model design which a competitor can bring to market at any time. Nevertheless, the Business Model Refinement Heuristic systematically considers the complex interdependencies of business model design in contrast to simplistic, one dimensional cooking recipes that are commonly used instead (if at all). To illustrate the heuristic with a real-world example, the following section contains an application to the design of General Electric's Ecomagination campaign.

9.4.3.4 Practical Example: Application to GE Ecomagination

The following is an attempt to reverse-engineer how the design process of General Electric's Ecomagination business model design process *could* have taken place (see chapter 7.5 for background information on Ecomagination). Obviously, the example is not entirely comparable to a real situation since it is done in retrospect – many uncertainties have vanished since then. However, the main goal is to demonstrate the process of reasoning in a comprehensible way.

This analysis is solely based on publicly available information including interviews, newspaper articles, websites, reports, etc. (sources include General Electric 2005; 2006; 2007; 2008; Fisher 2005; Kranhold 2007; Interbrand 2008). It is important to note that Ecomagination must be considered a meta-business model as it spans across many businesses, individual value propositions, revenue models, and so on. Hence, the focus is on implications on this meta-level, not on individual businesses that are included under the Ecomagination umbrella.

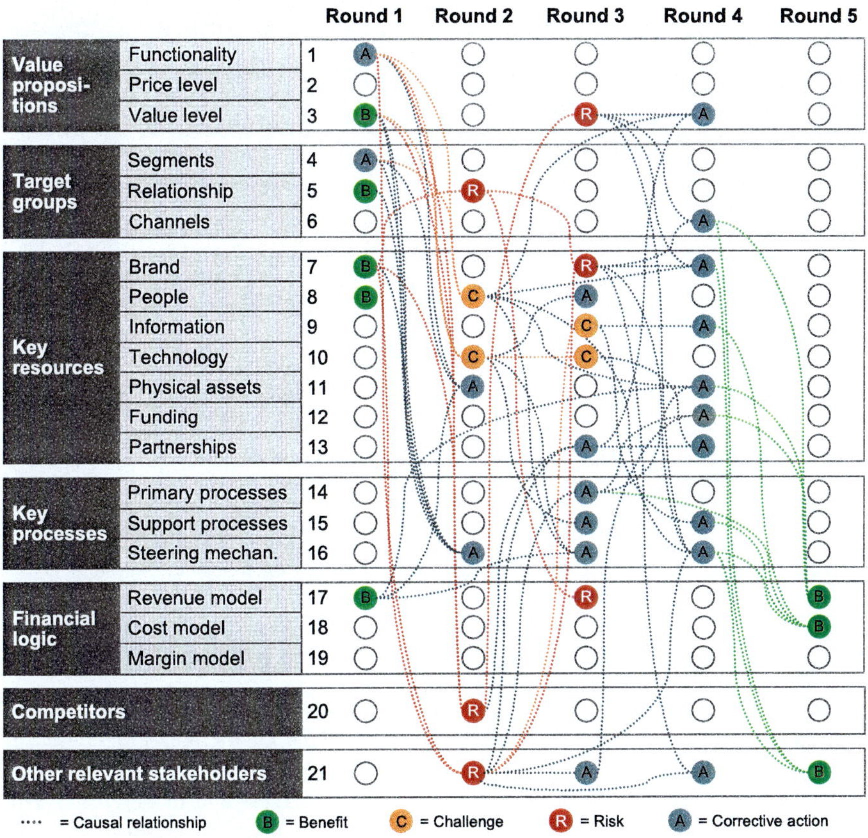

Fig. 9.17 The Business Model Refinement Heuristic applied to GE's Ecomagination campaign

Figure 9.17 presents an overview of the result of the Business Model Refinement Heuristic. The detailed rounds of analysis are explained in the following sections. What can be seen initially by looking at the whole picture is that there are plenty of interrelations (and there are certainly even more than the ones shown). Changes from round 1 lead to several challenges and risks. Round 3 reveals even more challenges and risks. In addition, first corrective actions are determined. Round 4 is fully dedicated to mitigating the identified risks and mastering the challenges. The last round reveals some additional benefits that may not have been expected initially.

Round 1: Initial goals and changes of the new business model

The first round includes the presumed initial goals and changes (see table 9.4). Presumed benefits include increased innovativeness and sales as well as a green reputation with customers, job applicants and the public at large:

Table 9.4 Round 1: Initial goals and changes of the new business model

Step #	Change type	Description	Main causes
1.1	Corrective action	Provide customers with products that help them meet their environmental challenges and improve operating performance	Initial change
1.3	Benefit	Improve value level of some products to "innovative green"	Initial goal
1.4	Corrective action	Address the (supposedly growing) segment of environmentally conscious customers	Initial change
1.5	Benefit	Become a trusted (and thus preferred) partner for these customer in meeting environmental challenges	Initial goal
1.7	Benefit	Be widely considered a sustainability leader	Initial goal
1.8	Benefit	A new green image would make it easier to recruit the best talents	Initial goal
1.17	Benefit	Increase revenue from environmentally-friendly products to US$ 20 billion by 2010	Initial goal

Round 2: Immediate implications

Round 2 is dominated by the immediately arising challenges and risks of the ambitious initiative that GE is pursuing (see table 9.5). Challenges are related to change management (2.8) and technology (2.10). Corrective actions need to be initiated with respect to production (2.11) and sales targets and incentives (2.16). Reactions of customers, competitors and NGOs to outspoken green aspirations represent formidable risks that need to be addressed (2.5, 2.20, 2.21).

Table 9.5 Round 2: Immediate implications

Step #	Change type	Description	Main causes
2.5	Risk	Not-so-green customers may oppose GE's efforts	1.7
2.8	Challenge	Create momentum internally and sustain the change	1.1, 1.4
2.10	Challenge	The aspired product functionality and value level require top-notch technology and R&D	1.1, 1.3
2.11	Corrective action	Production facilities are adapted to cope with changes in demand to avoid backlogs	1.1, 1.4, 1.17
2.16	Corrective action	Sales approaches and incentives are changed to be consistent with the goals of the green initiative	1.3, 1.4, 1.5, 1.7, 1.17
2.20	Risk	Competitors may try to copy or denounce the green initiative	1.3, 1.5, 1.7
2.21	Risk	NGOs will be alert, risk of greenwashing allegations	1.1, 1.7

Round 3: Major corrective actions and more challenges and risks

In round 3 (see table 9.6), a second wave of challenges (3.9, 3.10) and risks (3.3, 3.7, 3.17) are identified – many of which are related to the credibility of the initiative. However, an array of corrective actions can already be identified to mitigate this problem (3.13, 3.14, 3.21). Further corrective actions deal with technological challenges (3.8, 3.15, 3.16).

Table 9.6 Round 3: Major corrective actions, more challenges and risks

Step #	Change type	Description	Main causes
3.3	Risk	Smart competitors may be able to freeride on the green wave, thus negating the aspired innovative value level of GE's green products	2.20
3.7	Risk	Exposed GE brand may severely suffer if intended image change is not successful	2.5, 2.21
3.8	Corrective action	Open new research facilities (e.g., research centre in Munich, Germany) for green technology and hire researchers	2.10
3.9	Challenge	Need to know own environmental performance to reply to greenwashing allegations and become credible	2.21
3.10	Challenge	External technology needed in some fields to make fast progress	2.10
3.13	Corrective action	Hire external consultancy (GreenOrder) to make the green initiative less vulnerable against denunciations and harder to imitate Create advisory board with externals to help steer Ecomagination initiative	2.20, 2.21
3.14	Corrective action	Walk-the-talk: Improve environmental performance in own operations (e.g. "energy treasure hunts")	2.21
3.15	Corrective action	Increase investments in R&D for green products (target: US$ 1.5 billion by 2010)	2.10
3.16	Corrective action	Change incentives for research and product development to create more green products	1.17, 2.8, 2.10
3.17	Risk	"Eco laggard" customers may be lost to low-profile competitors	2.5
3.21	Corrective action	"Keep the public informed" commitment and public engagement to communicate progress and address suspicions of a shallow marketing campaign	2.21

Round 4: Overcoming remaining weak points of the business model design

In round 4, corrective actions for about half of the business model elements are added (see table 9.7). They eliminate remaining weak points of the initial business model and even provide additional competitive benefits, as discovered in round 5.

Table 9.7 Round 4: Overcoming remaining weak points of the business model design

Step #	Change type	Description	Main causes
4.3	Corrective action	Define criteria for an environmentally-friendly product portfolio that can be used to make progress explicit for target tracking and communication	2.8, 3.3, 3.13
4.6	Corrective action	Demonstrate green technology in-house to engage with potential customers and proof of concept	3.3, 3.7
4.7	Corrective action	Initiate a massive, global marketing campaign for several US$ 100 million (TV, print, dedicated website, online ads) that competitors can hardly match	3.3, 3.7
4.7	Corrective action	Use one strong claim ("Ecomagination"), integrated with the overall brand, that spans all efforts and multiplies its visibility internally and externally	2.8, 3.3, 3.7
4.9 4.15	Corrective action	Set up an environmental management system and track resource usage and emissions; initiate company-wide sustainability reporting	3.9
4.11	Corrective action	Leverage all divisions and their resources to contribute to Ecomagination and innovate	1.17, 2.8, 3.14
4.12	Corrective action	Use GE Financial Services to finance GHG reduction projects	3.14, 3.21
4.13	Corrective action	Enter partnerships for environmental key technologies, e.g. hydrogen and CCS (BP), integrated gasification combined cycle (IGCC) (Duke Energy), battery technology (A123 Systems), desalination (Texas Tech University)	3.10, 3.13
4.16	Corrective action	Set aggressive environmental targets (-30% / -1% GHG emissions relative to sales / absolute until 2008 / 2012; energy intensity -30% until 2012; announced in 2008: water usage -20% until 2012)	2.8, 3.3, 3.7, 2.21
4.21	Corrective action	Announce Ecomagination campaign in a joint press conference with a respected NGO (World Resources Institute)	2.21, 3.7

Round 5: Uncovering additional benefits from the business model design

Three more benefits can be expected from the transformed green business model (see table 9.8). The cumulative adaptations from rounds 1 through 5 add up to the final business model design.

Table 9.8 Round 5: Uncovering additional benefits from the business model design

Step #	Change type	Description	Main causes
5.17	Benefit	Additional revenues from synergies and cross sales possible (e.g., Green Credit Card that finances GE's GHG reduction projects)	4.6, 4.11, 4.12
5.18	Benefit	Increased expenses for systems and environmental performance improvements, but even higher savings due to environmental measures	3.14, 4.9, 4.15, 4.16
5.21	Benefit	Influence on political processes due to "green innovator" status	4.6, 4.7, 4.16

In reality, the various benefits, challenges, risks, and corrective actions obviously have not been conceived in the order or even causal chain that is described above. Some adaptations happened years after Ecomagination was launched. Nevertheless, all the stated implications fit into the logical structure generated with the Business Model Refinement Heuristic. As such, it offers a structure to augment business model designers' thought processes, but is not a guarantee for flawless business model designs (I doubt that this is principally possible anyway).

9.4.3.5 Common Considerations for Designing Green Business Models

There are patterns of implications that can be expected to occur frequently in Green Business Model Transformations. The potential benefits and levers for value creation have already been discussed from the angles of corporate sustainability, system dynamics, and Green Business Model Transformations (see chapters 3.2.3, 4.5.4, and 7.1.2 respectively) The following discussion of possible implications thus focuses on challenges and risks rather than benefits. It aims to provide a starting point for project teams that initiate a Business Model Refinement Heuristic. For each business model component, a checklist with probing questions is provided. This checklist aims to function as another quality gate: If the answer to one of the questions is unknown, this may indicate a hidden risk or opportunity. Obviously, the probing questions overlap to some extent with the questions presented in section 9.3.2.1 for the business model decomposition technique (although they are on a more detailed level).

The value propositions component

Changes to the value propositions component are principally subject to imitation by competitors. If the new value proposition is not based on unique capabilities, the firm needs to create other barriers that help to sustain its competitive advantage. Among other things, this may include branding, exclusive partnerships, or smart exploitation of a first mover advantage (see 8.3).

Companies should be cautious about basing a new business model solely on superior product functionality: It is not always the "best" product that wins the market (see Brynjolfsson & Urban 2001). Having the best technology or lowest environmental impact does not guarantee success. Conversely, objectively superior product features are not necessarily a precondition to establishing a "green innovation" value level. For example, the Toyota Prius hybrid represented no real environmental breakthrough in terms of CO_2 emissions. Nevertheless, Toyota managed to turn the Prius into an emotional, personal statement for its buyers and to achieve domination in the (newly created) global hybrid market (see 7.8.1).

However, not all products are principally suited to be marketed primarily as being green. In some cases, other product features are more important and green should only be used as an additional sales argument (Esty & Winston 2009, 129).

Moreover, business model designers should try to avoid green "discrimination": When a company splits its product portfolio into green and non-green products (like GE did), customers may suspect the former to be overpriced (or painted green), while the latter may become discredited as a non-ethical choice (Unruh & Ettenson 2010a, 96f.). One reason why the high-priced Prius did so well in the marketplace is that – unlike other Toyota models – there is no non-hybrid Prius (Maynard 2007).

Companies should also be aware that not all product categories allow asking for premiums. On the contrary, many consumers expect to pay the same or only a small premium at most (Manget et al. 2009, 14ff.). And indeed, green products do not always cost more to produce. For example, new materials may be more environmentally-friendly *and* cheaper. Conversely, companies should also be aware of the dynamics that a price reduction may set in motion in the market.

Table 9.9 includes sample probing questions to check assumptions and help with the business model design.

Table 9.9 Assumption checklist for the value propositions component

BM element	Probing questions
Functionality	▪ Would the intended change to functionality be appreciated / accepted by all customers? ▪ How will customers react? - Increased loyalty, more frequent purchases, higher value purchases, cross-selling effects, or general brand building effects?
Price level	▪ What is the price elasticity? ▪ Could an increase in price result in serious backlashes? ▪ Does the new business model allow for a cheap, no-frills version of the product? Could that damage the other product lines?
Value level	▪ Will a change in value level be recognised as such in the market? ▪ Are there other offerings in the portfolio that could discredit the green value proposition? ▪ Is the new value level hard to imitate by competition?

The target groups component

A product offering has to be in tune with the zeitgeist and needs of the addressed customer segments. This is true for socio-demographic as well as geographical segments. For example, Shell introduced a gasoline called "Pura" as a cleaner burning fuel in Thailand with great success. However, when Shell tried to introduce it in Holland, market success did not materialise. Reduced air pollution and longer engine life represented no worthwhile value proposition for Dutch customers given that the air was cleaner and the cars newer in Holland. Later, Shell more successfully marketed the fuel under the name "V-Power" to provide extra engine power (Esty & Winston 2009, 127f.).

Moreover, shifts in the customer base resulting from Green Business Model Transformations can change the relative importance of different customer needs and create unintended market dynamics. For example, while ethical-minded consumers may be willing to pay more than ordinary consumers, they may also be much more critical with regard to environmental product characteristics than ordinary consumers (Manget et al. 2009).

Another possible effect of shifts in customer segments is that the previous distribution channels may be no longer optimal. In a broad effort to get in touch with consumers, the public at large and other stakeholders, GE made "Keep the public informed" one of its central Ecomagination commitments (General Electric 2006) and systematically measures which channel is most effective in this respect (Manget et al. 2009, 22).

Table 9.10 lists respective probing questions:

Table 9.10 Assumption checklist for the target groups component

BM element	Probing questions
Segment	▪ Could a changed value proposition affect the addressable customer segments? ▪ Is there still a good fit between the needs profile and the value proposition? ▪ Are there stakeholder groups other than customers(e.g., politics, NGOs) that should be considered primary addressees of the value proposition?
Relationship	▪ Does a changed value proposition influence the customer relationship? ▪ Can the customer relationship be changed in the company's favour, e.g. by getting more feedback on products or by reorganising it to be less costly?
Channels	▪ Can the business model benefit from new or a different mix of channels? ▪ Are there channels (e.g., electronic ones) that can amplify the green aspects of the new business model?

The key resources component

Brand is probably the single business model element that is most often associated with Green Business Model Transformations (see Economist Intelligence Unit 2008b, 33). However, few companies have managed to capitalise on the green theme as successfully as GE has: According to Interbrand, GE's brand value has increased by almost US$ 9 billion (circa € 7 billion) between 2004 and 2008. This gain has been largely attributed to the launch of the Ecomagination campaign (Interbrand 2004; 2008).

Too many shallow green claims have created a subliminal scepticism within society regarding green branding (see Interbrand 2008, 13f.). Hence, green re-branding efforts will either be disregarded if efforts are too fainthearted, or prompt stakeholders (especially NGOs) to scrutinise the claims. NGOs may uncover sustainability issues that would otherwise have remained unnoticed (or at least wouldn't have been denounced publicly). In extreme cases this can even result in calls for a boycott. For example, once Dove's ethically-based "Campaign for Real Beauty" turned into a big commercial success, Greenpeace blamed the company for destroying rainforest through its palm oil business practices (palm oil is a raw material used for products in many industries, including personal care). The media attention garnered by the Greenpeace campaign forced Dove's parent company Unilever to react. Senior executives from Unilever met with Greenpeace activists in May 2008. In the meantime, the topic has even made it to the front page of the €40 billion company: Unilever prominently announced that it has discontinued business relations with one of its palm oil suppliers for sustainability reasons (Greenpeace 2008; Unilever 2009b). Moreover, Unilever committed itself in May 2008 to buy only certified sustainable palm oil by 2015, and promised not to pass on any potential extra cost to its customers (Unilever 2009a).

Regarding the people element, all large change initiatives have in common that building momentum and sustaining change represents a notable challenge. In some cases it may be sufficient to meet this challenge with conventional change management measures (compare chapter 6.3). However, some new business models may also require a new corporate culture that can be difficult to realise within the core company and with existing employees. Moreover, many firms have a competencies gap with regard to sustainability issues that may be rectified by hiring experts or conducting trainings.

Product development also needs attention if unfamiliar, sophisticated green technology is involved. This makes technology partnerships and alliances attractive means to realise fast results. This in turn requires managing these partnerships effectively. It should be kept in mind that reputational damage can also be inflicted by the misconduct of partners. This is true even more so for suppliers. For example, IKEA has strict ethical requirements for its suppliers. IKEA realises that it would be blamed by the media for any major scandal at one of its suppliers. However, IKEA also realises that it can never guarantee that all of its 1,400 suppliers follow the standards at all times, let alone the thousands of suppliers of the suppliers (see IKEA 2008, 10ff.). In order to be less vulnerable in the event of an incident, adherence to rules is strictly enforced – sometimes to an extent that may seem excessive (e.g., to avoid accusations of child labour IKEA employs no Chinese under the age of 16, even if they regularly finished school at the age of fifteen and a half).

Supplier management can also lead to new information requirements. New IT systems or process adaptations may be introduced to satisfy these. This also applies to other additional information needs arising from new business models. For instance, more sophisticated green consumers may demand more detailed product information. Potential measures to address this may include creating new product labels and packaging information or setting up a corporate blog and online forums (also see 7.2).

Through investments in new technologies, IT systems, equipment, staff and so on, new business models can burn a lot of money initially. Firms need to be prepared to secure sufficient funding until the business model becomes cash-flow positive. Contingencies should be considered to cope with slower-than-expected progress, especially if dynamic growth processes make anticipated payback periods highly uncertain. It is important, though, to clearly define measures of success up front and find ways to get a proof of concept early on (Johnson 2010, 141f.).

Like partnerships, borrowing existing assets from the old business model can be a good means of reducing investment needs, but can also be treacherous: Borrowing may seem like an obvious choice for written-off, underutilised assets that offer low marginal cost, but can at the same time commit the company to an inferior long-term cost position or technology vis-à-vis new entrants (Christensen et al. 2008, 101f.). Moreover, borrowing from the core business carries the risk of damaging interference, especially if the new and the old business model compete with each other (see also 8.4.3) (Govindarajan & Trimble 2005, 63ff.).

Probing questions for business model elements of the key resources component are listed in table 9.11:

Table 9.11 Assumption checklist for the key resources component

BM element	Probing questions
Brand	▪ Are there any branding risks stemming from the new business model? ▪ Does the new business model add brand value? ▪ Will a green brand appeal to customers (e.g., increase brand loyalty)? ▪ Does it require large investments in marketing to be effective? ▪ Is the existing branding strategy still adequate? Are there any contradictions with the new business model?
People	▪ Do the employees have the necessary skills and knowledge to operate the business model? ▪ Is the existing culture and employee mindset compatible with the new business model? ▪ Can the manpower requirements for implementation and operations be met? ▪ Is the organisational structure still suitable to support the new business model?

Table 9.11 *(continued)*

BM element	Probing questions
Information	▪ Does environmental management require additional information? ▪ Does the business model require any other information (e.g., about customers, suppliers, own operations)?
Technology	▪ Is new technology required (in manufacturing, logistics, etc.)? ▪ How much would it cost (one-time investment and running cost) to introduce the necessary changes?
Physical assets	▪ Can the physical assets from the old business model still be used for the new one? Would that be cost-effective (even when considering running cost only)? ▪ Do capacities need to be adjusted? Should new physical assets be provided or current assets be sold? ▪ How do other companies with similar business models leverage their assets?
Funding	▪ Does the business model require significant up front investments? ▪ When will the business model be cash-flow positive? Are there strong uncertainties in this respect (e.g., due to high fixed costs or exponential growth dynamics)?
Partnerships	▪ Are partnerships needed to acquire the needed competencies (quickly enough)? ▪ Could partners offer additional credibility to the venture? ▪ Are partnerships needed to share big business risks or gain sufficient scale for the business model to be profitable? ▪ What disadvantages and risks would these partnerships encompass? Is there a branding risk involved? ▪ What types of relationship with partners are desired? Are partnerships of strategic nature, tightly integrated, partly competitive? ▪ Can shared interests and lasting cooperation be ensured? ▪ Do the required type, number, or relationship to suppliers change? ▪ Need there be an additional code of conduct? How will it be monitored and enforced? ▪ Should some activities be out- or insourced?

The key processes component

Business model transformations will often require direct changes to one or more key processes. In addition, green business models may require specific further process changes that are less obvious. One important aspect, for instance, will usually be that companies which sell green are expected to be internally green themselves. This can entail significant changes to operations along the whole value chain. Corrective actions can range from green sourcing, to clean production methods and changed marketing and sales processes. In order to manage these changes consistently, many different corporate functions (procurement, production, communications, HR, etc.) need to cooperate closely. Additionally, sustainability is often related to complex, global issues and hence requires global management and potentially collaborating with external organisations (Senge et al. 2007).

Moreover, existing rules and business practices frequently hinder the success of new business models (Johnson et al. 2008, 58f.). For example, new incentive systems for R&D or procurement may be necessary to foster green product development and implement green sourcing.

Table 9.12 lists probing questions for key processes structured along primary and support processes as well as steering mechanisms (compare chapter 4.4.4):

Table 9.12 Assumption checklist for the key processes component

BM element	Probing questions
Primary processes	
Logistics (in- / outbound)	▪ Do logistics processes need to be changed to meet new customer needs or environmental performance criteria?
Operations	▪ Does the company "need to walk the talk" in order to make the business model credible with stakeholders? ▪ Are changes in manufacturing needed? ▪ Can waste be prevented, more products be reused or recycled?
Marketing & Sales	▪ Should marketing change its focus and processes? ▪ Does the sales force need to change its approach? Is training needed?
Service	▪ Does the new business model require more, better, or different service?
Support processes	
Procurement	▪ Do current procurement processes support the new business model (e.g., in order to source sustainable raw materials)?
R&D	▪ Does R&D have to be realigned to meet the new challenges of the transformed business model and make it sustainable in the long term? ▪ Are different incentives or additional budgets necessary?
HR	▪ Can HR benefit from the new business model in its recruiting efforts? ▪ Does the business model require different HR processes to generate critical competencies (e.g. additional recruiting activities, new development paths, training curricula, etc.)?
Communica-tions	▪ Should communication on certain issues be more open or more active in general? ▪ Are new interfaces necessary to collect the required information and coordinate communications throughout the company?
IT	▪ Do IT processes and systems need to be newly set up, automated, or be better integrated to support the business model?
Sustainability Mgmt	▪ Is sustainability management advanced enough to support the business model? ▪ Are sustainability risks adequately managed – also in the light of issues that will gain higher prominence in the future?
Others	▪ Is there a need to change any other processes: Legal, Finance, Accounting, M&A, etc.?
Steering mechanisms	▪ Do central business rules need to change, e.g. to allow for different types of investment ▪ Do behavioural norms need to be modified, e.g. to foster a new corporate culture or reward experimentation? ▪ Are different success metrics needed in order to adequately manage the new business model?

The financial logic component

The effects of the above-mentioned first- and second-order effects on revenues and the profitability of the new business model obviously need to be examined. Conversely, new revenue models can also influence other business model elements. For example, if products are leased instead of sold, this can facilitate a recycling-based business model. However, key logistics and production processes will then also need to be reconfigured accordingly.

Sample probing questions to scrutinise the financial logic are presented in table 9.13:

Table 9.13 Assumption checklist for the financial logic component

BM element	Probing questions
Revenue model	▪ Does the new business model offer additional revenue sources? ▪ Will some revenue sources no longer be available? ▪ Are there major shifts between the sources to be expected? ▪ Does this affect the volatility of the cash flow? ▪ Is there a risk of cannibalisation with other businesses? ▪ What implications does the new business model have on the number of units sold? ▪ How will prices develop? Is price erosion possible? ▪ Considering the change in volume and price: How will revenues develop overall? ▪ Do changes in the customer target segments have any pricing implications? ▪ Are new differential pricing methods expedient? Or, should pricing be simplified?
Cost model	▪ How will unit cost be affected by the new business model? ▪ What economies of scale and scope can be expected over time? ▪ How does the cost position compare to the market? ▪ Does the new business model require higher overhead cost or can they even be reduced? ▪ Are there any cost items that deserve special attention (e.g., because they are subject to uncertainty or strong volatility like certain raw materials)?
Margin model	▪ Does the new business model make some operations unprofitable? ▪ Are these operations necessary to run the profitable parts of the business? ▪ Does the new business model have implications on the overall profitability (e.g., is it more vulnerable to fluctuations in utilisation or does it affect intersegment sales)?

The business environment

The context in which the business model is embedded also needs to be thoroughly considered. This includes current boundary conditions (see 8.2), but also dynamic competitive patterns (see 8.3). Table 9.14 lists probing question about competitors and other relevant stakeholders:

Table 9.14 Assumption checklist for the external business environment and stakeholder reactions

BM context	Probing questions
Competitors	▪ Is the industry ripe for transformative changes? ▪ Does environmental sustainability play a significant role in the industry today? ▪ How are other relevant competitors currently positioned? Are they moving? ▪ Is the new business model beneficial to the whole industry or does it harm competitors? ▪ Does it build on the company's strengths or emphasise the competitors' weaknesses? ▪ Is there a risk of a price war, law suits, or any other means of defensive tactics that competitors might choose? Could this make the business model unprofitable (for all)? ▪ Is there a first mover advantage? ▪ Is it easy to imitate for competitors or do we possess unique and essential capabilities? ▪ Does the new business model increase the risk of new entrants? ▪ What additional entry barriers could be erected? ▪ Can we force competitors to move in a certain direction by introducing the new business model (e.g., by setting standards or destroying the profitability of the old one)? ▪ Would a collective strategy within the industry be beneficial (e.g., through alliances)? ▪ Can the new business model be used to ease competitive pressure?
Other relevant stakeholders	▪ How will the business model affect the nature and urgency of interests of other stakeholders (politics, NGOs, the media and public at large, etc.) ▪ Is the business model heavily dependant on regulatory conditions or public opinion? ▪ Are any stakeholders likely to oppose the new business models? Are there any natural allies, possibly in other adjacent industries or up- / downstream? ▪ Can the company gain political influence by introducing the new business model or use its existing influence to foster its success?

9.4.4 Change Management Challenges

Until now, Phase III was described as if it were merely an intellectual exercise. However, business model design necessarily requires the involvement of a larger group of people apart from top management and the core project team.

Regarding changes of this magnitude, these "outsiders" will likely take sides. They may either welcome the change or resist it, albeit not necessarily openly (see chapter 6.3.4 for a comprehensive discussion of resistance to change). Apart from those who see their own position threatened, resistance will mainly stem from diverging assessments or a lack of understanding during this phase. Although top management must make clear that it is serious about the transformation and that constructive behaviour is imperative, opponents should not be silenced at this stage: their concerns can provide valuable input. Apart from that, communication and education as well as participation and involvement need to be employed in order to gain commitment (instead of only compliance). Hence, the aim should be to build a small, powerful base of supporters that can later carry the change into the broader organisation. Business model design sessions are valuable not only because functional experts and line managers can provide the project team with their knowledge and experience, but also because these sessions offer an opportunity to educate the contributors about the intended change and gain their support. Senge (2006, 206f.) further suggests following three principles to generate commitment: To be committed oneself, to be straightforward about the benefits and downsides of the change, and to let the other person choose whether to enrol (or not).

However, participation and involvement need to be carefully managed. Business model transformation efforts are particularly susceptible to opposition that results from existing mental models: As Johnson (2010, 157ff.) points out, new business models challenge previous views on how the company should operate and make money – hence, the new business model may be rejected simply because it does not fit with the company's current way of doing business and its traditions ('This is against our corporate DNA!'). Another powerful source of resistance is when the new business model directly competes for resources internally or threatens to cannibalise existing revenue. Even if the new business model design is not outright rejected, there is a risk that various involved parties introduce moderate changes in order to make it more compatible with the current model (Johnson 2010, 159f.). For example, the marketing department may request to use a slightly higher pricing and production may want some small design changes to be more compatible with existing products. Seen in isolation, such changes may be relatively small; but in sum they can jeopardise the overall concept. Business model designers need to be very careful in choosing changes to trade for the support of certain internal stakeholders as opposed to insisting on the original design. Hence, it is crucial that the project team has a sufficiently robust management mandate – to say 'no' to high ranking managers, and to request cooperation from affected organisational units. Conversely, Osterwalder & Pigneur (2010, 254) warn that the project team itself needs to avoid "[f]alling in love with ideas too quickly". It should not commit to one particular business model design before other options have been explored adequately.

The decision on how many and which people to involve in the business model design phase contains trade-offs: If too many people are involved, premature ideas and designs may spread within the organisation. This can cause disturbance and unjustified resentment which is difficult to undo later on. On the other hand, if the involved group is too small, a lack of expertise may hurt the quality of business model designs and rumours may spread nevertheless.

Involved people should be picked both for professional and political reasons. Ultimately, the business model designs need to be of high quality in order to appeal to people, and information brokers and influencers are needed to promote them – so that eventually a shared vision can emerge (see Senge 2006, 191ff.). Initiating momentum is thus the main change management task during Phase III. Phases IV and V are about growing and sustaining this momentum.

9.5 Phase IV: Planning for Action

9.5.1 Problem Context

"Before beginning, plan carefully."

Attributed to Marcus Tullius Cicero (106-43 BC), Roman statesman (Weiss 2006, 132)

Highly-skilled chess players plan many steps ahead in order to gain a competitive advantage. A good plan thereby takes into account potential moves of the opponent. Professional players also make extensive use of chess software to study possible combinations and develop their strategy before the game. Despite the enormous advances in computing technology, chess computers were not worthwhile contenders for human world-class players until the late 1990s. While chess computers can access myriads of predetermined combinations and evaluate millions of options per second, strong human players can play out their ability to "see" opportunities and weaknesses in given positions. The real world is much more complicated than a 64-square board with 32 figures which are subject to very simple rules of movement.

In the business world, human ingenuity and luck will always continue to play a significant role. However, like in chess, careful planning can significantly increase prospects of success for Green Business Model Transformations. Analogously, the project team should rely on structured planning methods as well as the intuition of the organisation's skilful leaders.

The role of the project team is to perform what Mintzberg (1994) calls *strategic programming*: Strategies and business model designs are operationalised. Strategic programming translates broad visions and strategies into concrete goals. Means to achieve these goals are broken down into substrategies and detailed action plans. Both of these tasks have already been started during the business model design and need to be taken to the next level during Phase IV. This also involves adjusting budgets, performance controls, and certain policies and operating routines as necessary (Mintzberg 1994, 112).

Mintzberg remarks that strategic programming is not an advisable approach for every company under all circumstances. Relative stability and a need to coordinate "a myriad of intricate operations" (Mintzberg 1994, 112) are prerequisites for its meaningful application. Although stability will not always be present when companies undertake Green Business Model Transformations, the essential need for coordination nevertheless makes formal planning highly recommended. In cases that require maximum flexibility, it may not be the produced plan that yields the most benefits, but the process of planning itself.

Phase IV is divided into three major deliverables: the *economic model*, the *strategic plan*, and the *change plan*. The economic model provides the financial fact base for decisions on strategic choices and their implementation. These insights are fed into the strategic plan that contains a compelling business case with strategic goals and respective measures. The plans should complement each other, forming a feedback loop (Bierman & Smidt 1993, 5). Aggregate plans are then broken down into a detailed change (or implementation) plan.

9.5.2 Management Tool: Economic Model

Economic models have been commonly used for making corporate investment decisions for a long time. All established companies can be expected to have some standard procedure, at least for large investments. Hence, it is of little use for this publication to provide a detailed review of the theory and practice of capital budgeting (e.g., see instead Seitz & Ellison 2004; Shapiro 2005; Bierman & Smidt 2006; Bierman & Smidt 2007). Instead, this section will focus on the specifics of Green Business Model Transformations and present one generic procedure that works, including some suggestions on how the present practice of economic modelling may be enhanced. These include avoiding certain pitfalls as well as methods to deal with a high degree of uncertainty in assumptions and frequently occurring high levels of dynamic complexity of business model behaviour. Figure 9.18 presents an overview of what the procedure of building an economic model might look like:

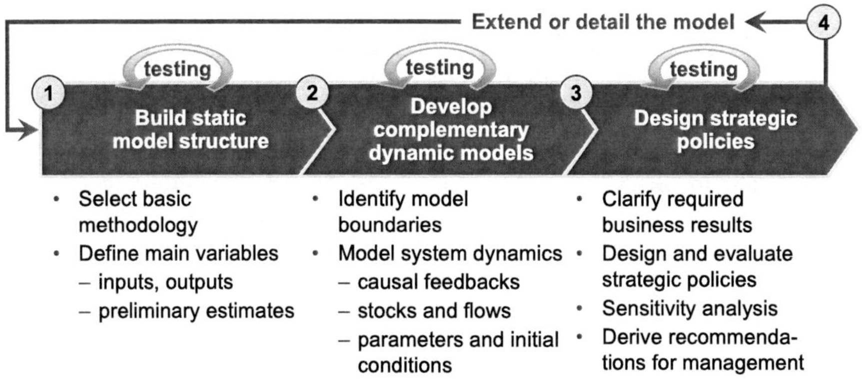

Fig. 9.18 Overview of economic model building

9.5.2.1 Step 1: Build Static Model Structure

A first important decision to be made by the project team is which basic methodology to use for economic modelling of the transformation. In its most simple form a net present value is calculated, possibly compared to alternative cash flows of a base case (e.g., continuation of the old business model).

The complexity of such calculations depends in part on whether or not the transformation can be represented as an independent investment in financial terms, i.e. if there are no or little effects on the remainder of the business: This would be the case if it aims at capturing a thus far unpenetrated geographic market. If the new green business model cannibalises the existing one, borrows its assets or has a general branding effect that benefits the whole company, however, the calculation is more difficult.

Estimating future cash flows is a very difficult task as they are rarely certain. A risk-adjusted discount rate may be used to factor in uncertainty. This can be done in more or less sophisticated ways. A more considerate approach, for example, would assign different discount rates to different cash flow series depending on their individual risk (Bierman & Smidt 2007). However, in the context of business model transformations deciders should not only be interested in the most accurate risk-adjusted return calculations, but also try to understand the causality behind them. Only then can management systematically minimise initial flaws of the business model, reliably evaluate its success in Phase VI and if necessary identify adequate adaptations. This is why the second step, i.e. complementing static evaluation models with dynamic elements, is explicitly mentioned in this framework: the uncertainties themselves are modelled instead of using only correlations and single best estimates. If key uncertainties are well-understood this knowledge can be used to determine a probability distribution of the project (e.g., via Monte Carlo simulation) and therefore give an answer to the question of how likely it is that the NPV exceeds a certain threshold (see Savvides 1994). As a precondition to this second step, however, important variables first need to be determined. Depending on the transformation path (see 8.4.3) this task is more or less straightforward. In case the transformation represents a direct modification of the existing business model most critical variables should be known and only the differences need to be accounted for. For example, Maersk Line had to pay special attention to engine maintenance cost in the course of the introduction of slow steaming as these costs were considered a major risk (see 7.7.2). For other transformation types this task may be more difficult. Osram did not fully understand important business model aspects before it actually started renting out its first off-grid lanterns (see 7.3).

In general it is preferable to have a simple model that covers only a few but important uncertainties rather than a very detailed but intransparent one: The weak points of economic models almost always lie in their structure of implied causalities or false assumptions, rarely in an insufficient level of detail. Once the structure of the model is confirmed to a reliable approximation of reality (through repeated testing), it can be broken down further as necessary. Regardless of whether the project team intends to use sophisticated dynamic models in a second step to understand uncertainties, attempting to create a first calculation with (rough) best estimates at this stage can be informative. The assumptions made in Phase III should provide all necessary information for this task and can be used to scrutinise both the model and the initial assumptions.

In terms of model outputs, it is often preferred to have only a single number, the NPV. However, although external benefits and cost are not part of conventional capital budgeting (Bierman & Smidt 1993, 348) these may be calculated as well. Especially if several alternative green business model designs are evaluated and found to have similar NPVs, an additional quantification of environmental (and social) value could be a highly useful piece of information for decision-making (and potentially even be reused for communication purposes).

The question of how many alternative business model designs are considered is non-trivial as well. Depending on the circumstances, only one particular green business model design may turn out to be relevant during completion of Phase III. The project team should be aware of cognitive biases, however, that can result in ignoring all but one alternative – particularly "in highly complex and uncertain decision environments" (Schwenk 1984, 119). Alternatively, Phase III may generate several variants of one basic green business model design that can then be evaluated financially. For example, alternatives could differ in their distribution model, but otherwise follow the same business logic. Or, Phase III may generate several very different designs. In an iterative manner, economic modelling and business model design may be used to narrow down promising designs efficiently without excluding interesting alternatives too early. The optimal timing and balance between focus and breadth of analysis cannot be generalised, though. Even when determining that only a single business model design is justified, the project team should be wary of the following pitfall: assuming that current profits of the old business model will continue to follow the trajectory from the past. Many business models show diminishing returns over time as competitors innovate or new entrants arrive (see Christensen et al. 2008). In addition, environmentally unsustainable business models can sometimes become economically unattractive very quickly, e.g. due to regulation that internalises environmental cost, declining support from ethically minded customers, or increasing prices for non-renewable raw materials.

Decisions for business model transformations do not always immediately fully determine the resulting path of the company. In a highly uncertain business environment it is obviously preferable to have the flexibility to delay some critical decisions (e.g., whether or not to use a certain technology) until a later point in time when more information is available. The value of flexibility of a business model design is also an aspect that can be incorporated in capital budgeting (Bierman & Smidt 1993, 463ff.). The so-called *real options* approach is thus useful to determine whether a company can create value by buying the option of making certain investments later on (e.g., Amram & Kulatilaka 1999). For instance, the decision to invest in technology for hybrid cars (see 7.8) can be evaluated this way (Avadikyan & Llerena 2010). However, due to certain shortcomings, the real options approach is not appropriate for all kinds of Green Business Model Business Transformations (see 6.5.2.2).

If the business model designs are characterised by high degrees of uncertainties, the project team can try to defang this problem through dynamic modelling. This will not necessarily allow predicting the economic success of a business model precisely, but it helps to understand why the business model behaves as it does, and what levers the firm may use to respond. Hence the first important task of the second step is to decide if and what to model respectively.

9.5.2.2 Step 2: Develop Complementary Dynamic Models

As business model transformations include changing two or more interrelated elements, the presence of dynamic effects is almost certain. The key question is whether these dynamics have significant influence within the problem scope and time frame of interest (Sterman 2000, 97ff.). One example from the industry survey that lends itself to dynamic modelling is the emerging electric car market with all its non-linear uncertainties (see 7.8).

Static economic models rely on estimated changes in inputs (e.g., a certain annual sales growth rate or rate of change in unit cost). These assumptions will usually be based on simple correlations using historic data, market research or expert judgement. In many cases, these approximations are adequate. For isolated actions like the launch of a new product version, the dynamics are usually going to be similar to those that were observed with its predecessors. However, in the case of novel, multifaceted changes the effects may become non-linear. Intuition and historic data can then be strongly misleading as the underlying causality is not understood. Even if the changes and conditions seem familiar this does not guarantee the result will be as expected. For example, certain limits to growth that cause sales to stall may be approached for the first time. In all such cases, explicit modelling of these system dynamics is advisable.

The basics on system dynamics and tools for modelling feedback structures have been laid out in chapter 4.5. There is a tradeoff between making a model comprehensive and keeping it manageable. Therefore, the project team has to carefully determine where the boundary should be drawn. Flows to and from outside the model boundary are implicitly assumed to have infinite capacity. It is critical to ensure that this simplification does not change model outputs in practice (Sterman 2000, 222ff.). For example, if sustainable raw materials could become scarce at some point, a volume-based strategy may have fatal medium-term consequences for the profitability of a business model once bottlenecks materialise and raw material prices rise. If the flow of raw materials comes from outside the model boundary this effect does not show. Instead, the supplier needs to be modelled endogenously in this case to draw the correct conclusions (also compare section 4.5.4 for the example of Bionade and its struggle to source large quantities of organic fruit).

After the project team has specified the model boundaries, the inner working of the model needs to be worked out as explained in chapter 4.5.2 (i.e., diagrams of causal feedback loops, stocks and flows as well as respective mathematical equations). Respective functional experts and experienced managers need to be involved to ensure that the developed causal structure is realistic. This procedure also ensures that relevant people support the results of the model even when it generates results that contradict previous beliefs (Sterman 2000, 84f.).

Sterman (2000) stresses that repeated testing is essential. Not only does this maximise technical and logical correctness, but it also helps the developers to understand the behaviour of the model including its limitations. The finished model should be a good representation of reality, which can usually be tested by comparing it to observable historic data. However, seeking a perfect match is not necessary and may even be counterproductive: A model can never reproduce reality without having the same level of complexity. Hence, bending model parameters and equations to fit historic data sets but without causal reasoning just produces black box models with hidden flaws and limitations to producing non-historic data (Sterman 2000, 328ff.). Such models are not useful for designing strategic policies (as in step 3).

While static models can conveniently be created with a spreadsheet, dynamic models are developed more effectively using special software (for a list of software see Wikipedia 2010b). The outline above is necessarily not sufficient to create good dynamic models in practice. Sterman (2000) provides a comprehensive resource that explains in great detail and with many examples how to create system dynamics models.

9.5.2.3 Step 3: Design Strategic Policies

In step 3, a working economic model has already been created and can now be used to derive insights. To this end, the project team first needs to clarify what financial results management expects so that the project survives the next go / no go decision. The project team can then critically examine assumptions and develop policies that satisfy these requirements. This approach, called *discovery based planning* (see section 6.5.2.2), applies a completely reversed logic from commonly used forward-calculating procedures: In discovery based planning the outcome is not a NPV or other financial measure, but assumptions under which the developed strategic policies fulfil financial requirements set by management. Simply calculating an NPV carries the risk that assumptions are tweaked until the project team believes that the business case seems attractive enough.

During the first two steps, "most likely" assumptions have already been made about parameter values and initial conditions for simulations. These values can be used to validate individual policies concerning issues like sales price, advertising, direct sales, sourcing, etc. (see section 4.5.4 for a summary of potential business model growth engines). The default strategic policy can be derived directly from the final business model design developed in Phase III. However, the economic model might reveal some flaws in the current business model design or the intended approach to its introduction that could not be detected without simulation. In any case, various alternative strategies need to be tried. As a positive side effect, some shortcomings of the current economic model structure may be noticed and corrected.

Often, alternative strategic policies will yield different benefits and drawbacks. One might lead to higher absolute profits, while the other promises higher profitability. Still others may produce positive cash flows faster or be more robust against changes in assumptions. Only a few of the most promising alternatives should be investigated in depth. For each of these alternatives, assumptions are then tightened (or relaxed if necessary) until the minimum requirements from management are met (in the sense of discovery based planning). The variations should be balanced, i.e. assumptions should have the same (estimated) probability. This means that assumptions which are believed to be very robust (e.g., because they are backed by extensive market research) should be varied less than assumptions that are associated with considerable uncertainty (e.g., because they solely depend on somewhat contradictory expert judgements). The project team should continue to monitor all critical assumptions in Phases V and VI.

The gathered insights form the basis of the next go / no go decision. If management is convinced that all minimum assumptions presented are conservative enough and sensitivities are acceptable, it can turn to the question of which strategic policies should be implemented. For this purpose, the project team needs to provide a more complete profile of each alternative. In addition to the "minimum" case, this includes the originally developed "most likely" case, and potentially an optimistic case as well. If the minimum case is close to the most likely case, a worst case can complement the risk profile.

However, in order to make these decisions, both the project team and management must be convinced that the model itself is a good enough approximation of reality. Hence, several iterations to extend and detail the economic model are usually necessary.

9.5.2.4 Step 4: Extend or Detail the Model

As Sterman (2000, 79ff.) emphasises, building models is an iterative procedure (especially if significant dynamic elements are included). The goal is to keep it as simple as possible and make it as complex as necessary to gain confidence in its adequacy. Variables and especially dynamic elements should only be incorporated if they are needed to solve a particular problem. However, it is important to realise that not only the model builder needs to be convinced of the model's usefulness, but also the people who are supposed to act based on its results. Hence, there will be a tradeoff between keeping the model simple and incorporating features considered relevant by other important interested parties. Nevertheless, the project team should seek broad feedback and be open about model assumptions and limitations. Involving critical individuals is the best way to make it more robust and preempt opposition in later phases when changes become difficult to justify.

As a general rule for system dynamics models, Sterman (2000, 81) states that it is more important to set boundaries wide enough to capture major feedbacks between components rather than modelling the components themselves in great detail. If the level of detail is increased, reliable data must be available to make it useful – otherwise the principle "garbage in, garbage out" applies. Moreover, developing useful models requires experienced personnel. A couple of expert modellers will suffice if they get strong support from people who carry the necessary knowledge about the new business model design.

9.5.2.5 Non-quantifiable Aspects

There are important factors that can hardly be expressed in dollar values. For instance, the required management attention and organisational paralysis that transformations entail also have an – albeit unknown – price tag. Conversely, transformations can conveniently be used to overcome outdated structures and inject the organisation with new energy (if successful). In addition, Green Business Model Transformations can increase a firm's social capital, e.g. resulting in increased loyalty of employees, an improved employer brand, greater political influence, or better community relations (see also 7.1.2).

The benefits and costs mentioned above should be made explicit alongside hard financial data. This is not only important to provide a balanced view on the real risks and benefits of the business model transformation, but also highlights the importance of some business model design features and measures to realise "soft" benefits. It is doubtful, for instance, that Osram would have initiated its solar-powered off-grid lighting initiative in Africa if it had offered no environmental benefits – even though these do not improve the financial prospects in any direct, quantifiable way (see 7.3).

Inclusion of hard-to-quantify aspects into the economic model is not advisable, however, unless they can be modelled in a credible (commonly accepted) way. Subjectiveness will usually be too great and provide an unnecessary opportunity for opponents of the project to discredit its economic justification. At the same time, it should never be forgotten that "hard" data may not be as reliable as is usually implied: All too often it turns out to be too limited in scope, too aggregated, outdated or not reliable enough to provide an accurate representation of the business opportunity under examination (Mintzberg et al. 2009, 74ff.).

9.5.3 Management Tool: Strategic Plan

While the economic model requires strong rigor on the analytical side, the strategic plan not only needs to provide a high quality of content, it also needs to become a pervasive, concise document (or set of documents) that captures all essential information on the business case and required (high-level) measures to make it work.

With the help of the strategic plan the addressed managers should be able to fully understand why the Green Business Model Transformation is pursued, how its goals are going to be achieved, what consequences it will have on the organisation and the business environment, and what role their area of responsibility will play in this context. Line managers who are supposed to implement the transformation thus need to gain an intuitive understanding that allows them to make decisions that advance the common goal. Within large and diversified companies, this high-level document can become critical in the light of likely scepticism and lack of knowledge related to green business models. It may also be broader in scope and integrate the Green Business Model Transformation with other current strategic endeavours (although this circumstance is excluded from the following outline).

However, it would be irresponsible of top management to rely on written communication alone. At GE for example, CEO Jeff Immelt heavily relies on personal strategic discussions with his management team to propagate and reinforce new strategic directions like that of Ecomagination. Osterwalder & Pigneur (2010, 170ff.) propose to use "storytelling" as a technique to introduce new business models to managers and engage people. In fact, at firms with strong entrepreneurial figureheads – where visionary leadership and informal communication dominates – a codified strategy may not even be necessary (Mintzberg et al. 2009, 149).

Even in its written form, the strategic plan does not yet contain detailed steps and timelines – these are reserved for the change plan. Figure 9.19 shows possible components of a strategic plan (its exact composition may vary depending on the individual circumstances):

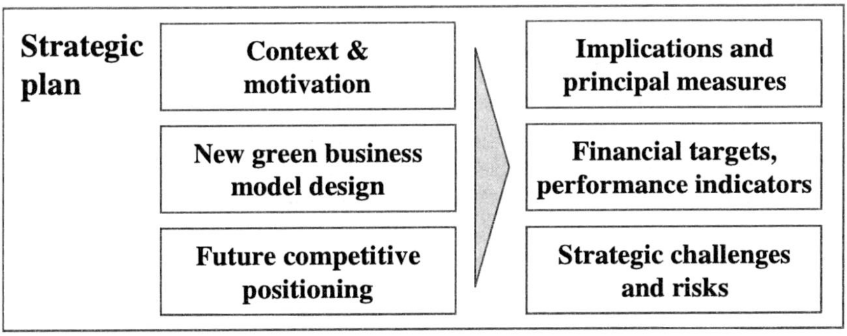

Fig. 9.19 Components of a strategic plan

The first part of the plan provides a brief summary of the context and motivation for the Green Business Model Transformation. This may or may not be limited to economic reasoning. This section should be followed by a concise explanation of the new business model design. The future competitive positioning and strategy is another component of the strategic plan. This can include expected reactions by competitors and other stakeholders (see 8.3) as well as respective contingencies (Casadesus-Masanell & Ricart 2010).

Principal means of reaching that positioning are presented next. This might include whether new capabilities will be developed internally or are intended to be acquired externally through partnerships. The principal transformation path (see 8.4.3) may also be outlined here – for instance, Siemens Building Technologies has identified strategic guidelines for the geographic expansion of its performance contracting business model (see 7.4). Next, the aspired financial results of the transformation over time (e.g., within 1-3-5-7 years) are laid out. Management should consider carefully how aggressive it wants to set these targets. While stretch goals are known to be a powerful source of growth in itself (see 4.5.4), continued underachievement can also paralyse the transformation effort. Conditional goals (if there is a significant exogenous success factor like future regulation) may be a sensible choice in this context. Regardless of how aggressive the communicated targets are, strategic challenges and risks need to be brought to the attention of responsible managers. They should ideally also be provided with the means to identify and counter these challenges and risks in practice. A sensible definition of performance indicators for business model monitoring should also be part of any strategic plan.

The strategic plan needs to be embedded into the regular annual planning process. If necessary, previous strategic plans need to be updated. Moreover, implications for the overall strategic direction are to be explored – even if the Green Business Model Transformation focuses only on one of several divisions.

9.5.4 Management Tool: Change Plan

The management tool called *change plan* in this book refers to a master plan in which everything that is needed for implementation is arranged. A change plan may be considered similar to what is sometimes referred to as a *programme plan*, i.e. a plan that coordinates several interrelated projects. However, in this context a change plan is defined as the entirety of plans on the lowest level of detail to be used during implementation. This terminology is not universal, nor is there a common agreement on what components a change plan should be composed of. The PMBOK and PRINCE2 are the two most widely recognised standards in project management (PMI 2008; OGC 2009a; 2009b). If a company uses one of these standards or a comparable one it should by all means do so for the Green Business Model Transformation. However, these project management frameworks are quite extensive and hence I will refrain from introducing them within this work. Instead, the broad planning tasks to prepare the implementation are outlined below.

Almost fifty years ago, the Stanford Research Institute proposed an extensive "system of plans" that encompasses all plans necessary to run a corporation (Stewart 1963; quoted in Mintzberg et al. 2009, 55f.), Meanwhile, management science has moved away from the temporary primacy of planning (Mintzberg et al. 2009, 366f.). Hence, the proposed change plan is meant as an optional tool for companies that are used to stringent central planning. The categorisation presented in figure 9.20 is used as it is comprehensive and yet comparatively simple. A change plan consists of several parts; the exact composition will depend on the company's management structure as well as the focus areas of the business model transformation.

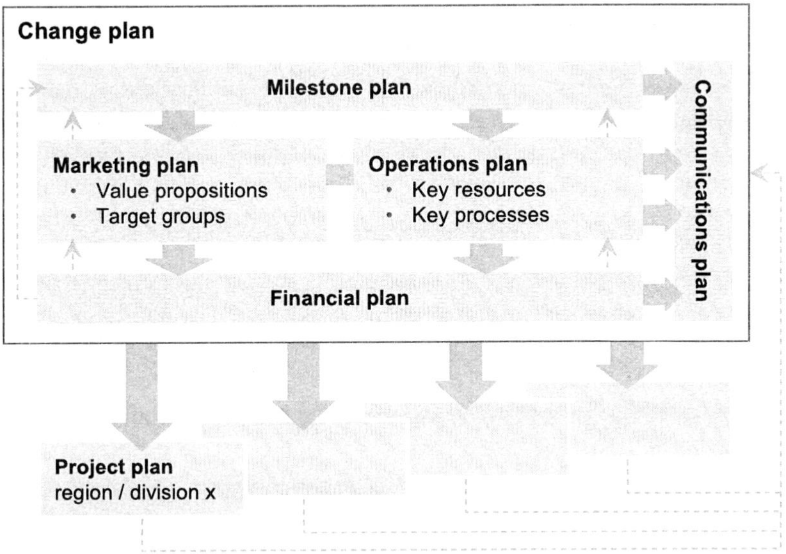

Fig. 9.20 A generic change plan and its planning modules

The milestone plan is a high level timeline that integrates all sub-plans and allows programme managers to identify critical interdependencies and delays quickly. It contains only the most important events for the purpose of clarity. The milestone plan further sets the pace for the *marketing plan*, the *operations plan*, and the *communications plan*.

The *marketing plan* is concerned with the successful managing of the value propositions and target groups components of the new business model. This includes an analysis of markets and customers (which is probably already available to a large extent by now), the marketing strategy used to capture these, and concrete measures for implementation (e.g., the marketing mix), including necessary budgets. In case the new business model requires extensive new product development, a separate *product development plan* may be appropriate.

The *operations plan* is similar to the marketing plan in its purpose with the difference that it covers the business model components of key processes and resources. If there is a critical resource or process that requires special attention, a separate plan may be sensible in this case, too. Examples could be an *organisational plan* for the arrangement and development of human resources (e.g., founding a separate incubation unit). Both the marketing plan and the operations plan feed back into the milestone plan (e.g., if detailed planning reveals that certain activities take longer than previously estimated). Analogous feedbacks need to be considered for all other plans.

The *financial plan* is informed by the economic model and consolidates the financial effects of the marketing and operations plan. It formalises the revenue, cost and margin model of the new business model and its implementation cost. It also specifies how the transformation is financed.

The *communications plan* coordinates all communication with internal and external stakeholders. Although it directly depends on the content of the other plans, it is listed as a separate component because the critical nature of communication is frequently underestimated in organisational transformations. The communications plan includes a systematic stakeholder analysis and a detailed schedule specifying who is to be informed when about what in which way.

All plans presented above should feature a risk management component. In addition to strategic risks that are explored in the strategic plan, the focus should be on implementation risks. For example, this may include potential delays in key process reconfigurations or difficulties in obtaining critical resources. A company may as well decide to consolidate all risk management topics into one separate *risk management* plan.

Not only do Green Business Model Transformations require changes to multiple business model elements. These changes may also need to be implemented in different business units, divisions, regions or countries. Hence, the change plan usually requires a whole portfolio of implementation projects. Respective individual project plans translate the overall change plan into detailed planning items like a work breakdown structure, project schedule, staff requirements and assignments, and cost.

Once individual budgets for the transformation get prepared, the undertaking may hit business reality: All resources of a company – not only capital – are limited. This may lead to bottlenecks and delays. After all, the Green Business Model Transformation will also compete for resources with unrelated projects – which underscores how important it is that the business case is substantiated. Phillips & Bana e Costa (2007) identified five key challenges for resource allocation that can be summarised as follows: (1) There are multiple (often conflicting) objectives; (2) information on available options is incomplete; (3) separate allocation for different organisational units rarely leads to the best possible solution overall; (4) personal interests of the many involved or affected individuals are conflicting; (5) those who disagree may defect, leading to hidden, counterproductive activities or projects.

While these five challenges seem to call for strictly centralised management, instructing responsible managers in detail on how to implement the new business model is a recipe for disaster: The risk that these managers get detached from the cause of the effort is too great. In addition, some decision autonomy is needed to account for local conditions. The appropriate balance between central and decentralised management of course depends on the company's structure and culture. Phillips and Bana e Costa (2007, 5) emphasise that ultimately human judgement is the dominant determinant for resource allocation decisions in practice, "with models informing that judgement". This implies that change management should always be paid great attention to and may often be more critical than finding the theoretically optimal solution.

9.5.5 Change Management Challenges

Placing a bet on the success of a major transformation is a difficult decision to make for a top management team. This is particularly true if there is no "burning platform" (see 8.4). However, the example of GE's Ecomagination campaign (see 7.5) demonstrates that it is possible to win over a sceptical management team for a seemingly daring transformation. BMW's project i provides another example (see 7.8.3): In order to gain support for the initiative, BMW used company events with former American Secretary of State Madeleine Albright and German Foreign Minister Joschka Fischer to make its employees question the old way of building fast, but polluting cars (Reiter 2010).

The required receptiveness within the top management team has hopefully already been created during the earlier phases. However, Phase IV puts dollar values on ideas, affects budgets and requires difficult decision around prioritisation. Conflicts will inevitably arise, and if high ranking managers do not demonstratively commit to the transformation or even oppose it openly, resistance may spread quickly. The CEO and other board members need to prevent this from happening, possibly even by means of displacement of certain resisters.

Ideally, measures like the ones mentioned above can be prevented altogether. At some point during Phase IV it is inevitable to inform a larger part of the organisation (and possibly even the public) about the planned changes. Since Green Business Model Transformations are capable of creating special enthusiasm due to their sustainability aspects, management should try to use that fact to build broad momentum.

However, large transformations are often approached in an overly technical way; symbolic actions like greening the headquarters, sponsoring certain events or launching a company-wide project in which employees work together to improve environmental performance can be strong signals. If employees start to feel emotionally attached to the project this can create very strong motivation. However, if these symbolic actions are not accompanied by the necessary "tough" decisions that may cost a lot of money or require sacrifices from top management, enthusiasm can turn into cynicism.

A related problem that occurs specifically in firms focused on analysis and planning is the assumption that Green Business Model Transformations can be planned exclusively from the desk of a central planning office. In some cases, deep practical knowledge is needed to prepare meaningful plans (and business model designs) (Mintzberg et al. 2009, 39ff.). Osram had to learn this lesson when it realised during implementation that some aspects of their business model did not create the response from customers they had hoped for (see 7.3). Respective learning is done in the field (by going to factories, meeting customers and suppliers, etc.) and can be transferred though the involvement of practitioners. It is critical to identify areas where pure analytical thinking would lead to "ivory tower results". Moreover, there is a risk that overly meticulous planning results in extensive control that destroys all entrepreneurial spirit at the lower levels of the organisation. On the other hand, Siemens Building Technologies made mistakes during its expansion because it followed an approach with too *little* structure. It is therefore the responsibility of top management to strike the right balance between analysis and experimentation, between thinking and action.

While current strategic planning tools and practices aim to avoid many pitfalls of misguided planning efforts in the past, some problems have not yet been eradicated within all organisations. Wilson (1994, 13) mentions various potential causes of failure, including the following: Project staff may take over the planning process, thereby excluding executives and line managers who need to implement the transformation. Or, the planning process itself becomes too dominant, so that analysis displaces true strategic thinking. Ever more sophisticated methods and systems only aggravate this trend. With respect to the development of system dynamics models Sterman (2000, 79ff.) suggests 12 principles designed to prevent inconsiderate use of this tool. Moreover, Wilson warns that planning may ignore cultural and organisational specifics: Some business model designs will be very successful as a slide show or spreadsheet, but very difficult to implement for a given company. It is a key change management task to prevent the issues described above from happening.

9.6 Phase V: Making it Happen

9.6.1 Problem Context

"Words on paper do not transform a company." (Mintzberg et al. 2009, 42)

Up to this point changes have been theoretical in nature. Phase V comprises the actual execution of the transformation – the business model now has to stand the test of reality. Corrections subsequently become much more difficult to make and design flaws cost real money.

Traditionally, design and implementation have been considered two separate tasks in strategic management. While some business model transformations will lend themselves to such a separation it is generally artificial: Within this framework the separation merely exists for the sake of conceptual clarity. Especially in unstable or complex business environments, however, this separation is not expedient in management practice (Mintzberg et al. 2009, 39ff.). Experience shows that often considerable experimentation is necessary to get the details of a new business model right (Sosna et al. 2010; McGrath 2010).

There are a large number of conceivable types of Green Business Model Transformations (see 5.5) which can entail very different implementation processes. Historic backgrounds of firms further add to this variety. It is thus not possible to cover this phase comprehensively on task-level as has been done for earlier phases. Considering that the scope of this work is focused on large, established companies this level of detail would be of limited value anyway: The vast majority of companies can be expected to have gained experience with other large-scale transformations in the past (although this does not mean they pose no challenge anymore). For this reason Phase V is kept intentionally short: Apart from some general points, the following discussion will pick up selected observations from the industry survey and illustrate the specifics that firms were facing depending on the kind of corporate journey.

9.6.2 General Considerations for Implementing Green Business Model Transformations

In chapter 6, supplemented by chapter 8, various perspectives on organisations and approaches to change have been presented. Whether a firm is best described as a machine organisation (see Mintzberg et al. 2009, 324) or whether it more closely resembles an organised anarchy (see Cohen et al. 1972); whether it is still relatively early in its lifecycle or has developed sophisticated means of collaboration over the years – such factors obviously play a role for the appropriate implementation approach.

The spectrum ranges from approaches that rely on highly detailed, prescriptive specifications versus ones that emphasise organisational learning and experimentation (in which case the distinction of implementation to earlier phases becomes blurry). Both ways will probably fail if pushed to the extreme. From a content perspective, central coordination and control can ensure consistency across large and complex organisations, but will inevitably lead to some unforeseeable, adverse consequences in the field. Instead of using thick handbooks and trying to determine every detail, management may rely on rough guidelines and rules to steer behaviour of middle managers and other employees implementing the transformation. Most advocates of complexity theory would probably argue that even these will have unpredictable outcomes (compare 6.2.7.1). Some scholars have thus proposed steering behaviour by establishing a shared vision instead (e.g., Senge 2006, 191ff.; also see Kanter 2008). This way, not the means but the ends are prescribed and implementers can decide themselves which course of action best serves this purpose. With its successful Ecomagination initiative, GE seems to have struck the right balance between decentralised autonomy and central direction. While the means of contributing to the initiative remained largely the responsibility of divisions, the strategic direction was personally communicated to responsible managers by the CEO (see 7.5).

Phase IV presented a rather rigid approach to planning that firms are supposed to relax as appropriate. However, even for organisations with formal planning procedures it is incorrect to assume that a Green Business Model Transformation can be a pure top-down initiative created at the drawing board. Every transformation of this kind relies on emergent learning that has occurred before. In some cases, this learning is deliberately exploited through Crowd Ideation Events (see 9.3.3). In other cases competencies generated in the past are needed to operate the new business model. Fostering this learning during the implementation phase is often critical (for general factors that foster learning within organisations see, for example, Senge 2006). Learning should usually focus on "testing the assumptions underlying what is being done" initially (DiBella et al. 1996, 374), and later move to improving respective practices (compare *learning focus* as one of seven learning orientations described by DiBella et al. 1996). Building necessary competencies can be facilitated through experimentation (like Osram did with its O-Hubs, see 7.3, and Maersk Line with slow steaming, see 7.7.2). Other learning orientations (e.g., product vs. process focus or the value chain focus) will depend on the nature of the business model and the firm. If the new business model requires significant unfamiliar competences, firms may use external knowledge sources and enter partnerships (like the technology partnerships for EVs in the automotive industry, see 7.8), or simply buy it on the market (like GE hired an environmental consultancy for Ecomagination).

Environmental issues often span across value chain segments. Combined with silo thinking the complexity of environmental issues can add an additional layer of complication to Green Business Model Transformations. This can lead to uncoordinated activities with highly undesirable effects, as Esty & Winston (2009, 244ff.) illustrate: For example, Toyota's government relations staff undermined the company's aspirations to be an environmental leader by lobbying against raising fuel-efficiency standards in California. Similarly, GE was criticised for advocating against banning incandescent light bulbs. GE subsequently sought to coordinate its business, regulatory and eco-marketing goals more actively (Kranhold 2007). Another common issue is if implementation lags or contradicts public environmental claims. When Ryanair's Michael O'Leary had to take back a previous claim that the company had reduced CO_2 emissions by 50 percent over five years (Shankleman 2007) that may have been just embarrassing (they have increased during that time). If a similar faux-pas happens in the context of a green business model this may be its death sentence. This also explains why Frosta put so much effort into ensuring it can actually adhere to its "purity command" (see 7.2.1). Like managing partnerships (see above), the ability for internal coordination and cooperation will be a crucial competency for many Green Business Model Transformations.

Companies with general deficiencies in managing environmental sustainability may put their new green business models at risk. Possible ways to mitigate such shortcomings include establishing formal sustainability leadership (e.g. with a Chief Sustainability Officer position), improvement of sustainability assessment and reporting as well as business integration of sustainability (e.g., through training, compensation, guidelines, etc.) (Lubin & Esty 2010, 47ff.).

9.6.3 Specific Considerations for Different Corporate Journeys

In chapter 8.4.3, different possible transformation paths have been presented, based on three patterns of progress and four mechanisms that represent the shifts in business activity. The pattern of progress is either *continuous*, *staged*, or represents a *"big bang"*. The mechanisms include *modifying* the existing business model, launching a new model that *cannibalises* the old one, introducing a *complementing* business model, or one that is *synergetic* with the existing (non-green) business model. Each combination entails a different implementation process and respective implications are briefly discussed below.

A big bang approach in which the business model is introduced in every applicable geographic market for all applicable divisions at once guarantees maximum attention from customers and the media, but also requires the most sophisticated project management and carries the greatest risks. Due to the great coordination challenge the probability of mistakes is high, and – if they affect the transformation in general – the consequences may be especially grave.

Frosta provides an example of a big bang approach (see 7.2.1). It was important to change all products under the Frosta brand at once. Otherwise, the repositioning could have been diluted and miss its intended effect. Conversely, Frosta had to endure massive losses in revenue initially that it did not see coming.

GE's Ecomagination (see 7.5) was designed as a global, company-wide initiative and thus also follows a big bang approach. The overall framework of the initiative could hardly have been limited for regions or divisions initially. However, the content of the initiative was far from completed on the date of its initial launch: For example, the certification process for so-called "Ecomagination products" was not completed: Only some of the existing products that were later included in the Ecomagination portfolio were featured initially (General Electric 2005, 14f.) – perhaps a deliberate choice in order to give the impression of greater progress in later years. Moreover, GE added a company-wide water target at a later stage (General Electric 2007, 2). Maybe the topic had not been deemed important enough at an earlier time. In any case, the decision to concentrate on energy and greenhouse gases reduced the complexity of the initial rollout.

Other transformations not only defer certain aspects of the transformation, but are intentionally designed to happen in several steps. Such a *staged approach* reduces planning and implementation complexity and allows for correcting initial flaws in later stages, but obviously takes longer. This may also dilute the competitive advantage vis-à-vis competitors, giving them more time to respond. In some cases, such as for utilities that want to switch to renewable energy generation or chemical companies that want to switch to green chemistry (see 7.6) a staged transformation is inevitable due to the nature of affected assets. However, acquisitions (like that of Cognis by BASF) can speed up the process.

Siemens Building Technologies (see 7.4) followed a staged approach with performance contracting and stretched geographical expansion over a period of many years. This way, learning from early mistakes and failures could be taken into account. The overall growth is therefore considered exponential: limited growth with lots of experimenting followed by scaling of proven implementation approaches.

Johnson (2010, 135ff.) describes a stage model consisting of *incubation*, *acceleration*, and *transition*. He emphasises that during incubation early testing of assumptions in the field is often more valuable than meticulous planning. Osram also learned this lesson with its first off-grid lighting operations in Africa (see 7.3). Moreover, interferences from the core business should be avoided during that time. Next, acceleration is about making processes of the now proven business model repeatable and efficient, leading to strong growth in revenues (and eventually profits). Although Osram has started to open additional branches it has yet to reach this stage. During the last stage of Johnson's model, transition, the new business model is either reintegrated into the core business or permanently established as an independent unit – depending on which option promises to be more conducive for long-term success (this does obviously not apply to transformations that modify the core business directly, like that of Frosta). Siemens Building Technologies had to recognise that reintegrating new performance contracting units failed if they had not reached a critical size and stability at the time.

The last pattern is continuous progress. Nestlé follows this approach (see 7.2.2). However, unless the breadth and speed of change is not substantially higher than that of rivals it is not in scope for this work (in the sense of a Green Business Model Transformation).

In terms of transformation mechanisms, substituting (i.e., modifying and cannibalising) transformations are the most risky. The example of Frosta has already been mentioned. Maersk Line first conducted extensive tests before it introduced slow steaming more broadly. A major concern here was that its assets (i.e. ship engines) could be damaged. However, going back to previous practices would be difficult at this stage should economic conditions change (e.g., a dropping oil price or increasing demand). Not only would this provoke criticism from environmentalists, but its investments in ships and equipment would lose their value (see 8.2.1.3). Lufthansa faces similar challenges in the context of switching to biofuels. Its new test route with a biofuel mix can be considered a technical as well as commercial test (see 7.7.1). Implementing a cannibalistic business model leads to various questions around timing its introduction and protecting it from the core business, e.g. by pursuing an "ambidextrous" organisational approach (see 8.4.3). Furthermore, these business model transformations may not be reversible if the economics turn out to be poor: Even if the firm discontinues business model operations and cuts its direct losses, a competitor or new entrant may still choose to pursue it – potentially destroying the profitability of the old business model this way (see 8.3.3).

Complementing business models like Osram's off-grid lighting business (see 7.3) or synergistic business models like the sustainability services of BASF (see 7.6) are less problematic as firms usually risk losing only their direct investments. However, these business models will sometimes have difficulties in securing the necessary resources for implementation. For synergistic business models, it is also crucial to ensure that customers are made aware that they also benefit from the synergy. Otherwise they may feel exploited. For example, BASF needs to ensure that customers do not feel too dependent as BASF may recommend its own products in the course of its environmental consulting services.

9.6.4 Change Management Challenges

The implementation phase to a certain extent will be a balancing act between realising the business model as it has been designed versus accommodating internal political pressures, external stakeholder demands, and design adaptations due to unanticipated business factors.

Change management is responsible for achieving the appropriate balance. As already seen during the design phase, affected parties may request changes to the business model design. Not all of them are easily assessed for their legitimacy: Some may seem small or make sense in isolation, but could still have adverse effects on the viability of the business model overall (see 9.4.4). In addition, some interventions may be less well-intended: Particularly for business models that cannibalise the old one (see above), resisters may try to damage the new business model (Johnson 2010, 160), e.g. by denying access to critical resources.

Change management is thus not only responsible for ensuring that new business model components are implemented correctly, but also needs to ensure that there is no organisational crisis. Internal champions that were already involved in the design phase can smooth transition between the design and implementation stage. Moreover, as mentioned before, testing assumptions as early as possible can help avoiding the need for corrections and guard against (justified) criticism from resisters. The emotional side of the change effort should also not be underestimated in its importance. If employees feel that management is being overly secretive, insincere or ruthless, even a striking business case is hardly sufficient to generate widespread support for the transformation. Management should try to prevent such developments through early, extensive, and honest communication and respond quickly if first warning signs appear. The full array of change management measures can be used to advance the transformation (see 6.3.4).

The type of corporate journey (see 8.4.2) also influences the nature and magnitude of change management challenges during implementation.

In the case of a "drowning man" type, the organisation will usually show a strong sense of urgency. The direction may be contested, but the principal need for change will usually be accepted. However, due to the pressure it is important to stay ahead of things and avoid entering firefighting mode, even if the business model needs adaptations (for a discussion of firefighting in the domain of product development see Repenning et al. 2001).

In contrast, creating broad acknowledgement of the need for change can be difficult for "tumbling giants". Even if top management has recognised the seriousness of the situation thanks to scenario planning exercises or other indications the feeling of invulnerability may still be prevalent within the larger organisation. Resistance from powerful divisional leaders can also threaten the success of the transformation.

For "game changers" and "opportunity seekers" Green Business Model Transformations happen under more positive circumstances, but usually require a deeper level of commitment as changes seem less urgent. When compliance is not enough, conducting trainings with whole management teams instead of only a few incubators may be advisable. GE – which is known for its execution capabilities – follows this approach. Not only does it reduce resistance, but the team can also develop a common understanding of the opportunities and challenges arising in the change, thereby increasing its speed and impact (Prokesch 2009; see also Senge et al. 2008, 143ff.).

"Do-gooders", the last of the described corporate journeys, can take advantage of the positive connotation of the change effort to create broad, intrinsic willingness to change. However, the transformation may not be taken seriously by some managers who insinuate that it follows a philanthropic (or public relations), and not a business rationale. Change agents thus need to ensure that the economic benefits are sufficiently emphasised. For example, GE always highlights that "green is green", Osram and Philips also make clear that their investments in off-grid lighting follow economic motives. After all, blind enthusiasm for a good cause (or any other) may not only lead to suboptimal business decisions, but can lead to a "true believers syndrome" that isolates innovators from the larger organisation (Senge 2006, 296ff.).

For the transformation to be successful it is not sufficient, however, to implement new business model elements: Management must also seek active discontinuation of previous business practices that are not adequate anymore for the new business model – the organisation needs to "unlearn" (Govindarajan & Trimble 2005, 60ff.). It has to be made very clear that doing things the old way is no longer acceptable. One frequent impediment to achieve this stems from conflicts of interest within middle management. Inadequate incentives or unclear responsibilities can hamper both environmental initiatives (Esty & Winston 2009, 242ff.) as well as business model transformations in general (Johnson 2010, 165f.). The transformation should also not be declared to be over too soon. Instead, management should put ongoing efforts into institutionalising the new business model (Kotter 1995, 66f.).

9.7 Phase VI: Finding New Signals

9.7.1 Problem Context

"Business success is less a function of grandiose predictions than it is a result of being able to respond rapidly to real changes as they occur."

Jack Welch, CEO of General Electric 1981-2003 (Welch 2001, 390)

Phase VI starts when the green business model is operational on a commercial scale. The firm now enters a state of stability as described by the theory of punctuated equilibrium (see 6.2.5). Yet especially in the beginning, some adaptations may still be necessary to increase efficiency or fine-tune value propositions. In the case of staged transformations, implementation and monitoring may run in parallel while the transformation is expanded.

Once business model operations are stable and successful, management attention should shift to finding new signals. As the above quote of Jack Welch indicates, many environmental changes will not be predicted (even if this was possible), but require fast and adequate responses. Depending on the magnitude of these changes, small or large adaptations may be necessary. In extreme cases, the prevailing business model paradigm may be challenged by tectonic shifts in the business model environment – as environmental sustainability has already done in many industries. Business model renewal can also be pursued independent of external stimulus. Firms should try to become active *before* they lose their competitive advantages.

9.7.2 Optimising the New Business Model

Once the firm has made the big jump (see 8.3.1), it needs to optimise the new green business model. The precondition for successful "hill-climbing" towards the local peak of the "fitness landscape" (see 6.1.4) is to gain a good understanding of what the fitness landscape looks like, where the business model is currently positioned, and how it can reach the peak. Now that informed guessing can be replaced by actual observation of business model performance, critical business model assumptions can be re-examined (although some firms may need to rectify deficiencies in reporting). This should start as early as possible, often already during implementation. Particularly for business models that offer large early-mover advantages (see 8.3.2) or that require reaching a certain tipping point (see Gladwell 2000) quick optimisation is essential.

If key performance indicators show problematic results, responsible managers need to identify the root causes and rectify them if possible. Even with relatively stable business environments, detailed business model designs and diligent planning, some problems may not surface until the actual implementation. For instance, while most formal rules and metrics may have been adjusted to work with the new business model, some may have been overlooked. In addition, implicit norms that at one time helped to make the old business model successful may now be thwarting the success of the new one (Johnson 2010, 161ff.). Consequently, "the forgetting curve is sometimes more important than the learning curve" (Prahalad 2010, 36; see also Govindarajan & Trimble 2005, 60ff.). Whatever the problem may be, when management recognises that the implemented business model does not perform satisfactorily, the obvious challenge is to find the appropriate response.

For example, if BMW realised in 2013 that desired sales figures of the i3 are short by a large margin – could modified pricing and additional advertisement rectify this, or should that be considered a symptom of a more fundamental problem (e.g. that the product does not meet customer requirements)? BMW may decide to be patient and work on improving sales, but it could also come to the conclusion that the EV segment is a dead end for BMW and discontinue its efforts. Such judgements are difficult – the Toyota Prius, for instance, was initially met with scepticism by many industry observers (Taylor 2006; Toyota 2010b).

Even in well-designed transformations firms may go through worse-before-better patterns during implementation (e.g., make a loss with the first generation of green products). Top management can easily lose patience in such cases as they start feeling the pressure from shareholders and internal opponents. Frosta's Green Business Model Transformation also caused serious financial troubles and even led to the replacement of its CEO. Frosta adhered to its strategy, however, and successfully introduced corrective measures to make its transformation a success (see 7.2.1). Yet if there are business model flaws that cannot be fixed, top management needs to ensure that this fact is acknowledged early enough. Although the industry survey in chapter 7 covered industries that have not produced true green business models so far, no example was provided of a Green Business Model Transformation that turned out to be an outright failure. Companies are obviously reluctant to publicise such failures and maybe that is why good examples are difficult to find. Nevertheless, the possibility of failure cannot be ignored so that no "good money is thrown after bad money".

Established companies often operate several business models in parallel. Hence, they should also optimise their portfolio of business models and try to generate synergies between them (Osterwalder & Pigneur 2010, 259). Some firms like GE have created green corporate platforms for growth for this purpose (see 7.5).

9.7.3 Finding New Signals

Even when a company successfully transforms a market through environmental sustainability, Lubin & Esty (2010, 47) caution that first movers should not "rest on [their] green laurels". Companies need to continue to question assumptions from time to time and seek new (green) ways to innovate (also see McGrath 2010, 256ff.).

As explained in the overview of this chapter, the management framework is designed to be iterative (compare figure 9.2). An iteration loop may be initiated based on a fixed schedule or be triggered by an event. The latter may include external stimuli like changes in the market environment or regulatory framework, KPIs that develop in unexpected ways, or stakeholders that change their attitude. Or, a new idea or technology is conceived within the organisation.

The magnitude of the event determines which iteration loop should be considered:

- Entering Phase I again can be justified if an unexpected, but potentially far-reaching event has happened. The recent financial crisis or the nuclear disaster triggered by the tsunami in Japan may be of such a scale, at least for some industries that were deeply affected. Otherwise, scenarios also become outdated after a while (circa five to ten years), and the benefits from the last scenario exercise may meanwhile have faded or been lost due to changes in management.
- If a particular topic emerges that will be of great importance in the future (e.g., eco-efficiency for IBM), the affected company may choose to rethink old certainties with Executive Ideation Workshops or Crowd Ideation Events (Phase II). IBM, for example, has conducted various Jams about different topics since 2001. Major ones include the two "InnovationJams" in 2006 and

2008 (Bjelland & Wood 2008), as well as the "Eco-efficiency Jam" in 2010 (IBM 2010c). Large Crowd Ideation Events are usually not undertaken too frequently as new learning within the audience has to occur first and the effort is substantial. Frequent Executive Ideation Events pose a challenge in terms of scheduling. Questioning business models too often on the executive level may also disturb the organisation.

- Going directly back to the drawing board (Phase III) will usually be the consequence of disappointing business results or a major shift in the competitive or regulatory landscape. Or, a definite opportunity has arisen, e.g. to capture an adjacent market with a similar green business model. In any case, a fundamental check of business model assumptions every few years does not hurt.
- Phase IV, the planning stage, will usually be part of an annual planning cycle. However, certain unusual circumstances that do not question the business model itself, but require strategic reconsiderations may justify unscheduled planning activities. For example, the financial crisis has hit many companies through the general downturn even though they were not directly involved in the financial industry. Such events may necessitate postponements of strategic endeavours (e.g., acquisitions) or changes to implementation plans.
- Minor adaptations related to business model optimisation may occur at any time, but should nevertheless be considered in aggregated terms. There is a risk that changes to accommodate certain customer groups or internal departments accumulate complexity that eventually damages the business model (see Gottfredson & Aspinall 2005).

Day & Schoemaker (2005) remark that important new developments will often happen at the periphery rather than the core of competition. As a result, initial signals may be weak, and KPIs may not deteriorate until an effective response becomes difficult. The extent to which firms have to develop their ability to scan their periphery depends on the stability of their industry. The authors suggest a test for "peripheral vision" that categorises a firm as *vulnerable* or *neurotic* (need-capability mismatch), or as *vigilant* or *focused* (appropriate capability).

Activities in Phases I and II (and to a certain extent Phase III) already provide specific frames for scanning the periphery. If the regular peripheral vision is underdeveloped, however, lessons from the past can help to catch up (i.e., considering past blind spots, benchmarking against more visionary companies). Listening more to outliers within the organisation may also be a valuable source of new signals. The same is true for external sources like non-ordinary customers that may provide early warning signs of the threat of an emerging trend (Day & Schoemaker 2005, 136ff.) (also compare the sources of innovation in 6.4.4).

9.7.4 Change Management Challenges

Following the theory of punctuated equilibrium (see 6.2.5), optimising a business model and finding new signals are somewhat antagonistic goals. On the one hand, inertia allows building up competencies during periods of equilibrium (i.e.,

Phase VI) which are necessary to execute the new business model effectively. On the other hand, the very same inertia hampers the ability of the organisation to reassess the environment or respond appropriately (see Tushman & Romanelli 1985).

At the beginning, change management may even call for a trial period until operating routines are sufficiently stable (Sastry 1997, 262ff.). However, once the immediate challenges of stabilising the business model are resolved, change management needs to ensure that the organisation remains receptive, and potentially foster pockets of divergence – without creating "neurotic" organisations, as Day & Schoemaker (2005) have warned against. In this context, Sastry (1997, 257ff.) proposes that decision-makers assess the organisation-environment fit in order to determine whether change is appropriate even if it impedes the process of further improving operating routines.

The inability to respond to environmental changes may in part originate in an increasing dependence on partners, which limits the option space. Yet alliances and partnerships can also be a potential source of important stimuli. Hence, successful change management may significantly depend on the effective management of alliances (see Kale & Singh 2009). The automotive industry provides an example: Decisions for alliances and partnerships today have considerable impact on future business model options.

As mentioned before, it is often new entrants who embrace new business models first – despite the superior resources that incumbent firms have at their disposal. The management framework effectively addresses this problem during early phases of Green Business Model Transformations, but does not necessarily help to sustain that agility later on. To accomplish this, firms may try to create better incentives for experimentation and anchor it permanently within the organisation (McGrath 2010, 260). Otherwise it is unlikely that decision-makers will follow-up on new ideas during ordinary times before their validity becomes obvious to everyone in the market.

9.8 Recapitulation of the Management Framework

Myriads of things can go wrong during endeavours as complex as Green Business Model Transformations – many of which I have warned against over the course of this chapter. To conclude the presentation of the management framework, the interplay of the six phases is reconsidered briefly. The challenge is to strike the right balance between *convergence* vs. *divergence* and *creativity and intuition* vs. *analytics and reasoning*. All these factors are needed to make Green Business Model Transformations successful, but each phase tends to require a particular mixture in order to produce the desired results. Figure 9.21 depicts the "wave shape" that I consider ideal to facilitate success:

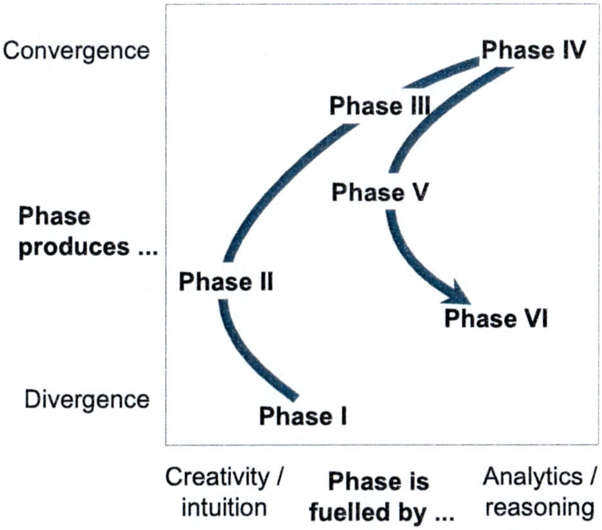

Fig. 9.21 Ideal mix of convergence vs. divergence and creativity/intuition vs. analytics/reasoning

- Phase I ("embracing ambiguity"): Maximum divergence needs to be reached by challenging mental models with different possible futures. Although this phase should be based on sound analysis and science, building credible and moving scenarios requires imagination and creativity. Most importantly, the right hemisphere of the brain of participants needs to be accessed; this phase will fall short if it becomes a mere intellectual exercise.
- Phase II ("rethinking old certainties"): Bold and possibly unorthodox ideas need to be generated. Hence, a maximum of creative energy is required. The ideas may create some additional divergence within the firm, but at the end of the process some convergence should be sought in the sense that ideas are narrowed down to the most promising ones.
- Phase III ("translating into business impact"): At some point, the trend towards more divergence has to be reversed and ideas need to be scrutinised for their economic potential and operational viability. This involves analytic rigor, but creating a sophisticated business model design also requires substantial creativity and business judgement. Towards the end of the phase, the project team needs to aim for basic agreement that the right ideas are pursued, and that the proposed business model design(s) represent the right means of doing so.
- Phase IV ("planning for action"): This phase makes extensive use of analytics and involves a high level of formalisation. The intended changes are operationalised: the path going forward must be clearly laid out, including tasks, resources, time schedule, and responsibilities.

- Phase V ("making it happen"): This is the moment of truth – the business model design now needs to prove its viability in reality and the wider organisation needs to become aligned. An overly mechanistic approach should be avoided, though. The "human factor" requires some flexibility and consideration. Furthermore, if plans are followed too rigidly, a poor fit with real-world conditions may be the result. Applying common sense and some creativity in ad-hoc problem solving may often lead to better results (not all complications can be foreseen). Conversely, too much divergence during this phase puts the overall concept at risk.
- Phase VI ("finding new signals"): This phase mostly relies on routine tasks as the business model is monitored and minor adaptations are made. The density of changes will obviously be low during this phase. However, if new signals are received this may justify questioning whether existing business model assumptions are still valid.

A smooth Green Business Model Transformation depends on the congruence of the six phases as described above. Disturbances may – but will not necessarily – severely impede the transformation. Below, I discuss six sample cases and their alleged consequences. For this purpose, a "surfer-wave" analogy is used (see figure 9.22 and figure 9.23) – although there is obviously no causal relationship between the shape of graphs that depict Green Business Model Transformations and conditions for surfing.

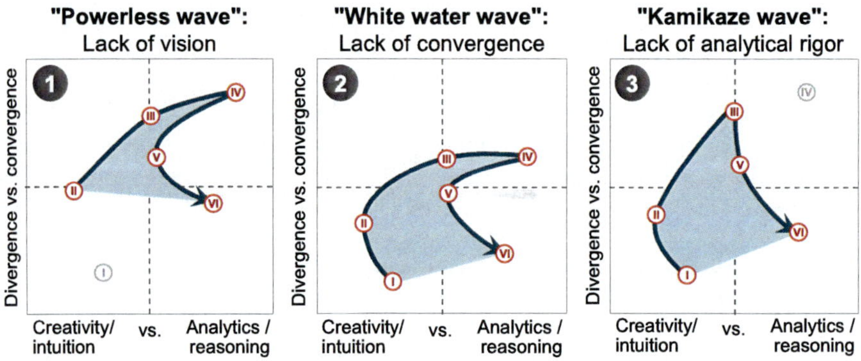

Fig. 9.22 Examples of unbalanced transformation approaches (1)

The first example, dubbed "powerless wave", is afflicted with a lack of vision as Phase I is omitted. Some companies may indeed not need Phase I because they have already achieved the respective outcomes beforehand. All other firms risk that the wave has too little energy due to the missing tension of divergence in the beginning – energy that is required to mobilise the organisation in later phases. Conversely, some strategic consulting projects I have participated in have suffered from the fact that divergent mental models of senior executives have never been made explicit, and hence there was no common ground for reaching convergence later when disputes on specific issues arose.

The "white water wave" in the second example relates to hard-to-control conditions resulting from a lack of convergence. Here, too little effort has been put into reaching consensus within important circles of power. The energy of the transformation is thus not aligned and may be wasted in power struggles with unknown outcomes. GE has avoided this problem through flat hierarchies and accountability of the management team (see 7.5).

The third example, "kamikaze wave", is also difficult to control. Instead of a lack of convergence, the problem in this case is the lack of analytical rigor. Phase IV has been omitted, and as a result many details remain unconsidered until they surface during implementation. Highly skilled implementation managers are needed to cope with the many ad-hoc challenges.

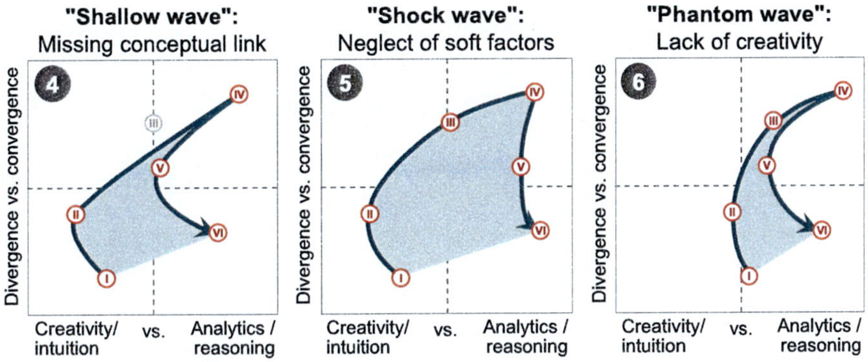

Fig. 9.23 Examples of unbalanced transformation approaches (2)

The fourth case, "shallow wave", suffers from a missing link between idea generation and detailed planning (see figure 9.23). Phase III has been omitted, either to save time or because it was not deemed necessary. As a result, the business model design lacks substance and some fundamental flaws in the business model may have been overlooked. Or, potential market responses have not been examined thoroughly. Depending on the severity of the deficiencies they may be irreparable. In the case of Osram, the business model design could have been tested earlier to avoid some misconceptions, although respective design flaws do not seem to be irreversible (see 7.3). Furthermore, the approach described in Phase III could have helped Frosta to uncover unrealistic assumptions and make its transformation less turbulent (see 7.2.1).

The fifth example, "shock wave", describes a scenario in which the organisation enters a state of crisis during implementation. The implementation approach is too mechanistic and ignores the fears and interests of those affected by the change. Poor communication may also contribute to the problem. Coercion may be used to enforce the changes, but no true commitment can be generated this way. Lasting adverse effects on motivation and distrust of those who decided on the transformation can be the consequence. Historically, many mergers have failed this way.

Finally, "phantom wave" describes an uninspiring initiative that is incapable of creating momentum. Causes may include overly rigid guidelines during the first, creative phases, or assignment of the wrong people (i.e., uninspired bureaucrats). An initial lack of originality can hardly be compensated for during later phases, but if the transformation is executed professionally, at least disastrous failures can be prevented. Half-hearted approaches in the electric car domain will likely produce some instructive negative examples in the future.

Other kinds of unbalanced transformation approaches are conceivable and the ones described above will not necessarily lead to failure of the green business model. Yet if the described pitfalls can be avoided the risks of the Green Business Model Transformation should be considerably lower.

Part IV

Conclusion and Outlook

Chapter 10
Conclusion and Outlook

10.1 Summary

Despite the efforts made towards sustainable development in recent decades, the current path of economic development still relies heavily on depleting resources and worsens environmental degradation. Incremental improvements and occasional radical green innovations by start-up companies seem to be insufficient to change that trajectory fast enough. Thus, the question of how large, established for-profit companies can significantly and quickly improve their environmental performance is of vital importance.

This work aims to contribute to the solution through the conceptualisation of Green Business Model Transformations and by providing a framework to manage them profitably in practice.

Green Business Model Transformations as envisioned in this book do not rely on wondrous moral enlightenment of decision-makers, but are based on economic reasoning that should principally appeal to even the most hard-nosed managers.

In this pursuit this monograph integrates diverse fields of research: it covers corporate sustainability and the many levers through which it can create economic value (chapter 3); it introduces the business model concept as a means to analyse complex, interconnected business and environmental issues (chapter 4); and it examines organisational change and innovation, highlighting the cognitive and organisational impediments that are associated with these green opportunities (chapter 6).

A systems view serves as the theoretical frame to integrate the diverse research domains. The final outcome is a comprehensive management framework for Green Business Model Transformations that allows firms to seize opportunities which they possibly did not even know existed (chapter 9). This systematic yet not simplistic six-phase approach facilitates radical green change beyond business-as-usual and at the same time limits associated risks.

It provides …

- … a holistic view on business model design, beyond product and process innovations

A. Sommer: Managing Green Business Model Transformations, SPLCEM, pp. 349–351.
springerlink.com

- … a holistic view on all aspects of change, including often neglected mental models
- … a holistic view on environmental sustainability with its systemic characteristics

This publication not only contributes to management practice, but also to theoretical research: The concept of business models has been refined as an analytical tool to better support issues arising from managing Green Business Model Transformations. Moreover, an initial taxonomy for green business models has been outlined in chapter 5. Illustrated with a broad survey of practical examples (chapter 7), general influencing factors have been derived on three levels in chapter 8: industry, business model, and company. The practical examples also enabled identification of various possible competitive dynamics related to Green Business Model Transformations.

10.2 Limitations

The greatest strength of this work is possibly also its greatest weakness: It aspires to give answers that are widely applicable, rather than limited to selected industries or specific conditions. Conclusions thus need to rely on a very broad foundation including diverse industries and a relatively large number of firms. This means that the presented case vignettes are based on a limited number of interviews in each case and may thus potentially lack some relevant details. Even using this broad approach, many interesting industries could not be covered. In addition, no adequate quantitative empirical data is available yet to support or reject the qualitative inferences that have been presented. Moreover, some of the described business models are still in an infant stage or only exist on paper, making respective analyses necessarily speculative at times. It may still take years until certain assessments can be conclusively validated (falsification may come much earlier, though). Despite the increased risk of getting exposed this way, however, the chosen research approach seems appropriate considering that the proposed management framework aims to guide decision-makers that themselves face situations of extreme ambiguity and uncertainty (it is certainly more honest than analysing only historic accounts).

With respect to the developed taxonomy of green business models, the level of abstraction is still high, in part owing to the broad scope and high complexity of environmental sustainability. It remains to be seen whether further operationalisations reveal as yet unidentified shortcomings of the taxonomy.

10.3 Future Research

Going forward, testing the management framework and its underlying theoretical assumptions may reveal important new insights both about the general nature of Green Business Model Transformations and about their specific occurrences.

As the empirical basis broadens and spreads to thus far untapped industries over time, the management framework may be enhanced. Exploring more specific circumstances in the framework may be worthwhile, particularly for the later phases that hitherto remain relatively generic. Moreover, some of the management situations and caveats described above may turn out to be less relevant in practice whereas other important ones may have been overlooked. Business model choices and dependencies between business model elements can probably be determined more precisely if the scope is limited to a given industry (and possibly geography). Ultimately, effective, industry-specific green business model configurations may be developed, although these would offer no competitive differentiation anymore, of course. Nevertheless, feasible green development paths of entire industries could be outlined this way from a business rather than a macroeconomic perspective.

Time will tell whether individual cognitive and organisational barriers to Green Business Model Transformations can be torn down across industries and whether enough firms are able to learn to manage these transformations successfully. A new breed of managers and realigned incentive systems may be needed in some companies to achieve this outcome.

Otherwise, as start-ups lack the leverage, the alternative paradigm of incremental corporate greening would need to gain considerable momentum in order to accomplish the desired reconciliation of environmental sustainability and economic prosperity. Should that fail also, then at least the few firms which have acted in the spirit of this work will be well-prepared for a grim, not-so-green future to unfold, as they will have considered such a scenario long ago.

Reference List

Adams, J.D.; Hayes, J. & Hopson, B. (1976): Transition: Understanding and Managing Personal Change. London: Robertson.

AFDC (2010): HEV Sales by Model. Trend of sales by HEV models from 1999-2009, http://www.afdc.energy.gov/afdc/data/docs/hev_sales.xls (Accessed: May 26, 2009).

Aftalion, F. (2001): A history of the international chemical industry. Philadelphia, PA: Chemical Heritage Foundation, 2nd edition.

Afuah, A. & Tucci, C. (2001): Internet Business Models and Strategies. New York, NY: McGraw Hill.

Ahlers, F. (2007): FRoSTA wird kleiner!, FRoSTA-Blog. http://www.frostablog.de/zutaten/frosta-wird-kleiner (Accessed: October 18, 2010).

Ahlers, F. (2008): FRoSTA ist für alle da., http://www.frostablog.de/nachhaltigkeit/was-soll-ich-zur-nachhaltigkeit-sagen (Accessed: October 1, 2010).

Alchian, A.A. & Demsetz, H. (1972): Production, Information Costs, and Economic Organization, American Economic Review, Vol. 62, No. 5, 777-795.

Al-Debei, M.M.; El-Haddadeh, R.H. & Avison, D. (2008): "Defining the business model in the new world of digital business", in: Proceedings of the Fourteenth Americas Conference on Information Systems. Toronto, Canada.

Aldrich, H.E. (2007): Organizations and Environments. Stanford, CA: Stanford University Press, Classic edition.

Ambec, S. & Lanoie, P. (2007): When and why does it pay to be green? Laboratoire d'Economie Appliquée de Grenoble. http://econpapers.repec.org/paper/gblwpaper/200704.htm (Accessed: April 7, 2010).

Ambec, S. & Lanoie, P. (2008): Does It Pay to Be Green? A Systematic Overview, Academy of Management Perspectives, Vol. 22, No. 4, 45-62.

Amit, R. & Schoemaker, P.J.H. (1993): Strategic Assets and Organizational Rent, Strategic Management Journal, Vol. 14, No. 1, 33-46.

Amit, R. & Zott, C. (2001): Value Creation in E-business, Strategic Management Journal, Vol. 22, No. 6-7, 493-520.

Amory, B.; Hunter, L.L. & Hawken, P. (1999): A Road Map for Natural Capitalism, Harvard Business Review, Vol. 77, No. 3, 145-158.

Amram, M. & Kulatilaka, N. (1999): Real options. Managing Strategic Investment in an Uncertain World. Boston, MA: Harvard Business School Press.

Anastas, P.T. (2010): 2020 Visions: Chemistry, Nature, Vol. 463, No. 7277, 26-32.

Anastas, P.T. & Beach, E.S. (2007): Green chemistry: The emergence of a transformative framework, Green Chemistry Letters and Reviews, Vol. 1, No. 1, 9-24.

Anastas, P.T. & Warner, J.C. (2000): Green Chemistry: Theory and Practice. Oxford: Oxford University Press.

Andersen, M.M. (2008): "Eco-innovation - towards a taxonomy and a theory", in: DRUID 25th Celebration Conference 2008 on Entrepreneurship and Innovation. Copenhagen.

Anderson, P. (1999): Complexity Theory and Organization Science, Organization Science, Vol. 10, No. 3, 216-232.

Anthony, S.D.; Eyring, M. & Gibson, L. (2006): Mapping Your Innovation Strategy, Harvard Business Review, Vol. 84, No. 5, 104-113.

Anthony, S.D.; Johnson, M.W. & Sinfield, J.V. (2008): Institutionalizing Innovation, MIT Sloan Management Review, Vol. 49, No. 2, 45-53.

AOL Time Warner Inc. (2001): SEC Info - Time Warner Inc - 10KT405 - For 12/31/00. http://www.secinfo.com/dsvRa.4hj.htm (Accessed: May 19, 2009).

Apple (2005): Apple to Use Intel Microprocessors Beginning in 2006, http://www.apple.com/pr/library/2005/jun/06intel.html (Accessed: January 4, 2010).

Applegate, L.M. (2001): "E-business Models: Making sense of the internet business landscape", in: Dickson, G.W. & DeSanctis, G. (Eds.): Information technology and the future enterprise: New models for managers. Upper Saddle River, NJ: Prentice Hall.

Aragón-Correa, J.A. & Sharma, S. (2003): A Contingent Resource-Based View of Proactive Corporate Environmental Strategy, Academy of Management Review, Vol. 28, No. 1, 71-88.

Argyris, C. & Schön, D.A. (1978): Organizational Learning: a Theory of Action Perspective. Reading, MA: Addison-Wesley.

Arisholm, E.; Gallis, H.; Dyba, T. & Sjoberg, D.I.K. (2007): Evaluating pair programming with respect to system complexity and programmer expertise, IEEE Transactions on Software Engineering, Vol. 33, No. 2, 65–86.

Armenakis, A.A. & Bedeian, A.G. (1999): Organizational Change: A Review of Theory and Research in the 1990s, Journal of Management, Vol. 25, No. 3, 293-315.

Arnann, S. (2009): Öko-Boom: Bionade stößt an Rohstoffgrenzen, Spiegel Online. http://www.spiegel.de/wirtschaft/0,1518,590002,00.html (Accessed: February 8, 2010).

Arnhold, U. (2010): User Generated Branding. Integrating User Generated Content into Brand Management. Wiesbaden: Gabler.

Arnold, H.; Kuhnert, F.; Kurtz, R. & Bauer, W. (2010): Elektromobilität. Herausforderungen für Industrie und öffentliche Hand. Frankfurt am Main: PricewaterhouseCoopers AG and FraunhoferInstitut für Arbeitswirtschaft und Organisation IAO.

Astley, W.G. (1984): Toward an Appreciation of Collective Strategy, Academy of Management Review, Vol. 9, No. 3, 526-535.

Astley, W.G. & Fombrun, C.J. (1983): Collective Strategy: Social Ecology of Organizational Environments, Academy of Management Review, Vol. 8, No. 4, 576-587.

Aston, A. (2009): Wal-Mart: Making Its Suppliers Go Green, BusinessWeek: Online Magazine. http://www.businessweek.com/magazine/content/09_21/b4132044814736.htm (Accessed: December 14, 2010).

Aston, A. (2010): BSR 2010: Maersk Slows Down to Go Greener, Greenbiz.com. http://www.greenbiz.com/blog/2010/11/05/bsr-2010-maersk-slows-down-go-greener?page=full (Accessed: November 25, 2010).

ATAG (2009): Beginner's Guide to Aviation Biofuels. Geneva: Air Transport Action Group (ATAG).

atmosfair (2011): atmosfair Airline Index 2011. Berlin: atmosfair.

Avadikyan, A. & Llerena, P. (2010): A real options reasoning approach to hybrid vehicle investments, Technological Forecasting and Social Change, Vol. 77, No. 4, 649-661.

Baden-Fuller, C. & Morgan, M.S. (2010): Business Models as Models, Long Range Planning, Vol. 43, No. 2/3, 156-171.

Bansal, P. & Roth, K. (2000): Why Companies Go Green: A Model of Ecological Responsiveness, Academy of Management Journal, Vol. 43, No. 4, 717-736.

Barnard, C.I. (1968): The functions of the executive. Cambridge, MA: Harvard University Press, anniversary edition.

Barnett, M.L. (2007): Stakeholder influence capacity and the variability of financial returns to corporate social responsibility, Academy of Management Review, Vol. 32, No. 3, 794-816.

Barney, J. (1986): Organizational Culture: Can It Be a Source of Sustained Competitive Advantage?, Academy of Management Review, Vol. 11, No. 3, 656-665.

Barney, J. (1991a): Firm Resources and Sustained Competitive Advantage, Journal of Management, Vol. 17, No. 1, 99 -120.

Barney, J. (1991b): Special Theory Forum: The Resource-Based Model of the Firm: Origins, Implications, and Prospects, Journal of Management, Vol. 17, No. 1, 97–98.

Baron, D.P. (2009): Business and its Environment. Upper Saddle River, NJ: Prentice Hall, 6th edition.

Baron, D.P. & Diermeier, D. (2007): Strategic Activism and Nonmarket Strategy, Journal of Economics & Management Strategy, Vol. 16, No. 3, 599-634.

BASF (2007): Brochure: Success - Added value through sustainability, http://www.basf.com/group/corporate/en/function/conversions:/publish/content/sustainability/management-and-instruments/success-added-value/images/BASF_Creating_Success.pdf (Accessed: May 21, 2010).

Baumgartner, R.J. & Ebner, D. (2010): Corporate sustainability strategies: sustainability profiles and maturity levels, Sustainable Development, Vol. 18, No. 2, 76-89.

Baumgartner, R.J. & Korhonen, J. (2010): Strategic thinking for sustainable development, Sustainable Development, Vol. 18, No. 2, 71-75.

Bayer (2010): Sustainable Development Report 2009. Leverkusen, Germany: Bayer AG.

BBC News (2000): AOL, Time Warner confirm merger, BBC News. http://news.bbc.co.uk/2/hi/business/597169.stm (Accessed: May 19, 2009).

BBC News (2007): UN warns on impacts of biofuels, BBC News. http://news.bbc.co.uk/2/hi/science/nature/6636467.stm (Accessed: November 24, 2010).

Beck, N.; Brüderl, J. & Woywode, M. (2008): Momentum or Deceleration? Theoretical and Methodological Reflections on the Analysis of Organizational Change, Academy of Management Journal, Vol. 51, No. 3, 413-435.

Bedsworth, L.W. & Taylor, M.R. (2007): Learning from California's Zero-Emission Vehicle Program, California Economic Policy, Vol. 3, No. 4, 1-20.

Beer, M.; Eisenstat, R.A. & Spector, B. (1990): Why Change Programs Don't Produce Change, Harvard Business Review, Vol. 68, No. 6, 158-166.

Berchicci, L. & King, A. (2007): Postcards from the Edge -- A Review of the Business and Environment Literature, Academy of Management Annals, Vol. 1, 513-547.

Berger, U. & Bernhard-Mehlich, I. (2002): "Die Verhaltenswissenschaftliche Entscheidungstheorie", in: Kieser, A. (Ed.): Organisationstheorien. Stuttgart: Kohlhammer, 133-168.

Berns, M.; Townend, A.; Zayna, K.; Balagopal, B.; Reeves, M.; Hopkins, M. & Kruschwitz, N. (2009a): The Business of Sustainability. Findings and Insights from the First Annual Business of Sustainability Survey and the Global Thought Leaders' Research Project. North Hollywood, CA: MIT Sloan Management Review.

Berns, M.; Townend, A.; Zayna, K.; Balagopal, B.; Reeves, M.; Hopkins, M. & Kruschwitz, N. (2009b): The Business of Sustainability. Imperatives, Advantages, and Actions. Boston, MA: The Boston Consulting Group.

Berst, J. (2009): Siemens: Will the Sleeping Giant Awaken?, Smart Grid News - Grid Modernization and the Smart Grid. http://www.smartgridnews.com/artman/publish/companies/Siemens_Will_the_Sleeping_Giant_Awaken-606.html (Accessed: August 15, 2010).

von Bertalanffy, L. (1950): An Outline of General System Theory, The British Journal for the Philosophy of Science, Vol. 1, No. 2, 134-165.

Better Place (2010a): Battery Switch Stations, http://www.betterplace.com/the-solution-switch-stations (Accessed: September 21, 2010).

Better Place (2010b): Global progress in the transition to electric vehicles, http://www.betterplace.com/global-progress (Accessed: September 21, 2010).

Better Place (2010c): Promoting international standards for electric vehicles, http://www.betterplace.com/the-solution-standards (Accessed: September 21, 2010).

Bierman, H. & Smidt, S. (1993): The Capital Budgeting Decision. Economic Analysis of Investment Projects. New York et al.: Macmillan, 8th edition.

Bierman, H. & Smidt, S. (2006): The Capital Budgeting Decision. Economic Analysis of Investment Projects. New York et al.: Routledge, 9th edition.

Bierman, H. & Smidt, S. (2007): Advanced Capital Budgeting: Refinements in the Economic Analysis of Investment Projects. New York et al.: Routledge.

Bjelland, O.M. & Wood, R.C. (2008): An Inside View of IBM's "Innovation Jam," MIT Sloan Management Review, Vol. 50, No. 1, 32-40.

Blanco, S. (2010): In depth: BMW Megacity Vehicle and Project I — Autoblog Green, autobloggreen. http://green.autoblog.com/2010/07/02/in-depth-bmw-megacity-vehicle-and-project-i/ (Accessed: November 16, 2010).

Blattberg, R.C.; Getz, G. & Thomas, J.S. (2001): Customer equity: Building and managing relationships as valuable assets. Boston, MA: Harvard Business School Press.

BMW Group (2010a): Leipzig: The future mobility plant, http://www.project-i.com/article/revolutionary-concept/leipzig-the-future-mobility-plant/ (Accessed: November 16, 2010).

BMW Group (2010b): LifeDrive Concept, http://www.project-i.com/article/revolutionary-concept/lifedrive-concept/ (Accessed: November 11, 2010).

BMW Group (2010c): project i - A future mobility think tank by the BMW Group, http://www.project-i.com/?WT.mc_id=bpi_s_ggl_phase1&WT.seg_1=BMW%20project%20i#home (Accessed: November 11, 2010).

BMW Group (2011): Experience BMW i - BMW i. Born Electric., http://www.bmw-i.com/en_ww/experience-bmw-i/ (Accessed: February 22, 2011).

BMW i (2011): Contact BMW i Ventures, http://www.bmw-i-usa.com/en_us/i-ventures/ (Accessed: March 15, 2011).

Bodie, M.T. (2006): AOL Time Warner and the False God of Shareholder Primacy, Journal of Corporation Law, Vol. 31, 975-1002.

Bonini, S. & Kaas, H.-W. (2010): Building a sustainable Ford Motor Company: An interview with Bill Ford, McKinsey Quarterly. http://www.mckinseyquarterly.com/The_next_environmental_issue_for_business_McKinsey_Global_Survey_results_2651 (Accessed: September 1, 2010).

de Bono, E. (1970): Lateral Thinking. A Textbook of Creativity. London: Ward Lock Educational.

Book, M.; Groll, M.; Mosquet, X.; Rizoulis, D. & Sticher, G. (2009): The Comeback of the Electric Car? How Real, How Soon, and What Must Happen Next. The Boston Consulting Group.

Boulding, K.E. (1956): General Systems Theory-The Skeleton of Science, Management Science, Vol. 2, No. 3, 197-208.

Bower, J.L. & Christensen, C.M. (1995): Disruptive Technologies: Catching the Wave, Harvard Business Review, Vol. 73, No. 1, 43-53.

BP (2009): Biofuels fact sheet. BP. http://www.bp.com/liveassets/bp_internet/globalbp/STAGING/global_assets/e_s_assets/e_s_assets_2008/downloads/bp_biofuels_biofuels_factsheet_BP_Final.pdf (Accessed: September 10, 2009).

de Brabandere, L. (2005): The forgotten half of change. Achieving greater creativity through changes in perception. Chicago, IL: Dearborn Trade Publishing.

de Brabandere, L. & Iny, A. (2010): Rethinking Scenarios. What a Difference a Day Makes. The Boston Consulting Group.

Bradsher, K. (2009): China Vies to Be World's Leader in Electric Cars, The New York Times. http://www.nytimes.com/2009/04/02/business/global/02electric.html (Accessed: October 23, 2010).

Brand Finance (2007): Global Intangible Tracker 2007. An annual review of the world's intangible value. http://www.brandfinance.com/Uploads/pdfs/BF_GIT_07_REPORT_Final%20Version%20Low%20Res.pdf (Accessed: June 7, 2009).

Brand, F. (2009): Critical natural capital revisited: Ecological resilience and sustainable development, Ecological Economics, Vol. 68, No. 3, 605-612.

Brandenburger, A.M. & Nalebuff, B.J. (1995): The right game: use game theory to shape strategy, Harvard Business Review, Vol. 73, No. 4, 57-71.

Brandenburger, A.M. & Nalebuff, B.J. (1996): Co-opetition: 1. A revolutionray mindset that combines competition and cooperation in the marketplace; 2. The game theory strategy that's changing the game of business. New York et al.: Doubleday.

Branigan, T. (2008): Chinese figures show fivefold rise in babies sick from contaminated milk, The Guardian. http://www.guardian.co.uk/world/2008/dec/02/china (Accessed: October 14, 2010).

Britannica Online Encyclopædia (2009): Strategy (military), http://www.britannica.com/EBchecked/topic/568259/strategy (Accessed: May 12, 2009).

Broberg, J. (2009): An Owner's View: A.P. Moller-Maersk, DNV Maritime Magazine, No. 1, 16-17.

Brown, S.L. & Eisenhardt, K.M. (1998): Competing on the edge: Strategy as structured chaos. Boston, MA: Harvard Business School Press.

Brugger, P. & Brugger, S. (1993): The Easter bunny in October: is it disguised as a duck?, Perceptual and Motor Skills, Vol. 76, No. 2, 577-578.

Brynjolfsson, E. & Urban, G.L. (2001): Strategies for e-business success. San Francisco, CA: Jossey-Bass.

Buchholz, T.G. (2007): New Ideas from Dead Economists: An Introduction to Modern Economic Thought. New York et al.: Plume, completely revised and updated edition.

Bundesregierung (2009): Nationaler Entwicklungsplan Elektromobilität der Bundesregierung. Berlin: Die Bundesregierung der Bundesrepublik Deutschland.

Bunker, B.B. & Alban, B.T. (1992): Conclusion: What Makes Large Group Interventions Effective?, The Journal of Applied Behavioral Science, Vol. 28, No. 4, 579-591.

Burke, C.S.; Stagl, K.C.; Klein, C.; Goodwin, G.F.; Salas, E. & Halpin, S.M. (2006): What type of leadership behaviors are functional in teams? A meta-analysis, The Leadership Quarterly, Vol. 17, No. 3, 288–307.

Burke, W.W. (2002): Organization change. Theory and Practice. Thousand Oaks et al.: Sage Publications.

Burmann, C. & Arnhold, U. (2008): User Generated Branding. State of the Art of Research. Münster: LIT.

Burns, T. & Stalker, G.M. (1994): The management of innovation. Oxford: Oxford University Press.

Burritt, R.L. & Schaltegger, S. (2010): Sustainability accounting and reporting: fad or trend?, Accounting, Auditing & Accountability Journal, Vol. 23, No. 7, 829-846.

Burritt, R.L.; Hahn, T. & Schaltegger, S. (2002): Towards a Comprehensive Framework for Environmental Management Accounting, Australian Accounting Review, Vol. 12, No. 27, 39-50.

Butner, K. & Gregory, J.J. (2009): Green and beyond. Getting smarter about the environment. Somers, NY: IBM Global Business Services.

Buysse, K. & Verbeke, A. (2003): Proactive Environmental Strategies. A Stakeholder Management Perspective, Strategic Management Journal, Vol. 24, No. 5, 453-470.

Buzby, J.C. (2003): International Trade and Food Safety: Economic Theory and Case Studies. United States Department of Agriculture (USDA).

BYD Auto (2010): BYD and RWE conclude cooperation on e-mobility, http://www.byd.com/press.php?id=102 (Accessed: September 18, 2010).

Cairns, S. & Newson, C. (2006): Predict and decide. Aviation, climate change and UK policy. Oxford: Environmental Change Institute, University of Oxford.

Callenbach, E. (1975): Ecotopia. Berkeley, CA: Heyday Books.

Carnall, C.A. (2007): Managing Change in Organizations. Harlow et al.: Financial Times/Prentice Hall, 5th edition.

Carson, R. (1962): Silent Spring. Cambridge, MA: Houghton Miffin.

Casadesus-Masanell, R. & Ricart, J.E. (2010): From Strategy to Business Models and onto Tactics, Long Range Planning, Vol. 43, No. 2/3, 195-215.

Chandy, R.K. & Tellis, G.J. (2000): The Incumbent's Curse? Incumbency, Size, and Radical Product Innovation, The Journal of Marketing, Vol. 64, No. 3, 1-17.

Chesbrough, H. (2003a): Open innovation: the new imperative for creating and profiting from technology. Boston, MA: Harvard Business School Press.

Chesbrough, H. (2003b): The Era of Open Innovation, MIT Sloan Management Review, Vol. 44, No. 3, 35-41.

Chesbrough, H. (2006): Open business models. Boston, MA: Harvard Business School Press.

Chesbrough, H. (2007): Why Companies Should Have Open Business Models, MIT Sloan Management Review, Vol. 48, No. 2, 22-28.

Chesbrough, H. (2010): Business Model Innovation: Opportunities and Barriers, Long Range Planning, Vol. 43, No. 2/3, 354-363.

Chesbrough, H. & Rosenbloom, R.S. (2002): The role of the business model in capturing value from innovation: evidence from Xerox Corporation's technology spin-off companies, Industrial and Corporate Change, Vol. 11, No. 3, 529-555.

Child, J. (1972): Organizational Structure, Environment and Performance: The Role of Strategic Choice, Sociology, Vol. 6, No. 1, 1-22.

Christensen, C.M. (1997): The innovator's dilemma. When new technologies cause great firms to fail. Boston, MA: Harvard Business School Press.

Christensen, C.M. & Raynor, M.E. (2003): The innovator's solution. Creating and sustainning successful growth. Boston, MA: Harvard Business School Press.

Christensen, C.M.; Anthony, S.D. & Roth, E.A. (2004): Seeing What's Next Using the Theories of Innovation to Predict Industry Change. Boston, MA: Harvard Business School Press.

Christensen, C.M.; Kaufman, S.P. & Shih, W.C. (2008): Innovation Killers. How Financial Tools Destroy Your Capacity to Do New Things, Harvard Business Review, Vol. 86, No. 1, 98-105.

von Clausewitz, C. (1832): Vom Kriege. Hinterlassenes Werk des Generals Carl von Clausewitz. Berlin: Ferdinand Dümmler.

CNET News (2005): Apple's Intel switch: Jobs' keynote transcript, CNET News. http://news.cnet.com/Apples-Intel-switch-Jobs-keynote-transcript/2100-1047_3-5748045.html?tag=mncol (Accessed: January 4, 2010).

Coase, R.H. (1937): The Nature of the Firm, Economica, Vol. 4, No. 16, 386-405.

Coenen, L. & Díaz López, F.J. (2010): Comparing systems approaches to innovation and technological change for sustainable and competitive economies: an explorative study into conceptual commonalities, differences and complementarities, Journal of Cleaner Production, Vol. 18, No. 12, 1149-1160.

Cognis (2010a): BASF completes acquisition of Cognis, http://www.cognis.com/company/Company/Owners/ (Accessed: January 4, 2011).

Cognis (2010b): Sustainability driven by innovation. Sustainability report 2009. Monheim, Germany: Cognis GmbH. http://www.cognis.de/NR/rdonlyres/8A24011D-1C40-4CDC-B58F-BE6747E5B92D/0/Cognis_SR_2009eng.pdf (Accessed: August 16, 2010).

Cognis (2010c): The 24 Principles as applied to Cognis' products, http://www.cognis.com/company/Sustainability/Key+Topics/24+Principles/The+24+principles+applied+to+Cognis+products.htm (Accessed: November 24, 2010).

Cohen, M.D.; March, J.G. & Olsen, J.P. (1972): A Garbage Can Model of Organizational Choice, Administrative Science Quarterly, Vol. 17, No. 1, 1-25.

Cohen, S.G. & Bailey, D.E. (1997): What Makes Teams Work: Group Effectiveness Research from the Shop Floor to the Executive Suite, Journal of Management, Vol. 23, No. 3, 239-290.

Cohen, W.M. & Levinthal, D.A. (1990): Absorptive Capacity: A New Perspective on Learning and Innovation, Administrative Science Quarterly, Vol. 35, No. 1, 128-152.

Cohn, J.; Katzenbach, J. & Vlak, G. (2008): Finding and Grooming Breakthrough Innovators, Harvard Business Review, Vol. 86, No. 12, 62-69.

Collins, D. (1998): Organizational Change: Sociological Perspectives. London et al.: Routledge.

Collins, J.C. & Porras, J.I. (1996): Building Your Company's Vision, Harvard Business Review, Vol. 74, No. 5, 65-77.

Conner, K.R. (1988): Strategies for Product Cannibalism, Strategic Management Journal, Vol. 9, 9-26.

Consolandi, C.; Jaiswal-Dale, A.; Poggiani, E. & Vercelli, A. (2008): Global Standards and Ethical Stock Indexes: The Case of the Dow Jones Sustainability Stoxx Index, Journal of Business Ethics, Vol. 87, 1-13.

Cooper, M. & Sussman, D. (2011): Nuclear Power Loses Public Support in New Poll, The New York Times. http://www.nytimes.com/2011/03/23/us/23poll.html (Accessed: April 12, 2011).

Cornelius, P.; Van de Putte, A. & Romani, M. (2005): Three decades of scenario planning in shell: experience and possible extensions in the future, California Management Review, Vol. 48, No. 1, 92-109.

Costanza, R. et al. (1998): The value of the world's ecosystem services and natural capital, Ecological Economics, Vol. 25, No. 1, 3-15.

Coyne, K.P. & Horn, J. (2009): Predicting Your Competitor's Reaction, Harvard Business Review, Vol. 87, No. 4, 90-97.

Coyne, K.P.; Clifford, P.G. & Dye, R. (2007): Breakthrough Thinking from Inside the Box, Harvard Business Review, Vol. 85, No. 12, 70-78.

Crossan, M.M. & Berdrow, I. (2003): Organizational Learning and Strategic Renewal, Strategic Management Journal, Vol. 24, No. 11, 1087-1105.

Crossan, M.M.; Lane, H.W. & White, R.E. (1999): An Organizational Learning Framework: From Intuition to Institution, Academy of Management Review, Vol. 24, No. 3, 522-537.

Cyert, R.M. & March, J.G. (1963): A behavioral theory of the firm. Englewood Cliffs, NJ: Prentice Hall.

Daft, R.L. & Lewin, A.Y. (1990): Can Organization Studies Begin to Break out of the Normal Science Straitjacket? An Editorial Essay, Organization Science, Vol. 1, No. 1, 1-9.

Dahan, N.M.; Doh, J.P.; Oetzel, J. & Yaziji, M. (2010): Corporate-NGO Collaboration: Co-creating New Business Models for Developing Markets, Long Range Planning, Vol. 43, No. 2/3, 326-342.

Daimler (2008): Evonik and Daimler Establish Strategic Alliance for the Development and Production of Lithium-Ion Batteries, http://www.daimler.com/dccom/0-5-7153-1-1160670-1-0-0-0-0-0-8-7145-0-0-0-0-0-0-0.html (Accessed: September 21, 2010).

Daimler (2010a): BYD Company Limited and Daimler AG Sign Joint Venture Contract to Develop Electric Vehicles in China, http://www.daimler.com/dccom/0-5-7153-1-1298502-1-0-0-0-0-0-8-7145-0-0-0-0-0-0-0.html (Accessed: November 11, 2010).

Daimler (2010b): OPTIRESOURCE: Well-to-wheel Analysis of the Energy Efficiency of Passenger Car Drivetrains, http://www.daimler.com/dccom/0-5-920736-1-932626-1-0-0-925197-0-0-8-876574-0-0-0-0-0-0-0.html (Accessed: September 14, 2010).

Daley, B. (2010): Air Transport and the Environment. Farnham/Burlington: Ashgate Publishing.

Daly, H.E. (1992): Steady-State Economics. London, UK: Earthscan, 2nd edition.

Daly, H.E. & Cobb, J.B. (1989): For the Common Good: Redirecting the Economy Toward Community, the Environment and a Sustainable Future. Boston, MA: Beacon.

Damanpour, F. (1991): Organizational Innovation: A Meta-Analysis of Effects of Determinants and Moderators, The Academy of Management Journal, Vol. 34, No. 3, 555-590.

Damanpour, F. (1992): Organizational Size and Innovation, Organization Studies, Vol. 13, No. 3, 375-402.

Damanpour, F. (1996): Organizational Complexity and Innovation: Developing and Testing Multiple Contingency Models, Management Science, Vol. 42, No. 5, 693-716.

Damanpour, F. & Wischnevsky, J.D. (2006): Research on innovation in organizations: Distinguishing innovation-generating from innovation-adopting organizations, Journal of Engineering & Technology Management, Vol. 23, No. 4, 269-291.

Davenport, T.H. (1993): Process innovation: reengineering work through information technology. Boston, MA: Harvard Business Press.

David, P.A. (1985): Clio and the Economics of QWERTY, American Economic Review, Vol. 75, No. 2, 332-337.

Davies, N. (2007): The inconvenient truth about the carbon offset industry, The Guardian. http://www.guardian.co.uk/environment/2007/jun/16/climatechange.climatechange (Accessed: March 13, 2011).

Day, G.S. & Schoemaker, P.J.H. (2005): Scanning the Periphery, Harvard Business Review, Vol. 83, No. 11, 135-148.

DCTI (2010): CleanTech Studienreihe. Band 4 eMobilität. Bonn: Deutsches CleanTech Institut (DCTI). http://www.dcti.de/studien/emobilitaet/de/ (Accessed: October 13, 2010).

Dechant, K. et al. (1994): Environmental Leadership: From Compliance to Competitive Advantage, Academy of Management Executive, Vol. 8, No. 3, 7-27.

Dell (2010): IdeaStorm, http://www.ideastorm.com/ideaAbout?pt=About+IdeaStorm (Accessed: January 4, 2010).

Demil, B. & Lecocq, X. (2010): Business Model Evolution: In Search of Dynamic Consistency, Long Range Planning, Vol. 43, No. 2/3, 227-246.

Dengel, B. (2008): Der neue Kampf von Bionade, The Financial Times Deutschland. http://www.ftd.de/unternehmen/industrie/:agenda-der-neue-kampf-von-bionade/401887.html (Accessed: February 7, 2010).

DESERTEC (2009): Clean Power from Deserts. The DESERTEC Concept for Energy, Water and Climate Security. Bonn: Protext Verlag, 4th edition.

Diamond, J. (2005): Collapse: How Societies Choose to Fail or Succeed. New York et al.: Viking.

DiBella, A.J.; Nevis, E.C. & Gould, J.M. (1996): Understanding Organizational Learning Capability, Journal of Management Studies, Vol. 33, No. 3, 361-379.

DiMaggio, P.J. & Powell, W.W. (1983): The Iron Cage Revisited: Institutional Isomorphism and Collective Rationality in Organizational Fields, American Sociological Review, Vol. 48, No. 2, 147-160.

Dimitri, C. & Greene, C. (2002): Recent Growth Patterns in the U.S. Organic Foods Market. Washington, DC: U.S. Department of Agriculture (USDA).

Dinger, A.; Martin, R.; Mosquet, X.; Rabl, M.; Rizoulis, D.; Russo, M. & Sticher, G. (2010): Batteries for Electric Cars. Challenges, Opportunities, and the Outlook to 2020. The Boston Consulting Group.

von Ditfurth, H. (1985): So laßt uns denn ein Apfelbäumchen pflanzen: Es ist soweit. Hamburg: Rasch & Röhring.

DOE (2008): The Smart Grid: An Introduction. U.S. Department of Energy. http://www.oe.energy.gov/DocumentsandMedia/DOE_SG_Book_Single_Pages%281%29.pdf (Accessed: August 13, 2010).

Døhlie, K.A. (2009): The future of the ultra large container ship, DNV Container Ship Update, No. 3. http://www.dnv.com/industry/maritime/publicationsanddownloads/publications/dnvcontainershipupdate/2009/3-2009/Thefutureoftheultralarge containership.asp (Accessed: September 20, 2010).

Donaldson, T. & Preston, L.E. (1995): The stakeholder theory of the corporation: Concepts, evidence, and implications, Academy of Management Review, Vol. 20, No. 1, 65-91.

Dooley, K.J. & Van de Ven, A.H. (1999): Explaining Complex Organizational Dynamics, Organization Science, Vol. 10, No. 3, 358-372.

Dosi, G. (1982): Technological paradigms and technological trajectories: A suggested interpretation of the determinants and directions of technical change, Research Policy, Vol. 11, No. 3, 147-162.

Dougherty, D. (1992): Interpretive Barriers to Successful Product Innovation in Large Firms, Organization Science, Vol. 3, No. 2, 179-202.

Doz, Y.L. & Hamel, G. (1998): Alliance advantage: The art of creating value through partnering. Boston, MA: Harvard Business School Press.

Droste-Franke, B. (2004): Quantifizierung von Umweltschäden als Beitrag zu Umweltökonomischen Gesamtrechnungen. Stuttgart: Universität Stuttgart, Institute of Energy Economics and the Rational Use of Energy (IER): Ph.D. thesis.

Drucker, P.F. (1963): Managing for Business Effectiveness, Harvard Business Review, Vol. 41, No. 3, 53–60.

Dunphy, D.C. (1996): Organizational change in corporate settings, Human Relations, Vol. 49, No. 5, 541-552.

Dunphy, D.C. & Stace, D.A. (1988): Transformational and Coercive Strategies for Planned Organizational Change: Beyond the O.D. Model, Organization Studies, Vol. 9, No. 3, 317-334.

Dunphy, D.C. & Stace, D.A. (2000): "Strategies for Organizational Transition", in: Golembiewski, R.T. (Ed.): Handbook of organizational consultation. New York/Basel: CRC Press.

Economist (2007): Toyota: A wobble on the road to the top, The Economist. http://www.economist.com/node/10097827?story_id=10097827 (Accessed: December 8, 2010).

Economist (2009): Copenhagen climate talks: Better than nothing, The Economist. http://www.economist.com/node/15124802 (Accessed: December 29, 2010).

Economist Intelligence Unit (2008a): Corporate Citizenship: Profiting from a Sustainable Business. London et al.: Economist Intelligence Unit. http://a330. g. akamai. net/7/330/25828/20081114204729/graphics. eiu. com/upload/Corporate_ Citizens. pdf (Accessed: March 26, 2009).

Economist Intelligence Unit (2008b): Doing good: Business and the sustainability challenge. London et al.: Economist Intelligence Unit. http://a330.g.akamai.net/7/330/25828/20080208191823/graphics.eiu.com/upload/ Sustainability_allsponsors.pdf (Accessed: March 26, 2009).

Eesley, C. & Lenox, M.J. (2006): Firm Responses to Secondary Stakeholder Action, Strategic Management Journal, Vol. 27, No. 8, 765-781.

Ehrenfeld, J.R. (2005): The roots of sustainability, MIT Sloan Management Review, Vol. 46, No. 2, 23-25.

Ehrenfeld, J.R. (2008): Sustainability by Design: A Subversive Strategy for Transforming Our Consumer Culture. New Haven/London: Yale University Press.

EIA (2010): International Energy Statistics. Total Electricity Net Consumption, http://tonto.eia.doe.gov/cfapps/ipdbproject/IEDIndex3.cfm?tid=2&pid=2&aid=2 (Accessed: September 21, 2010).

Eisenhardt, K. (1989): Agency Theory: An Assessment and Review, Academy of Management Review, Vol. 14, No. 1, 57-74.

Eisenhardt, K.M. & Martin, J.A. (2000): Dynamic capabilities: what are they?, Strategic Management Journal, Vol. 21, No. 10-11, 1105-1121.

Eisenhardt, K.M. & Schoonhoven, C.B. (1996): Resource-Based View of Strategic Alliance Formation: Strategic and Social Effects in Entrepreneurial Firms, Organization Science, Vol. 7, No. 2, 136-150.

Ekins, P. (1992): "A four-capital model of wealth creation", in: Ekins, P. & Max-Neef, M. (Eds.): Real-Life Economics: Understanding Wealth Creation. London/New York: Routledge, 147-155.

Ekins, P.; Simon, S.; Deutsch, L.; Folke, C. & De Groot, R. (2003): A framework for the practical application of the concepts of critical natural capital and strong sustainability, Ecological Economics, Vol. 44, No. 2-3, 165-185.

Elkington, J. (1994): Towards the Sustainable Corporation: Win-Win-Win Business Strategies for Sustainable Development, California Management Review, Vol. 36, No. 2, 90-100.

Elkington, J. (1999): Cannibals with forks. The triple bottom line of 21st century business. Oxford: Capstone.

Endres, A. & Radke, V. (1998): Indikatoren einer nachhaltigen Entwicklung: Elemente ihrer wirtschaftstheoretischen Fundierung. Berlin: Duncker & Humblot.

Equator Principles Association (2010): The Equator Principles. A benchmark for the financial industry to manage social and environmental issues in project financing, http://www.equator-principles.com/index.shtml (Accessed: November 22, 2010).

Esders, M. (2008): Conceptualising the Assessment of Eco-Innovation Performance. A Theory Based Framework for Deriving Eco-Innovation Key Performance Indicators and Drivers (EI-KPIs). Lueneburg: Centre for Sustainability Management (CSM), Leuphana University of Lueneburg.

Esty, D.C. & Winston, A.S. (2009): Green to gold. How smart companies use environmental strategy to innovate, create value, and build competitive advantage. Hoboken, NJ: Wiley, 2nd edition.

ETP (2010): SmartGrids. Strategic Deployment Document for Europe's Electricity Networks of the Future. Brussels, Belgium: European Technology Platform.

Ettlie, J.E.; Bridges, W.P. & O'Keefe, R.D. (1984): Organization strategy and structural differences for radical versus incremental innovation, Management Science, Vol. 30, No. 6, 682-695.

EU (2008): DIRECTIVE 2008/101/EC OF THE EUROPEAN PARLIAMENT AND OF THE COUNCIL, Official Journal of the European Union, Vol. L, No. 8, 3-21.

European Commission (2005): Organic Farming in the European Union. Facts and Figures. Brussels.

European Parliament (2010): MEPs set out clearer and more consistent food labelling rules, http://www.europarl.europa.eu/en/pressroom/content/20100615IPR76127 (Accessed: October 19, 2010).

Ewing, J. (2010): BMW Offers Preview of Battery-Powered Car, The New York Times. http://www.nytimes.com/2010/07/02/business/global/02bmw.html (Accessed: November 16, 2010).

Fargione, J.; Hill, J.; Tilman, D.; Polasky, S. & Hawthorne, P. (2008): Land Clearing and the Biofuel Carbon Debt, Science, Vol. 319, No. 5867, 1235-1238.

Fayol, H. (1929): Allgemeine und industrielle Verwaltung (Administration industrielle et générale). München et al.: Oldenbourg.

Fedrigo, D. & Hontelez, J. (2010): Sustainable Consumption and Production, Journal of Industrial Ecology, Vol. 14, No. 1, 10-12.

FHWA (2009): Highway Statistics 2008. Annual Vehicle Distance Traveled in Miles and Related Data - 2008 1/ By Highway Category and Vehicle Type, http://www.fhwa.dot.gov/policyinformation/statistics/2008/vm1.cfm (Accessed: September 21, 2010).

Fichter, K.; Noack, T.; Beucker, S.; Bierter, W. & Springer, S. (2006): Nachhaltigkeitskonzepte für Innovationsprozesse. Stuttgart: nova-net Konsortium.

de Figueiredo, J.M. (2000): Finding sustainable profitability in electronic commerce, MIT Sloan Management Review, Vol. 41, No. 4, 41-54.

Fine, C.H. (1998): Clockspeed. Winning Industry Control in the Age of Temporary Advantage. New York, NY: Perseus Books.

Fiol, C.M. & Lyles, M.A. (1985): Organizational Learning, Academy of Management Review, Vol. 10, No. 4, 803-813.

Fisher, D. (2005): GE Turns Green, Forbes. http://www.forbes.com/forbes/2005/0815/080.html (Accessed: September 11, 2009).

Fjeldstad, Ø.D. & Haanæs, K. (2001): Strategy Tradeoffs in the Knowledge and Network Economy, Business Strategy Review, Vol. 12, No. 1, 1-10.

Florida (1996): Lean and Green: The Move to Environmentally Conscious Manufacturing, California Management Review, Vol. 39, No. 1, 80-105.

Foodwatch (2010): foodwatch fights for consumers' rights, http://foodwatch.de/english/index_ger.html (Accessed: October 14, 2010).

Ford, J.D. & Ford, L.W. (2009): Decoding Resistance to Change, Harvard Business Review, Vol. 87, No. 4, 99-103.

Forrester, J.W. (1958): Industrial dynamics, Harvard Business Review, Vol. 36, No. 4, 37-66.

Forrester, J.W. (2003): Dynamic models of economic systems and industrial organizations, System Dynamics Review, Vol. 19, No. 4, 329-345.

Forstmoser, P. (2006): "Foreword", in: Schaltegger, S. & Wagner, M. (Eds.): Managing the business case for sustainability. The integration of social, environmental and economic performance. Sheffield: Greenleaf Publishing, 1-27.

Freeman, C. & Soete, L. (1997): The economics of industrial innovation. Boston, MA: MIT Press.

Freeman, R.E. (1984): Strategic management. A stakeholder approach. Boston, MA: Pitman Publishing.

Freeman, R.E.; Wicks, A.C. & Parmar, B. (2004): Stakeholder Theory and "The Corporate Objective Revisited," Organization Science, Vol. 15, No. 3, 364-369.

Friedman, A.L.; Miles, S.; House, S. & Bristol, B.S. (2002): Developing stakeholder theory, Journal of Management Studies, Vol. 39, No. 1, 1-21.

Friedman, M. (1970): The social responsibility of business is to increase its profits, New York Times Magazine, Vol. 32, No. 13, 122-126.

Frontier Economics (2006): Economic consideration of extending the EU ETS to include aviation. London: The European Low Fares Airline Association (ELFAA).

Frosta AG (2001): Annual Report 2001. Bremerhaven: Frosta AG.

Frosta AG (2002): Annual Report 2002. Bremerhaven: Frosta AG.

Frosta AG (2003a): Annual Report 2003. Bremerhaven: Frosta AG.

Frosta AG (2003b): Presseinformation. Zwischenbericht über den Geschäftsverlauf vom 1.1. – 30.9. 2003, http://www.frosta-ag.com/uploads/tx_smsttnewserw/ZB101003.pdf (Accessed: September 30, 2010).

Frosta AG (2005): Annual Report 2005. Bremerhaven: Frosta AG.

Frosta AG (2008): Annual Report 2008. Bremerhaven: Frosta AG.

Frynas, J.G.; Mellahi, K. & Pigman, G.A. (2006): First mover advantages in international business and firm-specific political resources, Strategic Management Journal, Vol. 27, No. 4, 321-345.

Fussler, C. (1996): Driving Eco-Innovation: A breakthrough discipline for innovation and sustainability. London: Pitman.

Galbraith, J.R. (1973): Designing complex organizations. Boston, MA: Addison-Wesley Longman Publishing.

Galpin, T.J. (1996): The Human Side of Change a Practical Guide to Organization Redesign. San Francisco: Jossey-Bass Publishers.

Gao, P.; Wang, A. & Wu, A. (2008): China Charges Up: The Electric Vehicle Opportunity. Shanghai: McKinsey&Company.

GE Ecomagination (2010): GE ecomagination: Products, Technologies, Solutions, http://ge.ecomagination.com/products/index.html (Accessed: May 19, 2010).

Geels, F.W. & Schot, J. (2007): Typology of sociotechnical transition pathways, Research Policy, Vol. 36, No. 3, 399-417.

General Electric (2005): GE 2005 ecomagination Report: Taking on Big Challenges. General Electric. http://ge.ecomagination.com/_files/downloads/reports/ge_2005_ecomagination_report.pdf (Accessed: March 25, 2009).

General Electric (2006): GE 2006 ecomagination Report: Delivering on ecomagination. General Electric. http://ge.ecomagination.com/_files/downloads/reports/ge_2006_ecomagination_report.pdf (Accessed: March 25, 2009).

General Electric (2007): GE ecomagination Report 2007: Investing and Delivering on ecomagination. General Electric. http://ge.ecomagination.com/site/downloads/news/2007ecoreport.pdf (Accessed: March 25, 2009).

General Electric (2008): 2008 ecomagination Annual Report: ecomagination is GE. General Electric. http://ge.ecomagination.com/_files/downloads/reports/ge_2008_ecomagination_report.pdf (Accessed: March 25, 2009).

General Electric (2010a): ecomagination 2009 Annual Report. General Electric. http://ge.ecomagination.com/_files/downloads/reports/ge_2008_ecomagination_report.pdf (Accessed: March 25, 2009).

General Electric (2010b): GE & Partners Announce $200 Million Global Commitment to Accelerate Power Grid Technology Through Open Collaboration, http://www.genewscenter.com/content/detail.aspx?ReleaseID=10580&NewsAreaID=2 (Accessed: September 8, 2010).

Genus, A. & Coles, A.-M. (2007): "A Critique of Geels' Multi-level Perspective of Technological Transition", in: International Summer Academy on Technology Studies - Transforming the Energy System. Deutschlandsberg, Austria.

Gersick, C.J.G. (1988): Time and Transition in Work Teams: Toward a New Model of Group Development, Academy of Management Journal, Vol. 31, No. 1, 9-41.

Gersick, C.J.G. (1991): Revolutionary Change Theories: A Multilevel Exploration of the Punctuated Equilibrium Paradigm, Academy of Management Review, Vol. 16, No. 1, 10-36.

Ghemawat, P. (1985): Building strategy on the experience curve, Harvard Business Review, Vol. 63, No. 2, 143-149.

Ghobadian, A.; Viney, H.; Liu, J. & James, P. (1998): Extending linear approaches to mapping corporate environmental behaviour, Business Strategy and the Environment, Vol. 7, No. 1, 13-23.

von Ghyczy, T. (2003): The Fruitful Flaws of Strategy Metaphors, Harvard Business Review, Vol. 81, No. 9, 86-94.

Gibson, C.B. & Gibbs, J.L. (2006): Unpacking the Concept of Virtuality: The Effects of Geographic Dispersion, Electronic Dependence, Dynamic Structure, and National Diversity on Team Innovation, Administrative Science Quarterly, Vol. 51, No. 3, 451-495.

Gladwell, M. (2000): The tipping point: How little things can make a big difference. Boston, MA: Little Brown.

Gladwin, T.N.; Kennelly, J.J. & Krause, T.-S. (1995): Shifting Paradigms for Sustainable Development: Implications for Management Theory and Research, Academy of Management Review, Vol. 20, No. 4, 874-907.

GM (2010): 2011 Chevy Volt | Electric Car | Chevrolet, http://www.chevrolet.com/volt/ (Accessed: December 20, 2010).

GNF (2010): UMEME KWA WOTE. Eco-friendly light for developing countries. Global Nature Fund supports worldwide unique solar project at Lake Victoria in Kenya, GNF - Solar Lamps at Lake Victoria. http://www.globalnature.org/21032/Projects/Climate-Protection/Solar-Lamps/02_vorlage.asp (Accessed: August 18, 2010).

Goddard, T. (2006): "Do Social Objectives Integrate with Core Corporate Objectives? The Future of Social Auditing", in: Schaltegger, S. & Wagner, M. (Eds.): Managing the Business Case for Sustainability. Sheffield: Greenleaf Publishing, 64-81.

GoodGuide (2010): iPhone App + Barcode Scanning, http://www.goodguide.com/about/mobile (Accessed: October 19, 2010).

Goodman, P.S. & Finn, P. (2007): Corruption Stains Timber Trade. Forests Destroyed in China's Race to Feed Global Wood-Processing Industry, The Washington Post. http://www.washingtonpost.com/wp-dyn/content/article/2007/03/31/AR2007033101287.html?hpid=topnews (Accessed: November 23, 2010).

Goodward, J. & Kelly, A. (2010): The Bottom Line on Offsets. Washington, DC: World Resources Institute (WRI).

Google (2010a): Google.org - Projects - Technology-Driven Solutions that Address Global Challenges, http://www.google.org/projects.html (Accessed: December 31, 2010).

Google (2010b): What is Google.org?, http://www.google.org/about.html (Accessed: December 31, 2010).

Gordijn, J. (2002): Value-based Requirements Engineering: Exploring Innovative e-Commerce Ideas. Vrije Universiteit Amsterdam: Ph.D. thesis.

Gordon, W.J. (1961): Synectics: The development of creative capacity. New York, NY: Harper.

Gottfredson, M. & Aspinall, K. (2005): Innovation vs Complexity. What Is Too Much of a Good Thing?, Harvard Business Review, Vol. 83, No. 11, 62-71.

Gould, S.J. & Eldredge, N. (1972): "Punctuated equilibria: an alternative to phyletic gradualism", in: Schopf, T.J. (Ed.): Models in paleobiology. San Francisco, CA: Freeman Cooper & Co, 82-115.

Gould, S.J. & Eldredge, N. (1977): Punctuated Equilibria: The Tempo and Mode of Evolution Reconsidered, Paleobiology, Vol. 3, No. 2, 115-151.

Govindarajan, V. & Trimble, C. (2005): Building Breakthrough Businesses Within Established Organizations, Harvard Business Review, Vol. 83, No. 5, 58-68.

Greenhouse Gas Services (2007): Press Release: GE, AES Plan Partnership To Lead US Market In Offsetting Greenhouse Gas Emissions. http://investor.aes.com/phoenix.zhtml?c=76149&p=irol-newsArticle&ID=950476&highlight= (Accessed: May 19, 2010).

Greenpeace (2008): Public pressure for Indonesia's forests works, Ask Unilever, http://www.greenpeace.org/international/campaigns/forests/asia-pacific/dove-palmoil-action (Accessed: January 15, 2010).

Greenpeace (2010): Sweet success for Kit Kat campaign: you asked, Nestlé has answered, http://www.greenpeace.org/international/en/news/features/Sweet-success-for-Kit-Kat-campaign/ (Accessed: October 19, 2010).

Greiner, L.E. (1998): Evolution and revolution as organizations grow, Harvard Business Review, Vol. 76, No. 3, 55-68.

Grunwald, M. (2008a): Nuclear's Comeback: Still No Energy Panacea, Time. http://www.time.com/time/magazine/article/0,9171,1869203,00.html (Accessed: April 12, 2011).

Grunwald, M. (2008b): The Clean Energy Scam, Time. http://www.time.com/time/magazine/article/0,9171,1725975,00.html (Accessed: April 22, 2010).

Guggenheim, D. (2006): An Inconvenient Truth. Paramount Vantage et al.

Gulati, R.; Nohria, N. & Zaheer, A. (2000): Strategic networks, Strategic Management Journal, Vol. 21, No. 3, 203-215.

Günther, E. (2009): Know-how und Projekterfahrung mit Smart Grids, http://www.siemens.com/press/pool/de/events/corporate/2009-08-pk/praesentation__guenther_d.pdf (Accessed: August 15, 2010).

Gunther, M. (2006): The green machine, Fortune Magazine, Vol. 154, No. 3. http://money.cnn.com/magazines/fortune/fortune_archive/2006/08/07/8382593/index.htm (Accessed: December 29, 2010).

Haanœs, K. et al. (2011): Sustainability: The "Embracers" Seize Advantage. North Hollywood, CA: MIT and BCG.

Hackman, J.R. (1987): "The design of work teams", in: Lorsch, J.W. (Ed.): Handbook of organizational behavior. Englewood Cliffs, NJ: Prentice-Hall, 315-342.

Hage, J. (1999): Organizational Innovation and Organizational Change, Annual Review of Sociology, Vol. 25, 597-622.

Hage, J. & Aiken, M. (1970): Social Change in Complex Organizations. New York, NY: Random House.

Halme, M. & Laurila, J. (2009): Philanthropy, Integration or Innovation? Exploring the Financial and Societal Outcomes of Different Types of Corporate Responsibility, Journal of Business Ethics, Vol. 84, No. 3, 325-339.

Hamel, G. (2002): Leading the Revolution. Boston, MA: Harvard Business School Press, 2nd edition.

Hamprecht, J. (2005): Sustainable Purchasing Strategy in the Food Industry. University of St. Gallen: Ph.D. thesis.

Hannan, M.T. & Freeman, J. (1977): The Population Ecology of Organizations, American Journal of Sociology, Vol. 82, No. 5, 929-964.

Hannan, M.T. & Freeman, J. (1984): Structural Inertia and Organizational Change, American Sociological Review, Vol. 49, No. 2, 149-164.

Harrison, J.S. & Freeman, R.E. (1999): Stakeholders, Social Responsibility, and Performance: Empirical Evidence and Theoretical Perspectives, Academy of Management Journal, Vol. 42, No. 5, 479-485.

Harrison, J.S.; Bosse, D.A. & Phillips, R.A. (2010): Managing for Stakeholders, Stakeholder Utility Functions, and Competitive Advantage, Strategic Management Journal, Vol. 31, No. 1, 58-74.

Hart, S.L. (1995): A natural-resource-based view of the firm, Academy of Management Review, Vol. 20, No. 4, 986-1014.

Hart, S.L. & Ahuja, G. (1996): Does it pay to be green? An empirical examination of the relationship between emission reduction and firm performance, Business Strategy and the Environment, Vol. 5, No. 1, 30-37.

Hart, S.L. & Milstein, M.B. (1999): Global sustainability and the creative destruction of industries, MIT Sloan Management Review, Vol. 41, No. 1, 23-33.

Hartwick, J.M. (1977): Intergenerational Equity and the Investing of Rents from Exhaustible Resources, The American Economic Review, Vol. 67, No. 5, 972-974.

Hawken, P.; Lovins, A.B. & Lovins, L.H. (1999): Natural capitalism: the next industrial revolution. London, UK: Earthscan.

Heiskanen, E. & Jalas, M. (2003): Can services lead to radical eco-efficiency improvements? - a review of the debate and evidence, Corporate Social Responsibility and Environmental Management, Vol. 10, No. 4, 186-198.

Henderson, R.M. & Clark, K.B. (1990): Architectural innovation: the reconfiguration of existing product technologies and the failure of established firms, Administrative Science Quarterly, Vol. 35, No. 1, 9-30.

Henriques, I. & Sadorsky, P. (1999): The Relationship Between Environmental Commitment and Managerial Perceptions of Stakeholder Importance, Academy of Management Journal, Vol. 42, No. 1, 87-99.

Hensley, R.; Knupfer, S. & Pinner, D. (2009): Electrifying cars: How three industries will evolve, McKinsey Quarterly. http://www.mckinseyquarterly.com/Electrifying_cars_How_three_industries_will_evolve_2370 (Accessed: September 1, 2010).

Herzberg, F. (1968): One more time: How do you motivate employees?, Harvard Business Review, Vol. 46, No. 1, 53-62.

Hickman, M. (2010): Global warming? It doesn't exist, says Ryanair boss O'Leary, The Independent. http://www.independent.co.uk/environment/climate-change/global-warming-it-doesnt-exist-says-ryanair-boss-oleary-2075420.html (Accessed: March 15, 2011).

von Hippel, E. (1988): The sources of innovation. New York, NY: Oxford University Press.

Hockerts, K. & Wüstenhagen, R. (2010): Greening Goliaths versus emerging Davids - Theorizing about the role of incumbents and new entrants in sustainable entrepreneurship, Journal of Business Venturing, Vol. 25, No. 5, 481-492.

Hoffert, M.I. et al. (2002): Advanced Technology Paths to Global Climate Stability: Energy for a Greenhouse Planet, Science, Vol. 298, No. 5595, 981-987.

Hoffman, A.J. (1999): Institutional Evolution and Change: Environmentalism and the U.S. Chemical Industry, Academy of Management Journal, Vol. 42, No. 4, 351-371.

Holloway, H. (2010): BMW Megacity: in-depth analysis, Autocar.co.uk. http://www.autocar.co.uk/News/NewsArticle/AllCars/252076/ (Accessed: November 16, 2010).

Holweg, M. & Pil, F.K. (2006): Evolving from Value Chain to Value Grid, MIT Sloan Management Review, Vol. 47, No. 4, 72-80.

Horst, J.; Frey, G. & Leprich, U. (2009): Auswirkungen von Elektroautos auf den Kraftwerkspark und die CO2-Emissionen in Deutschland. Frankfurt am Main: WWF Deutschland.

HSBC (2010): Sustainability Report 2009. London, UK: HSBC Holdings plc.

Huber, G.P. (1984): The Nature and Design of Post-Industrial Organizations, Management Science, Vol. 30, No. 8, 928-951.

Hunt, C.B. & Auster, E.R. (1990): Proactive Environmental Management: Avoiding the Toxic Trap, Sloan Management Review, Vol. 31, No. 2, 7-18.

IATA (2008): Building a greener future. International Air Transport Association (IATA), 3rd edition. http://www.iata.org/SiteCollectionDocuments/building_greener_future_oct08.pdf (Accessed: May 17, 2010).

IATA (2010): IATA 2010 Report on Alternative Fuels. Montreal/Geneva: International Air Transport Association (IATA), 5th edition. http://www.iata.org/ps/publications/Documents/IATA%202010%20Report%20on%20Alternative%20Fuels.pdf (Accessed: December 15, 2010).

IBM (2008a): IBM - InnovationJam 2008, http://www.ibm.com/ibm/jam/ (Accessed: September 8, 2009).

IBM (2008b): IBM India: Facts and History, http://www.ibm.com/news/in/en/2008/01/23/v141107r67598o89.html (Accessed: March 2, 2011).

IBM (2009): 22 opportunities for a smarter planet. IBM Corp. http://www.ibm.com/smarterplanet/global/files/us__en_us__overview__com03003usen.pdf (Accessed: January 5, 2011).

IBM (2010a): IBM Establishes New Corporate Responsibility and Environmental Requirements to Advance Sustainability Across its Global Supply Chain, IBM Press room. http://www-03.ibm.com/press/us/en/pressrelease/29953.wss (Accessed: May 8, 2010).

IBM (2010b): IBM Jam Events, https://www.collaborationjam.com/ (Accessed: January 5, 2010).

IBM (2010c): The emergence of the eco-efficient economy. The IBM Eco-efficiency Jam 2010: Key messages from a global online discussion on eco-efficiency. IBM Global Business Services.

IBM (2010d): Welcome to the decade of smart. IBM Corp. http://www.ibm.com/smarterplanet/global/files/us__en_us__overview__decade_of_smart_011310.pdf (Accessed: January 5, 2011).

IBM (2011a): IBM Archives: 1990s, http://www-03.ibm.com/ibm/history/history/decade_1990.html (Accessed: March 2, 2011).

IBM (2011b): Welcome to The Smarter City, http://www-03.ibm.com/innovation/us/thesmartercity/index.shtml (Accessed: March 24, 2011).

IEA (2009a): CO2 Emissions from Fuel Combustion 2009 Edition. Highlights. Paris: International Energy Agency (IEA).

IEA (2009b): Key World Energy Statistics 2009. Paris: International Energy Agency (IEA).

IEA (2009c): Technology Roadmap. Electric and plug-in hybrid electric vehicles. International Energy Agency (IEA).

IKEA (2008): Sustainability Report 08. IKEA Group. http://www.ikea.com/ms/en_US/about_ikea/pdf/Sustainability_report_2008.pdf (Accessed: September 17, 2010).

IKEA (2009): The Never Ending Job. Sustainability Report 09. IKEA Group. http://www.ikea.com/ms/en_US/about_ikea/pdf/sustainability_Report_2009.pdf (Accessed: November 23, 2010).

IKEA (2010): People and the environment. Responsibility beyond home furnishing, http://www.ikea.com/ms/en_US/the_ikea_story/people_and_the_environment/index.html (Accessed: November 23, 2010).

Ilinitch, A.Y. & Schaltegger, S. (1995): Developing a green business portfolio, Long Range Planning, Vol. 28, No. 2, 29-38.

IMO (2009): Second IMO GHG Study 2009. London, UK: International Maritime Organization (IMO).

Interbrand (2001): Best Global Brands: 2001 rankings, http://www.interbrand.com/best_global_brands.aspx?year=2001&langid=1000 (Accessed: August 18, 2009).
Interbrand (2004): Best Global Brands: 2004 rankings, http://www.interbrand.com/best_global_brands.aspx?year=2004&langid=1000 (Accessed: January 21, 2010).
Interbrand (2008): Best Global Brands 2008. Interbrand. http://www.interbrand.com/Libraries/Branding_Studies/Best_Global_Brands_2008.sflb.ashx (Accessed: August 18, 2009).
Interbrand (2009): Best Global Brands 2009. Interbrand. http://www.interbrand.com/Libraries/Branding_Studies/Best_Global_Brands_2009.sflb.ashx (Accessed: January 21, 2010).
IPCC (2007): "Summary for Policymakers", in: Metz, B., Davidson, O.R., Bosch, P.R., Dave, R. & Meyer, L.A. (Eds.): Climate Change 2007: Mitigation. Contribution of Working Group III to the Fourth Assessment Report of the Intergovernmental Panel on Climate Change. Cambridge/New York: Cambridge University Press.
ISO (2010): ISO 14040:2006. Environmental management - Life cycle assessment - Principles and framework, http://www.iso.org/iso/iso_catalogue/catalogue_tc/catalogue_detail.htm?csnumber=37456 (Accessed: January 26, 2011).
Jackson, T. (2009): Prosperity Without Growth. London, UK: Earthscan.
Jaffe, A.B.; Peterson, S.R.; Portney, P.R. & Stavins, R.N. (1995): Environmental Regulation and the Competitiveness of U.S. Manufacturing: What Does the Evidence Tell Us?, Journal of Economic Literature, Vol. 33, No. 1, 132-163.
Jameson, J.K. (1999): Toward a comprehensive model for the assessment and management of intraorganizational conflict: developing the framework, International Journal of Conflict Management, Vol. 10, No. 3, 268-294.
Jarvis, J. (2009): What Would Google Do? New York, NY: HarperCollins.
Jastrow, J. (1899): The mind's eye, Popular Science Monthly, No. 54, 299-312.
Jennings, P.D. & Zandbergen, P.A. (1995): Ecologically Sustainable Organizations: An Institutional Approach, Academy of Management Review, Vol. 20, No. 4, 1015-1052.
Jensen, M.C. & Meckling, W.H. (1976): Theory of the firm: Managerial behavior, agency costs and ownership structure, Journal of Financial Economics, Vol. 3, No. 4, 305-360.
Johnson, M.W. (2010): Seizing the White Space. Business Model Innovation for Growth and Renewal. Boston, MA: Harvard Business Press.
Johnson, M.W.; Christensen, C.M. & Kagermann, H. (2008): Reinventing Your Business Model, Harvard Business Review, Vol. 86, No. 12, 50-59.
Jones, T.M. (1995): Instrumental Stakeholder Theory: A Synthesis of Ethics and Economics, Academy of Management Review, Vol. 20, No. 2, 404-437.
Kachaner, N.; Deimler, M.S. & Saussois, C. (2008): Does Your Strategy Need Stretching? Adapting Your Strategy-Development Approach to Fit Today's Rapidly Changing Competitive Environment. Boston, MA: The Boston Consulting Group.
Kaenzig, J. & Wüstenhagen, R. (2010): The Effect of Life Cycle Cost Information on Consumer Investment Decisions Regarding Eco-Innovation, Journal of Industrial Ecology, Vol. 14, No. 1, 121-136.
Kale, P. & Singh, H. (2009): Managing Strategic Alliances: What Do We Know Now, and Where Do We Go From Here?, Academy of Management Perspectives, Vol. 23, No. 4, 45-62.
Kambil, A.; Ginsberg, A. & Bloch, M. (1996): Re-Inventing Value Propositions. Stern School of Business, New York University: Working Paper.
Kanellos, M. (2007): Father of the compact fluorescent bulb looks back, CNET News. http://news.cnet.com/Father-of-the-compact-fluorescent-bulb-looks-back/2100-11392_3-6202996.html (Accessed: March 25, 2011).
Kanter, R.M. (2008): Transforming Giants, Harvard Business Review, Vol. 86, No. 1, 43-52.

Kanter, R.M.; Stein, B.A. & Jick, T.D. (1992): The challenge of organizational change: how companies experience it and leaders guide it. New York et al.: Free Press.

Kaplan, R.S. & Norton, D.P. (2000): Having Trouble with Your Strategy? Then Map It, Harvard Business Review, Vol. 78, No. 5, 168-177.

Katzensteiner, T. (2009): BMW: Ökoautoprojekt "Project i" vom Sparzwang ausgeschlossen, WirtschaftsWoche. http://www.wiwo.de/unternehmen-maerkte/bmw-oekoautoprojekt-project-i-vom-sparzwang-ausgeschlossen-389848/ (Accessed: November 11, 2010).

Katzev, R. (2003): Car Sharing: A New Approach to Urban Transportation Problems, Analyses of Social Issues and Public Policy, Vol. 3, No. 1, 65-86.

Kauffman, S.A. (1993): The origins of order: self-organzation and selection in evolution. New York et al.: Oxford University Press.

Kaufman, L. (2011): Japan Crisis Could Rekindle U.S. Antinuclear Movement, The New York Times. http://www.nytimes.com/2011/03/19/science/earth/19antinuke.html (Accessed: April 12, 2011).

Kelly, D. & Amburgey, T.L. (1991): Organizational Inertia and Momentum: A Dynamic Model of Strategic Change, Academy of Management Journal, Vol. 34, No. 3, 591-612.

Kets de Vries, M.F.R. & Miller, D. (1984): The Neurotic Organization. Diagnosing and Changing Counterproductive Styles of Management. San Francisco et al.: Jossey-Bass.

Kieser, A. (2002): "Der Situative Ansatz", in: Kieser, A. (Ed.): Organisationstheorien. Stuttgart: Kohlhammer, 169-198.

Kieser, A. & Woywode, M. (2002): "Evolutionstheoretische Ansätze", in: Kieser, A. (Ed.): Organisationstheorien. Stuttgart: Kohlhammer, 253-285.

Kilduff, M.; Angelmar, R. & Mehra, A. (2000): Top Management-Team Diversity and Firm Performance: Examining the Role of Cognitions, Organization Science, Vol. 11, No. 1, 21-34.

King, A.A. & Lenox, M.J. (2002): Exploring the Locus of Profitable Pollution Reduction, Management Science, Vol. 48, No. 2, 289-299.

Klose, F.; Kofluk, M.; Lehrke, S. & Rubner, H. (2010): Toward a Distributed-Power World. Renewables and Smart Grids Will Reshape the Energy Sector. The Boston Consulting Group.

Knaut, A. (2010): Panel discussion: "Social Business and Social Entrepreneurship." Berlin: 4th International Conference on Corporate Social Responsibility.

Köhler, W. (1971): Die Aufgabe der Gestaltpsychologie. Berlin et al.: de Gruyter.

Kolk, A. (2002): The evolution of environmental management: from stage models to performance evaluation, Business Strategy and the Environment, Vol. 11, No. 1, 14-31.

Kotter, J.P. (1995): Leading Change: Why Transformation Efforts Fail, Harvard Business Review, Vol. 73, No. 2, 59-67.

Kotter, J.P. & Schlesinger, L.A. (2008): Choosing Strategies for Change, Harvard Business Review, Vol. 86, No. 7/8, 130-139.

Kozlowski, S.W.J. & Bell, B.S. (2003): "Work groups and teams in organizations", in: Borman, W.C., Ilgen, D.R. & Klimoski, R.J. (Eds.): Handbook of psychology: Industrial and organizational psychology. London, UK: Wiley, 333-375.

KPMG (2008): KPMG International Survey of Corporate Responsibility Reporting 2008. KPMG International. http://www.kpmg.com/SiteCollectionDocuments/International-corporate-responsibility-survey-2008.pdf (Accessed: March 26, 2009).

Kranhold, K. (2007): GE's Environment Push Hits Business Realities, The Wall Street Journal. http://online.wsj.com/article/SB118973485406827339.html (Accessed: April 2, 2009).

Krauss, C. (2010): The Lithium Chase, The New York Times. http://www.nytimes.com/2010/03/10/business/energy-environment/10lithium.html (Accessed: September 21, 2010).

Krugman, P. (2009): How Did Economists Get It So Wrong?, The New York Times. http://www.nytimes.com/2009/09/06/magazine/06Economic-t.html?_r=1&pagewanted=all (Accessed: January 2, 2010).

Krugman, P. (2010): Building a Green Economy, The New York Times. http://www.nytimes.com/2010/04/11/magazine/11Economy-t.html?ref=homepage&src=me&pagewanted=all (Accessed: April 19, 2010).

Kuhn, T.S. (1996): The structure of scientific revolutions. Chicago et al.: Chicago University Press, 3rd edition.

Lado, A.A.; Boyd, N.G. & Hanlon, S.C. (1997): Competition, Cooperation, and the Search for Economic Rents: A Syncretic Model, Academy of Management Review, Vol. 22, No. 1, 110-141.

Lai, R.; Weill, P. & Malone, T. (2006): Do Business Models Matter? MIT Sloan School of Management. http://seeit.mit.edu/Publications/DoBMsMatter7.pdf (Accessed: March 26, 2009).

Lambert, S. (2003): A Review of the Electronic Commerce Literature to Determine the Meaning of the Term "Business Model." Flinders University of South Australia: School of Commerce research paper series.

Lambert, S. (2006): "A business model research schema", in: 19th Bled eConference. Bled, Slovenia. http://www.bledconference.org/proceedings.nsf/Proceedings/C6CC94943A84A8A4C125718000319598/$File/20_Lambert.pdf (Accessed: March 26, 2009).

Landler, M. (2007): Germans See Imitation in Chinese Cars, The New York Times. http://www.nytimes.com/2007/09/12/business/worldbusiness/12auto.html (Accessed: November 11, 2010).

Lankoski, L. (2000): Determinants of environmental profit: An analysis of the firm-level relationship between environmental performance and economic performance. Helsinki University of Technology: Ph.D. thesis.

Lankoski, L. (2006): "Environmental and Economic Performance. The Basic Links", in: Schaltegger, S. & Wagner, M. (Eds.): Managing the Business Case for Sustainability. Sheffield: Greenleaf Publishing, 32-46.

Lash, J. & Wellington, F. (2007): Competitive advantage on a warming planet, Harvard Business Review, Vol. 85, No. 3, 94-102.

Laverty, K.J. (1996): Economic "Short-Termism": The Debate, the Unresolved Issues, and the Implications for Management Practice and Research, Academy of Management Review, Vol. 21, No. 3, 825-860.

Lawrence, P.R. & Lorsch, J.W. (1967): Differentiation and Integration in Complex Organizations, Administrative Science Quarterly, Vol. 12, No. 1, 1-47.

Lee, K.H. & Ball, R. (2006): "Achieving Sustainable Corporate Competitiveness. The Strategic Link between Top Management's (Green) Commitment and Corporate Environmental Strategy", in: Schaltegger, S. & Wagner, M. (Eds.): Managing the Business Case for Sustainability. Sheffield: Greenleaf Publishing, 378-397.

Lee, M.D.P. (2008): A review of the theories of corporate social responsibility: Its evolutionary path and the road ahead, International Journal of Management Reviews, Vol. 10, No. 1, 53-73.

Leeds, D.J. (2009): The Smart Grid in 2010: Market Segments, Applications and Industry Players. GTM Research.

Leetmaa, S.E.; Arnade, C. & Kelch, D. (2004): A Comparison of U.S. and EU Agricultural Productivity With Implications for EU Enlargement. Economic Research Service, USDA.

Levinthal, D.A. (1997): Adaptation on rugged landscapes, Management Science, Vol. 43, No. 7, 934-950.

Levinthal, D.A. & Warglien, M. (1999): Landscape Design: Designing for Local Action in Complex Worlds, Organization Science, Vol. 10, No. 3, 342-357.

Lewin, K. (1951): Field theory in social science: selected theoretical papers. New York, NY: Harper.

Lichtenthaler, U. (2009): Absorptive capacity, environmental turbulence, and the complementarity of organizational learning processes, Academy of Management Journal, Vol. 52, No. 4, 822-846.

Lieberman, M.B. & Montgomery, D.B. (1988): First-mover advantages, Strategic Management Journal, Vol. 9, Special Issue: Strategy Content Research, 41-58.

Lieberman, M.B. & Montgomery, D.B. (1998): First-mover (dis)advantages: retrospective and link with the resource-based view, Strategic Management Journal, Vol. 19, No. 12, 1111-1125.

Lighting Africa (2010): Solar Lighting for the Base of the Pyramid - Overview of an Emerging Market. IFC/World Bank/Dalberg Global Development Advisors.

Linder, J.C. & Cantrell, S. (2000): Changing Business Models: Surveying the Landscape. Accenture Institute for Strategic Change.

Linder, J.C. & Cantrell, S. (2001): Cast in Concrete or Carved in Water: Five Business Myths That Hold Companies Back, Strategy & Leadership, Vol. 29, No. 6, 13-18.

Linder, J.C.; Jarvenpaa, S. & Davenport, T.H. (2003): Toward an Innovation Sourcing Strategy, MIT Sloan Management Review, Vol. 44, No. 4, 43-50.

Lindgardt, Z.; Reeves, M.; Stalker, G.M. & Deimler, M.S. (2009): Business Model Innovation. When the Game Gets Tough, Change the Game. The Boston Consulting Group.

von Linné, C. (1806): A general system of nature. Vol. 1. London: Lackington, Allen, and Co.

Linstone, H.A. & Turoff, M. (1975): The Delphi Method. Techniques and Applications. Reading, MA: Addison-Wesley.

Linthorst, J.A. (2009): An overview: origins and development of green chemistry, Foundations of Chemistry, Vol. 12, No. 1, 55-68.

Lomborg, B. (2001): The Skeptical Environmentalist. Measuring the Real State of the World. Cambridge et al.: Cambridge University Press, revised and updated edition.

London, T. (2009): Making Better Investments at the Base of the Pyramid, Harvard Business Review, Vol. 87, No. 5, 106-113.

Longva, T. (2009): Cost-Benefit Assessment of Energy Efficiency, DNV Maritime Magazine, No. 1, 44-47.

Lovelace, K.; Shapiro, D.L. & Weingart, L.R. (2001): Maximizing Cross-Functional New Product Teams' Innovativeness and Constraint Adherence: A Conflict Communications Perspective, Academy of Management Journal, Vol. 44, No. 4, 779-793.

Lowder, B.T. (2006): Masters of Change: Transformational Growth Strategies at General Electric from 1982 to 2006, SSRN eLibrary. http://papers.ssrn.com/sol3/papers.cfm?abstract_id=918546 (Accessed: May 19, 2010).

Lowe, J. (1998): Bill Gates speaks. New York, NY: John Wiley and Sons.

Lubin, D.A. & Esty, D.C. (2010): The Sustainability Imperative, Harvard Business Review, Vol. 88, No. 5, 42-50.

Lüdeke-Freund, F. (2009): Business Model Concepts in Corporate Sustainability Contexts. From Rhetoric to a Generic Template for "Business Models for Sustainability." Centre for Sustainability Management, Universität Lüneburg.

Luehrman, T.A. (1998a): Investment Opportunities as Real Options: Getting Started on the Numbers, Harvard Business Review, Vol. 76, No. 4, 51-67.

Luehrman, T.A. (1998b): Strategy as a portfolio of real options, Harvard Business Review, Vol. 76, No. 5, 87-99.

Lufthansa (2009): Balance. Company/Social Responsibility/Environment/Corporate Citizenship. Deutsche Lufthansa AG.
Lufthansa (2010): World premiere: Lufthansa first airline to use biofuel on commercial flights, http://presse.lufthansa.com/en/news-releases/singleview/archive/2010/november/29/article/1828.html (Accessed: November 30, 2010).
MacMillan, D. (2008): The Issue: Immelt's Unpopular Idea, BusinessWeek Online. http://www.businessweek.com/managing/content/mar2008/ca2008034_906295.htm (Accessed: December 31, 2010).
Macve, R. & Chen, X. (2010): The "equator principles": a success for voluntary codes?, Accounting, Auditing & Accountability Journal, Vol. 23, No. 7, 890-919.
Maersk (2010): Slow steaming here to stay, http://www.maersk.com/AboutMaersk/News/Pages/20100901-145240.aspx (Accessed: September 3, 2010).
Maersk Line (2009): Maersk Line wins award for super slow steaming initiative, http://www.maerskline.com/link/?page=news&path=/news/story_page/09/initiative (Accessed: September 6, 2010).
Magnusson, T. & Berggren, C. (2011): Entering an era of ferment - radical vs. incrementalist strategies in automotive power train development, Technology Analysis and Strategic Management, Vol. 23, No. 3, 313-330.
Magretta, J. (2002): Why business models matter, Harvard Business Review, Vol. 80, No. 5, 86-93.
Makower, J. (2009): Strategies for the Green Economy. New York, NY: McGraw-Hill.
Maloni, M.J. & Brown, M.E. (2006): Corporate Social Responsibility in the Supply Chain: An Application in the Food Industry, Journal of Business Ethics, Vol. 68, No. 1, 35-52.
Maney, K. (2005): Tech titans wish we wouldn't quote them on this baloney, USA Today. http://www.usatoday.com/money/industries/technology/maney/2005-07-05-famous-quotes_x.htm (Accessed: January 1, 2010).
Manget, J.; Münnich, F. & Roche, C. (2009): Capturing the Green Advantage for Consumer Companies. The Boston Consulting Group.
March, J.G. (1991): Exploration and Exploitation in Organizational Learning, Organization Science, Vol. 2, No. 1, 71-87.
March, J.G. & Olsen, J.P. (1975): The Uncertainty of the Past: Organizational Learning under Ambiguity, European Journal of Political Research, Vol. 3, No. 2, 147-171.
March, J.G. & Olsen, J.P. (1976): Ambiguity and choice in organizations. Bergen: Universitetsforlaget.
March, J.G. & Simon, H. (1958): Organizations. New York, NY: Wiley.
Marcus, A.A. & Fremeth, A.R. (2009): Green Management Matters Regardless, Academy of Management Perspectives, Vol. 23, No. 4, 17-26.
Marginson, D. & McAulay, L. (2008): Exploring the debate on short-termism: a theoretical and empirical analysis, Strategic Management Journal, Vol. 29, No. 3, 273-292.
Marine Log (2010): Maersk orders slow steaming kits for 34 more vessels, Marine Log. http://www.marinelog.com/DOCS/NEWSMMIX/2010jul00121.html (Accessed: December 1, 2010).
Markard, J. & Truffer, B. (2008): Technological innovation systems and the multi-level perspective: Towards an integrated framework, Research Policy, Vol. 37, No. 4, 596-615.
Markides, C. & Charitou, C.D. (2004): Competing with Dual Business Models: A Contingency Approach, The Academy of Management Executive, Vol. 18, No. 3, 22-36.
Markides, C. & Geroski, P. (2005): Fast second. How smart companies bypass radical innovation to enter and dominate new markets. San Francisco, CA: Jossey-Bass.
Markus, M.L. & Benjamin, R.I. (1997): The magic bullet theory in IT-enabled transformation, MIT Sloan Management Review, Vol. 38, No. 2, 55-68.

Mathieu, J.; Maynard, M.T.; Rapp, T. & Gilson, L. (2008): Team Effectiveness 1997-2007: A Review of Recent Advancements and a Glimpse Into the Future, Journal of Management, Vol. 34, No. 3, 410-476.

Maynard, M. (2007): Say "Hybrid" and Many People Will Hear "Prius," The New York Times. http://www.nytimes.com/2007/07/04/business/04hybrid.html (Accessed: May 26, 2009).

Mayo, E. (1945): The Social Problems of an Industrial Civilization. Boston, MA: Graduate School of Business Administration, Harvard University.

McGrath, R.G. (2010): Business Models: A Discovery Driven Approach, Long Range Planning, Vol. 43, No. 2/3, 247-261.

McGrath, R.G. & MacMillan, I.C. (1995): Discovery-driven planning, Harvard Business Review, Vol. 73, No. 4, 44-54.

McGrath, R.G.; Tsai, M.-H.; Venkataraman, S. & MacMillan, I.C. (1996): Innovation, Competitive Advantage and Rent: A Model and Test, Management Science, Vol. 42, No. 3, 389-403.

McGregor, D. (1960): The Human Side of Enterprise. New York, NY: McGraw-Hill.

McKelvey, B. (1999): Avoiding Complexity Catastrophe in Coevolutionary Pockets: Strategies for Rugged Landscapes, Organization Science, Vol. 10, No. 3, 294-321.

McKinsey (2008): The state of corporate philanthropy: A McKinsey Global Survey, McKinsey Quarterly. http://www.mckinseyquarterly.com/Strategy/Strategic_Thinking/The_state_of_corporate_philanthropy_A_McKinsey_Global_Survey_2106?pagenum=2 (Accessed: December 3, 2010).

McWilliams, A.; Van Fleet, D.D. & Cory, K.D. (2002): Raising rivals' costs through political strategy: an extension of resource-based theory, Journal of Management Studies, Vol. 39, No. 5, 707-723.

Meadows, D.H.; Meadows, D.L.; Randers, J. & Behrens, W.W. (1972): The Limits to Growth. A report for the Club of Rome's Project on the Predicament of Mankind. New York, NY: Universe Books.

Meijkamp, R. (1998): Changing consumer behaviour through eco-efficient services: an empirical study of car sharing in the Netherlands, Business Strategy and the Environment, Vol. 7, No. 4, 234-244.

Mercedes-Benz (2010): BlueEFFICIENCY. Fast forward to tomorrow, http://www3.mercedes-benz.com/mbcom_v4/xx/blueefficiency/en.html (Accessed: January 5, 2011).

Meyer, J. & Rowan, B. (1977): Institutionalized Organizations: Formal Structure as Myth and Ceremony, The American Journal of Sociology, Vol. 83, No. 2, 340-363.

Miles, M.P. & Covin, J.G. (2000): Environmental Marketing: A Source of Reputational, Competitive, and Financial Advantage, Journal of Business Ethics, Vol. 23, No. 3, 299-311.

Miles, R.E.; Snow, C.C.; Meyer, A.D. & Coleman, H.J. (1978): Organizational Strategy, Structure, and Process, The Academy of Management Review, Vol. 3, No. 3, 546-562.

Millennium Ecosystem Assessment (2005): Ecosystems and Human Well-being: Synthesis. Washington, DC: Island Press.

Miller, D. (1986): Configurations of Strategy and Structure: Towards a Synthesis, Strategic Management Journal, Vol. 7, No. 3, 233-249.

Miller, D. (1996): Configurations Revisited, Strategic Management Journal, Vol. 17, No. 7, 505-512.

Miller, D. & Chen, M.-J. (1994): Sources and Consequences of Competitive Inertia: A Study of the U.S. Airline Industry, Administrative Science Quarterly, Vol. 39, No. 1, 1-23.

Miller, J.W. (2010): Trade Slump Swamps Maersk, The Wall Street Journal. http://online.wsj.com/article/SB10001424052748704187204575100932350914988.html (Accessed: November 4, 2010).

Mintzberg, H. (1994): The Fall and Rise of Strategic Planning, Harvard Business Review, Vol. 72, No. 1, 107-114.

Mintzberg, H. & Westley, F. (1992): Cycles of Organizational Change, Strategic Management Journal, Vol. 13, 39-59.
Mintzberg, H.; Ahlstrand, B.W. & Lampel, J. (2009): Strategy Safari. Your Complete Guide Through the Wilds of Strategic Management. Harlow, UK: Pearson Education Limited, 2nd edition.
Mitchell, R.K.; Agle, B.R. & Wood, D.J. (1997): Toward a theory of stakeholder identification and salience: Defining the principle of who and what really counts, Academy of Management Review, Vol. 22, No. 4, 853-886.
Mitroff, I.I. & Emshoff, J.R. (1979): On Strategic Assumption-Making: A Dialectical Approach to Policy and Planning, Academy of Management Review, Vol. 4, No. 1, 1-12.
Mitroff, S.R.; Sobel, D.M. & Gopnik, A. (2006): Reversing how to think about ambiguous figure reversals: Spontaneous alternating by uninformed observers, Perception, Vol. 35, No. 5, 709-715.
Molina-Azorin, J.F.; Claver-Cortes, E.; Lopez-Gamero, M.D. & Tari, J.J. (2009): Green management and financial performance: a literature review, Management Decision, Vol. 47, No. 7, 1080-1100.
Moore, G.A. (2005): Strategy and your stronger hand, Harvard Business Review, Vol. 83, No. 12, 62-72.
Moriarty, R.T. & Moran, U. (1990): Managing Hybrid Marketing Systems, Harvard Business Review, Vol. 68, No. 6, 146-155.
Morris, M.; Schindehutte, M. & Allen, J. (2005): The entrepreneur's business model: toward a unified perspective, Journal of Business Research, Vol. 58, No. 6, 726-735.
Mui, Y.Q. & Rosenwald, M.S. (2007): Wal-Mart's New Tack: Show Em the Payoff, The Washington Post. http://www.washingtonpost.com/wp-dyn/content/article/2007/09/12/AR2007091202513.html (Accessed: March 26, 2009).
Murillo-Luna, J.L.; Garcés-Ayerbe, C. & Rivera-Torres, P. (2008): Why do patterns of environmental response differ? A stakeholders' pressure approach, Strategic Management Journal, Vol. 29, No. 11, 1225-1240.
myclimate (2010): Calculate CO2 emissions, http://lufthansa.myclimate.org/EN (Accessed: May 18, 2010).
Nadler, D.A. & Tushman, M.L. (1990): Beyond the Charismatic Leader: Leadership and Organizational Change, California Management Review, Vol. 32, No. 2, 77-97.
Nakao, Y.; Amano, A.; Matsumura, K.; Genba, K. & Nakano, M. (2007): Relationship between environmental performance and financial performance: an empirical analysis of japanese corporations, Business Strategy and the Environment, Vol. 16, No. 2, 106-118.
Nelson, R.R. (1995): Recent Evolutionary Theorizing About Economic Change, Journal of Economic Literature, Vol. 33, No. 1, 48-90.
Nelson, R.R. & Winter, S.G. (1974): Neoclassical vs. Evolutionary Theories of Economic Growth: Critique and Prospectus, The Economic Journal, Vol. 84, No. 336, 886-905.
Nestlé (2010a): About us - Frequently Asked Questions - Current Issues, http://www.nestle.com/AllAbout/FAQs/CurrentIssues/FAQs.htm (Accessed: November 18, 2010).
Nestlé (2010b): Annual Report 2009. Cham and Vevey: Nestlé S.A. http://www.nestle.com/Resource.axd?Id=93839A79-943E-4942-81B3-86461C0F2AAB (Accessed: November 17, 2010).
Nestlé (2010c): Nestlé Creating Shared Value Report 2009. Nestlé. http://www.nestle.com/Common/NestleDocuments/Documents/Reports/CSV%20reports/Global%20report%202009/Global_report_2009_GB.pdf (Accessed: October 13, 2010).
von Neumann, J. & Morgenstern, O. (1944): Theory of Games and Economic Behavior. Princeton: Princeton University Press.

Neumayer, E. (2003): Weak Versus Strong Sustainability. Exploring the Limits of Two Opposing Paradigms. Cheltenham et al.: Elgar, 2nd edition.

New Energy Finance (2009): Global Trends in Clean Energy Investment. Journalist Fact-pack - WEF Meeting, http://www.newenergymatters.com/download.php?n=NEF_2009_28_01_Davos_Fact_pact.pdf&f=pdffile&t=pressreleases (Accessed: May 12, 2009).

Nidumolu, R.; Prahalad, C.K. & Rangaswami, M.R. (2009): Why sustainability is now the key driver of innovation, Harvard Business Review, Vol. 87, No. 9, 56-64.

Nielsen (2007): Global Consumers Vote Al Gore, Oprah Winfrey and Kofi Annan Most Influential to Champion Global Warming Cause: Nielsen Survey, http://nz.nielsen.com/news/GlobalWarming_Jul07.shtml (Accessed: January 17, 2011).

Nike (2010): Nike, Inc. Corporate Responsibility Report FY07-09. Nike, Inc.

Nonaka, I. & Takeuchi, H. (1995): The knowledge creating company: How Japanese companies create the dynamics of innovation. New York et al.: Oxford University Press.

Nordhaus, W.D. (2008): A Question of Balance. Weighing the Options on Global Warming Policies. New Haven/London: Yale University Press.

Nordhaus, W.D. & Tobin, J. (1972): "Is Growth Obsolete?", in: Economic Research: Retrospect and Prospect Vol 5: Economic Growth. Cambridge, MA: National Bureau of Economic Research, 1-80.

O'Reilly III, C.A. & Tushman, M.L. (2004): The ambidextrous organization, Harvard Business Review, Vol. 82, No. 4, 74-81.

OECD & Eurostat (1999): The Environmental Goods & Services Industry. Manual for Data Collection and Analysis. Paris, France: Organisation for Economic Co-operation and Development (OECD) / Statistical Office of the European Communities (Eurostat).

von Oetinger, B.; von Ghyczy, T. & Bassford, C. (2003): Clausewitz. Strategie denken. München: Deutscher Taschenbuch Verlag, 5th edition.

OGC (2009a): Directing Successful Projects with PRINCE2 2009 Edition Manual, Office of Government Commerce (GGC). London: Stationery Office Books.

OGC (2009b): Managing Successful Projects with PRINCE2 2009 Edition Manual, Office of Government Commerce (GGC). London: Stationery Office Books.

Olsen, J.P. (2001): Garbage Cans, New Institutionalism, and the Study of Politics, American Political Science Review, Vol. 95, No. 1, 191-198.

Osram (2010): Details on the OSRAM Off-Grid Lighting Solution, http://www.osram.com/osram_com/About_Us/Society_and_the_Environment_-_Global_Care/Products_and_the_environment/Off-Grid_Lighting/Details/index.html (Accessed: August 18, 2010).

Osterwalder, A. (2004): The Business Model Ontology - a proposition in a design science approach. Universite de Lausanne, Ecole des Hautes Etudes Commerciales: Ph.D. thesis.

Osterwalder, A. & Pigneur, Y. (2010): Business Model Generation. A Handbook for Visionaries, Game Changers, and Challengers. Hoboken, NJ: John Wiley & Sons.

Osterwalder, A.; Pigneur, Y. & Tucci, C.L. (2005): Clarifying business models: Origins, present, and future of the concept, Communications of the Association for Information Systems, Vol. 16, No. 1, 1-25.

Ottaway, R.N. (1983): The Change Agent: A Taxonomy in Relation to the Change Process, Human Relations, Vol. 36, No. 4, 361-392.

Otto Group (2009): Sustainability Report 2009. Otto Group.

P&G (2010): Products & Packaging: product lifecycle, renewable materials, http://www.pg.com/en_US/sustainability/environmental_sustainability/products_packaging/index.shtml (Accessed: October 3, 2010).

Paine, C. (2006): Who Killed the Electric Car? Sony Pictures Classics.

Pais, A. (2005): Subtle is the lord. The science and the life of Albert Einstein. Oxford: Oxford University Press.

Palmisano, S.J. (2010): A Smarter Planet for a Sustainable Future. London, UK: The IBM Summit at Start.
Papakiriakopoulos, D.; Poulymenakou, A. & Doukidis, G. (2001): "Building e-business models. An analytical framework and development guidelines", in: 14th Bled Electronic Commerce Conference. Bled, Slovenia.
Pearson, D.; Moffett, S. & Landauro, I. (2011): Electric Cars Focus of Renault Spying, The Wall Street Journal. http://online.wsj.com/article/SB10001424052748704415104576065281541459652.html?mod=googlenews_wsj (Accessed: January 6, 2011).
Peloza, J. & Yachnin, R. (2008): Valuing Business Sustainability: A Systematic Review. London, Canada: Research Network for Business Sustainability.
Penrose, E.T. (1959): The Theory of the Growth of the Firm. Oxford: Blackwell.
Peteraf, M.A. (1993): The Cornerstones of Competitive Advantage: A Resource-Based View, Strategic Management Journal, Vol. 14, No. 3, 179-191.
Petrovic, O.; Kittl, C. & Teksten, R.D. (2001): "Developing business models for eBusiness", in: International Conference on Electronic Commerce. Vienna.
Pettigrew, A.M. & Whipp, R. (1991): Managing Change for Competitive Success. Oxford: Blackwell.
Pettigrew, A.M.; Woodman, R.W. & Cameron, K.S. (2001): Studying Organizational Change and Development: Challenges for Future Research, Academy of Management Journal, Vol. 44, No. 4, 697-713.
Pezzey, J. (1994): "The Optimal Sustainable Depletion of Non-renewable Resources", in: Association of Environmental and Resource Economists Workshop. Boulder, CO.
Pfeffer, J. (1997): New directions for organization theory. New York, NY: Oxford University Press.
Pfeffer, J. & Salancik, G.R. (1978): The external control of organizations. A resource dependence perspective. New York, NY: Harper and Row.
Philips (2008): Philips Sustainable Energy Solution for Africa (SESA), http://www.newscenter.philips.com/main/standard/about/news/press/20080707_sesa_backgrounder.wpd (Accessed: August 19, 2010).
Philips (2010a): Off-grid lighting solutions in Africa, http://www.lighting.philips.com/af_en/solar_lighting/index.php?main=af_en&parent=af_en&id=af_en_solar_lighting&lang (Accessed: August 19, 2010).
Philips (2010b): Philips provides 300 households in Rwanda with solar lighting, http://www.newscenter.philips.com/main/standard/news/press/2010/20100605_rwanda.wpd (Accessed: August 19, 2010).
Phillips, L.D. (1982): Requisite Decision Modelling: A Case Study, The Journal of the Operational Research Society, Vol. 33, No. 4, 303-311.
Phillips, L.D. & Bana e Costa, C.A. (2007): Transparent prioritisation, budgeting and resource allocation with multi-criteria decision analysis and decision conferencing, Annals of Operations Research, Vol. 154, No. 1, 51-68.
Piet, J. (1994): The Challenge of Going Green, Harvard Business Review, Vol. 72, No. 4, 43.
Pinch, T.J. & Bijker, W.E. (1989): "The Social Construction of Facts and Artifacts: Or How the Sociology of Science and the Sociology of Technology Might Benefit Each Other", in: Bijker, W.E., Hughes, T.P. & Pinch, T.J. (Eds.): The Social Construction of Technological Systems: New Directions in the Sociology and History of Technology. Cambridge/London: MIT Press, 17-50.
Pinker, E.J.; Seidmann, A. & Foster, R.C. (2002): Strategies for Transitioning “Old Economy” Firms to e-Business, Communications of the ACM, Vol. 45, No. 5, 76-83.
PMI (2008): A Guide to the Project Management Body of Knowledge (PMBOK Guide). Project Management Institute (PMI), 4th edition.

Poliakoff, M.; Fitzpatrick, J.M.; Farren, T.R. & Anastas, P.T. (2002): Green Chemistry: Science and Politics of Change, Science, Vol. 297, No. 5582, 807-810.

Porter, M.E. (1980): Competitive Strategy: Techniques for Analyzing Industries and Competitors. New York, NY: Free Press, first edition with a new introduction.

Porter, M.E. (1985): Competitive advantage. Creating and sustaining superior performance. New York, NY: Free Press.

Porter, M.E. (1996): What is strategy?, Harvard Business Review, Vol. 74, No. 6, 61-78.

Porter, M.E. (2008a): On Competition. Boston, MA: Harvard Business School Press, updated and expanded edition.

Porter, M.E. (2008b): The five competitive forces that shape strategy, Harvard Business Review, Vol. 86, No. 1, 78-93.

Porter, M.E. & Kramer, M.R. (2006): Strategy and society. The link between competitive advantage and corporate social responsibility, Harvard Business Review, Vol. 84, No. 12, 78-92.

Porter, M.E. & Kramer, M.R. (2011): Creating Shared Value, Harvard Business Review, Vol. 89, No. 1/2, 62-77.

Porter, M.E. & van der Linde, C. (1995a): Green and Competitive: Ending the Stalemate, Harvard Business Review, Vol. 73, No. 5, 120-134.

Porter, M.E. & van der Linde, C. (1995b): Toward a New Conception of the Environment-Competitiveness Relationship, The Journal of Economic Perspectives, Vol. 9, No. 4, 97-118.

Porter, M.E. et al. (2007): Climate Business/Business Climate, Harvard Business Review, Forethought special report.

Portney, P.R. & Weyant, J.P. (Eds.) (1999): Discounting and intergenerational equity. Washington, DC: Resources for the Future.

Post, J.E. & Altman, B.W. (1994): Managing the Environmental Change Process: Barriers and Opportunities, Journal of Organizational Change Management, Vol. 7, No. 4, 64-81.

Power, S. (2008): If a Tree Falls in the Forest, Are Biofuels To Blame? It's Not Easy Being Green, The Wall Street Journal. http://online.wsj.com/article/SB122636711059015989 .html (Accessed: November 24, 2010).

Prahalad, C.K. (2010): Why Is It So Hard to Tackle the Obvious?, Harvard Business Review, Vol. 88, No. 6, 36.

Prein, H. (1987): Strategies for Third Party Intervention, Human Relations, Vol. 40, No. 11, 699-719.

Prochaska, J.O.; DiClemente, C.C. & Norcross, J.C. (1992): In Search of How People Change. Applications to Addictive Behaviors, American Psychologist, Vol. 47, No. 9, 1102-1114.

Prokesch, S. (2009): How GE Teaches Teams to Lead Change, Harvard Business Review, Vol. 87, No. 1, 99-106.

Quinn, R.E. & Cameron, K. (1983): Organizational Life Cycles and Shifting Criteria of Effectiveness: Some Preliminary Evidence, Management Science, Vol. 29, No. 1, 33-51.

Rajala, R. & Westerlund, M. (2005): "Business Models: A new perspective on knowledge-intensive services in the software industry", in: 18th Bled eCommerce Conference eIntegration in Action. Bled, Slovenia.

Ramzy, A. (2008): China's Tainted Milk Scandal of 2008, Time Magazine. http://www.time.com/time/world/article/0,8599,1844750,00.html (Accessed: October 14, 2010).

Rappa, M. (2001): Managing the digital enterprise. Business models on the Web, North Carolina State University. http://digitalenterprise.org/models/models.html (Accessed: May 28, 2009).

Rawls, J. (1999): A theory of justice. Cambridge, MA: Harvard University Press, revised edition.

Reinhardt, F.L. (1999): Bringing the Environment Down to Earth, Harvard Business Review, Vol. 77, No. 4, 149-158.

Reiter, C. (2010): BMW's Ultimate Driving Machine Is a Tiny Little Electric Car, BusinessWeek: Online Magazine. http://www.businessweek.com/magazine/content/10_40/b4197080555679.htm (Accessed: November 16, 2010).

Renault (2009): Better Place focus, http://www.renault.com/en/groupe/developpement-durable/environnement/pages/focus-better-place.aspx (Accessed: September 18, 2010).

Renault (2010): Kangoo be bop Z.E., http://www.renault.com/en/innovation/vehicule-electrique/pages/kangoo-be-bop-ze.aspx (Accessed: September 21, 2010).

Repenning, N.P.; Goncalves, P. & Black, L.J. (2001): Past The Tipping Point: The Persistence of Firefighting in Product Development, California Management Review, Vol. 43, No. 4, 44-63.

Revkin, A.C. (2007): Carbon-Neutral Is Hip, but Is It Green?, The New York Times. http://www.nytimes.com/2007/04/29/weekinreview/29revkin.html?ex=1335499200&en=d9e2407e4f1a20f0&ei=5124 (Accessed: March 13, 2011).

Richardson, G.P. (1996): Problems for the future of system dynamics, System Dynamics Review, Vol. 12, No. 2, 141-157.

Robinson, W.T. & Min, S. (2002): Is the First to Market the First to Fail? Empirical Evidence for Industrial Goods Businesses, Journal of Marketing Research, Vol. 39, No. 1, 120-128.

Robinson, W.T.; Kalyanaram, G. & Urban, G.L. (1994): First-mover advantages from pioneering new markets: A survey of empirical evidence, Review of Industrial Organization, Vol. 9, No. 1, 1-23.

Rockwood, K. (2010): Will Walmart's "Sustainability Index" Actually Work?, Fast Company. http://www.fastcompany.com/magazine/142/attention-walmart-shoppers-clean-up-in-aisle-nine.html (Accessed: October 24, 2010).

Rogers, E.M. (1962): Diffusion of Innovations. New York, NY: Free Press.

Romanelli, E. & Tushman, M.L. (1994): Organizational Transformation as Punctuated Equilibrium: An Empirical Test, Academy of Management Journal, Vol. 37, No. 5, 1141-1166.

Roome, N. (1992): Developing environmental management strategies, Business Strategy and the Environment, Vol. 1, No. 1, 11-24.

Rosenbloom, S. (2009): At Wal-Mart, Labeling to Reflect Green Intent, The New York Times. http://www.nytimes.com/2009/07/16/business/energy-environment/16walmart.html (Accessed: October 24, 2010).

Rosenbloom, S. (2010): Wal-Mart Unveils Plan to Make Supply Chain Greener, The New York Times. http://www.nytimes.com/2010/02/26/business/energy-environment/26walmart.html (Accessed: December 14, 2010).

Rosenthal, E. (2010): Slow Trip Across Sea Aids Profit and Environment, The New York Times. http://www.nytimes.com/2010/02/17/business/energy-environment/17speed.html (Accessed: September 6, 2010).

Rothenberg, S. (2007): Sustainability through servicizing, MIT Sloan Management Review, Vol. 48, No. 2, 83-89.

Rothwell, R. (1994): Towards the Fifth-generation Innovation Process, International Marketing Review, Vol. 11, No. 1, 7-31.

Rubel, H. & Sommer, A. (2007): "Green" Competitive Advantages. How companies succeed in integrating sustainability goals with growth strategies. The Boston Consulting Group.

Rugman, A.M. & Verbeke, A. (1998): Corporate Strategies and Environmental Regulations: An Organizing Framework, Strategic Management Journal, Vol. 19, No. 4, 363-375.

Russo, M.V. & Fouts, P.A. (1997): A Resource-Based Perspective on Corporate Environmental Performance and Profitability, Academy of Management Journal, Vol. 40, No. 3, 534-559.

RWE (2010): e-mobility - RWE Mobility, http://www.rwe-mobility.com/web/cms/de/236726/rwemobility/ (Accessed: September 21, 2010).

Ryanair (2008): Ryanair. Europe's Greenest Airline. http://www.ryanair.com/doc/about/ryanair_greenairline_2008.pdf (Accessed: March 9, 2011).

Ryanair (2010): The World's Favourite Airline. Annual Report & Financial Statements 2010. Ryanair Holdings plc.

Rybak, A. (2009): Osram - Kundenfang in Afrika, Financial Times Deutschland. http://www.ftd.de/karriere-management/management/:enable-osram-kundenfang-in-afrika/458840.html (Accessed: August 18, 2010).

Salesforce.com (2010): Ideas - salesforce.com Europe, http://www.salesforce.com/eu/crm/customer-service-support/ideation/ (Accessed: January 6, 2010).

Salzmann, O.; Ionescu-Somers, A. & Steger, U. (2005): The business case for corporate sustainability. Literature review and research options, European Management Journal, Vol. 23, No. 1, 27-36.

SAM & PwC (2010): The Sustainability Yearbook 2010. Sustainable Asset Management (SAM) and PricewaterhouseCoopers (PwC).

SAP (2009): The Business of Sustainability. Information Technology as a Catalyst for Short- and Long-Term Profitability. SAP AG.

Sastry, M.A. (1997): Problems and Paradoxes in a Model of Punctuated Organizational Change, Administrative Science Quarterly, Vol. 42, No. 2, 237-275.

Sauer, H.D. (2009): Mehr Licht am Victoriasee, Neue Zürcher Zeitung. http://www.nzz.ch/nachrichten/forschung_und_technik/mehr_licht_am_victoriasee_1.1695229.html (Accessed: August 18, 2010).

Savvides, S.C. (1994): Risk Analysis in Investment Appraisal, Project Appraisal Journal, Vol. 9, No. 1, 3-18.

Sawhney, M.; Wolcott, R.C. & Arroniz, I. (2007): The 12 different ways for companies to innovate, IEEE Engineering Management Review, Vol. 35, No. 1, 45-52.

Schaltegger, S. (1997): Economics of life cycle assessment: Inefficiency of the present approach, Business Strategy and the Environment, Vol. 6, No. 1, 1-8.

Schaltegger, S. (2002): A framework for ecopreneurship: Leading bioneers and environmental managers to ecopreneurship, Greener Management International, No. 38, 45-58.

Schaltegger, S. & Burritt, R.L. (2000): Contemporary Environmental Accounting Issues, Concepts, and Practice. Sheffield: Greenleaf Publishing.

Schaltegger, S. & Burritt, R.L. (2005): "Corporate sustainability", in: Folmer, H. & Tietenberg, T. (Eds.): The International Yearbook of Environmental and Resource Economics 2005/2006. A Survey of Current Issues. Cheltenham: Edward Elgar, 185-222.

Schaltegger, S. & Burritt, R.L. (2010): Sustainability accounting for companies: Catchphrase or decision support for business leaders?, Journal of World Business, Vol. 45, No. 4, 375-384.

Schaltegger, S. & Hasenmüller, P. (2005): Nachhaltiges Wirtschaften aus Sicht des "Business Case of Sustainability". Ergebnispapier zum Fachdialog des Bundesumweltministeriums (BMU) am 17. November 2005. Lüneburg: Centre for Sustainability Management (CSM).

Schaltegger, S. & Sturm, A. (1990): Ökologische Rationalität. Ansatzpunkte zur Ausgestaltung von ökologieorientierten Managementinstrumenten, Die Unternehmung, No. 4, 273-290.

Schaltegger, S. & Synnestvedt, T. (2002): The link between "green" and economic success: environmental management as the crucial trigger between environmental and economic performance, Journal of Environmental Management, Vol. 65, No. 4, 339-346.

Schaltegger, S. & Wagner, M. (2006a): "Managing and measuring the business case for sustainability", in: Schaltegger, S. & Wagner, M. (Eds.): Managing the business case for sustainability. The integration of social, environmental and economic performance. Sheffield: Greenleaf Publishing, 1-27.

Schaltegger, S. & Wagner, M. (2008): "Types of Sustainable Entrepreneurship and Conditions for Sustainability Innovation: from the Administration of a Technical Challenge to the Management of an Entrepreneurial Opportunity", in: Wüstenhagen, R., Hamschmidt, J., Sharma, S. & Starik, M. (Eds.): Sustainable Innovation and Entrepreneurship. Cheltenham: Edward Elgar, 27-48.

Schaltegger, S. & Wagner, M. (Eds.) (2006b): Managing the business case for sustainability. The integration of social, environmental and economic performance. Sheffield: Greenleaf Publishing.

Schaltegger, S.; Bennett, M. & Burritt, R.L. (Eds.) (2006): Sustainability Accounting and Reporting. Dordrecht: Springer.

Schaltegger, S.; Herzig, C.; Kleiber, O.; Klinke, T. & Müller, J. (2007): Nachhaltigkeitsmanagement in Unternehmen. Von der Idee zur Praxis: Managementansätze zur Umsetzung von Corporate Social Responsibility und Corporate Sustainability. Berlin/Lüneburg: BMU, econsense & CSM.

Scherer, A.G. (2002): "Kritik der Organisation oder Organisation der Kritik? - Wissenschaftstheoretische Bemerkungen zum kritischen Umgang mit Organisationstheorien", in: Kieser, A. (Ed.): Organisationstheorien. Stuttgart: Kohlhammer, 1-37.

Schlich, E.H. (2008): "Energy Economics and the Ecology of Scale in the Food Business", in: Caldwell, P.G. & Taylor, E.V. (Eds.): New Research on Energy Economics. Nova Science Publishers, 67-106.

Schmalensee, R. (1982): Product Differentiation Advantages of Pioneering Brands, The American Economic Review, Vol. 72, No. 3, 349-365.

Schmidheiny, S. (1992): Changing course: a global business perspective on development and the environment. Cambridge, MA: MIT Press.

Schneidewind, U. (1995): Chemie zwischen Wettbewerb und Umwelt. Perspektiven für eine wettbewerbsfähige und nachhaltige Chemieindustrie. Marburg: Metropolis-Verlag.

Schneidewind, U. (1998): Die Unternehmung als strukturpolitischer Akteur. Marburg: Metropolis-Verlag.

Schneidewind, U. & Rehm, A. (2010): "Vom Corporate Citizen zum strukturpolitischen Akteur – Gesellschaftliche Effekte und betrieblicher Nutzen unternehmerischen Engagements", in: Braun, S. (Ed.): Gesellschaftliches Engagement von Unternehmen. Wiesbaden: VS Verlag für Sozialwissenschaften.

Schoemaker, P.J.H. (1993): Multiple Scenario Development: Its Conceptual and Behavioral Foundation, Strategic Management Journal, Vol. 14, No. 3, 193-213.

Schoemaker, P.J.H. (1995): Scenario planning: a tool for strategic thinking, MIT Sloan Management Review, Vol. 36, 25-40.

Schön, D.A. (1971): Beyond the Stable State. New York, NY: Random House.

Schumpeter, J.A. (1912): Theorie der wirtschaftlichen Entwicklung. Leipzig: Duncker & Humblot.

Schweizer, L. (2005): Concept and evolution of business models, Journal of General Management, Vol. 31, No. 2, 37-56.

Schwenk, C.R. (1984): Cognitive Simplification Processes in Strategic Decision-Making, Strategic Management Journal, Vol. 5, No. 2, 111-128.

Schwenk, C.R. & Cosier, R.A. (1980): Effects of the Expert, Devil's Advocate, and Dialectical Inquiry Methods on Prediction Performance, Organizational Behavior & Human Performance, Vol. 26, No. 3, 409-424.

Scott, W.R. (1992): Organizations: Rational, Natural, and Open Systems. Englewood Cliffs, NJ: Prentice Hall, 3rd edition.

Scott, W.R. (2001): Institutions and organizations. Thousand Oaks et al.: Sage Publications, 2nd edition.

Scott, W.R. (2002): Organizations: Rational, Natural, and Open Systems. Prentice Hall, 5th edition.

Seelos, C. & Mair, J. (2007): Profitable Business Models and Market Creation in the Context of Deep Poverty: A Strategic View, Academy of Management Perspectives, Vol. 21, No. 4, 49-63.

Seitz, N. & Ellison, M. (2004): Capital Budgeting and Long-Term Financing Decisions. Mason, OH et al.: Thomson/South-Western, 4th edition.

Sells, B. (1994): What asbestos taught me about managing risk, Harvard Business Review, Vol. 72, No. 2, 76-90.

Selznick, P. (1948): Foundations of the Theory of Organization, American Sociological Review, Vol. 13, No. 1, 25-35.

Selznick, P. (1957): Leadership in Administration. New York, NY: Harper & Row.

Senge, P.M. (1990): The fifth discipline: the art and practice of the learning organization. New York et al.: Doubleday.

Senge, P.M. (1994): The fifth discipline fieldbook: strategies and tools for building a learning organization. New York, NY: Doubleday.

Senge, P.M. (2006): The fifth discipline: the art and practice of the learning organization. New York et al.: Currency/Doubleday, 2nd edition.

Senge, P.M.; Kleiner, A.; Roberts, C.; Ross, R.; Roth, G. & Smith, B. (1999): The Dance of Change. The challenges of sustaining momentum in learning organizations. New York et al.: Currency/Doubleday.

Senge, P.M.; Linchtenstein, B.B.; Kaeufer, K.; Bradbury, H. & Carroll, J. (2007): Collaborating for systemic change, MIT Sloan Management Review, Vol. 48, No. 2, 44-53.

Senge, P.M.; Smith, B.; Kruschwitz, N.; Laur, J. & Schley, S. (2008): The necessary revolution. How individuals and organizations are working together to create a sustainable world. London/Boston: Nicholas Brealey Publishing.

Sethi, S.P. (1994): Multinational corporations and the impact of public advocacy on corporate strategy: Nestlé and the infant formula controversy. Boston, MA: Kluwer Academic.

Shankar, V. (1997): Pioneers' Marketing Mix Reactions to Entry in Different Competitive Game Structures: Theoretical Analysis and Empirical Illustration, Marketing Science, Vol. 16, No. 3, 271-293.

Shankleman, M. (2007): Ryanair retracts emissions claim, BBC News. http://news.bbc.co.uk/2/hi/business/6310571.stm (Accessed: April 1, 2011).

Shapiro, A.C. (2005): Capital Budgeting and Investment Analysis. Upper Saddle River, NJ: Pearson/Prentice Hall.

Shapiro, S.M. (2001): 24/7 Innovation: A Blueprint for Surviving and Thriving in an Age of Change. New York, NY: McGraw-Hill.

Shapiro, S.M. (2008): Why Statistics Kill Innovation...And What You Can Learn From Vegas Card Counters. http://www.steveshapiro.com/pdfs/whystatisticskillinnovation.pdf (Accessed: September 29, 2009).

Sharfman, M.P. & Fernando, C.S. (2008): Environmental Risk Management and the Cost of Capital, Strategic Management Journal, Vol. 29, No. 6, 569-592.

Sharma, S. (2000): Managerial Interpretations and Organizational Context as Predictors of Corporate Choice of Environmental Strategy, Academy of Management Journal, Vol. 43, No. 4, 681-697.

Shell (2008): Scenarios: An Explorer's Guide. The Hague: Shell International BV.

Shell, E.R. (2009): Cheap: The High Cost of Discount Culture. New York, NY: The Penguin Press.

Shrivastava, P. (1994): Castrated Environment: Greening Organizational Studies, Organization Studies, Vol. 15, No. 5, 705-726.

Shrivastava, P. (1995a): Environmental Technologies and Competitive Advantage, Strategic Management Journal, Vol. 16, Special Issue: Technological Transformation and the New Competitive Landscape, 183-200.

Shrivastava, P. (1995b): The Role of Corporations in Achieving Ecological Sustainability, Academy of Management Review, Vol. 20, No. 4, 936-960.

Siegel, D.S. (2009): Green Management Matters Only If It Yields More Green: An Economic/Strategic Perspective, Academy of Management Perspectives, Vol. 23, No. 4, 5-16.

Siemens (2007): Performance Based Solutions: Putting energy savings back into your facilities. Zug, Switzerland et al.: Siemens Building Technologies. http://www.hqs.sbt.siemens.com/gip/general/dlc/data/assets/hq/A6V10084358_031_0-91841_en_hq-en.pdf (Accessed: September 2, 2010).

Siemens (2009): Sustainability Report 2008. Berlin/Munich: Siemens AG.

Siemens (2010a): Building Technologies: Performance Contracting, http://www.siemens.com/sustainability/en/environment/portfolio/building_technologies.htm#toc-1 (Accessed: September 2, 2010).

Siemens (2010b): Our Environmental Portfolio. Excerpt from the Siemens Sustainability Report 2009. Berlin/Munich: Siemens AG.

Siemens (2010c): Richard Hausmann to manage the Siemens Smart Grid Applications project, http://www.siemens.com/press/en/pressrelease/?press=/en/pressrelease/2010/corporate_communication/AXX20100453.htm (Accessed: August 14, 2010).

Da Silveira, G.; Borenstein, D. & Fogliatto, F.S. (2001): Mass customization: Literature review and research directions, International Journal of Production Economics, Vol. 72, No. 1, 1-13.

Simanis, E. & Hart, S. (2006): Expanding Possibilities at the Base of the Pyramid, Innovations: Technology, Governance, Globalization, Vol. 1, No. 1, 43-51.

Simon, H.A. (1947): Administrative Behavior. A Study of Decision-Making Processes in Administrative Organization. New York, NY: The Macmillan Company.

Simon, H.A. (1993): Strategy and Organizational Evolution, Strategic Management Journal, Vol. 14, 131-142.

Smartmeters.com (2010): Siemens to play catch-up, http://www.smartmeters.com/the-news/785-siemens-to-play-catch-up.html (Accessed: August 15, 2010).

Smith, A. (1863): An inquiry into the nature and causes of the wealth of nations. Edinburgh: Adam and Charles Black, new edition, McCulloch, J.R. (Ed.).

Smith, A. (2007): Translating Sustainabilities between Green Niches and Socio-Technical Regimes, Technology Analysis and Strategic Management, Vol. 19, No. 4, 427-450.

Smith, A.; Voß, J.-P. & Grin, J. (2010): Innovation studies and sustainability transitions: The allure of the multi-level perspective and its challenges, Research Policy, Vol. 39, No. 4, 435-448.

Solow, R.M. (1974): Intergenerational Equity and Exhaustible Resources, The Review of Economic Studies, Vol. 41, 29-45.

Sosna, M.; Trevinyo-Rodríguez, R.N. & Velamuri, S.R. (2010): Business Model Innovation through Trial-and-Error Learning: The Naturhouse Case, Long Range Planning, Vol. 43, No. 2/3, 383-407.

Sperling, C.W.; Millner, C. & Warner, J. (1998): Hollywood Be Thy Name. The Warner Brothers Story. Lexington, KY: University Press of Kentucky.

Stabell, C.B. & Fjeldstad, Ø.D. (1998): Configuring value for competitive advantage: On chains, shops, and networks, Strategic Management Journal, Vol. 19, No. 5, 413-437.

Stacey, R.D. (1995): The Science of Complexity: An Alternative Perspective for Strategic Change Processes, Strategic Management Journal, Vol. 16, No. 6, 477-495.

Stacey, R.D.; Griffin, D. & Shaw, P. (2002): Complexity and management: fad or radical challenge to systems thinking? London et al.: Routledge.

Steel, K.D. (2007): "The choice between grid and off-grid electrification in Kenya and its impact on system development", in: The 2007 International Conference of the System Dynamics Society and 50th Anniversay Celebration. Boston, MA: The System Dynamics Society.

Steel, K.D. (2008): Energy System Development in Africa: The case of grid and off-grid power in Kenya. Massachusetts Institute of Technology: Ph.D. thesis.

Steger, U. (2006): "Building a Business Case for Corporate Sustainability", in: Schaltegger, S. & Wagner, M. (Eds.): Managing the Business Case for Sustainability. Sheffield: Greenleaf Publishing, 412-443.

Steiners, D. (2010): Klimamärkte der Zukunft. Berlin: Bayer AG. http://www.econsense.de/_VERANSTALTUNGEN/images/2010_7_SWP-WS/Bayer%20Klimamaerkte%20der%20Zukunft%20econsense%20SWP.pdf (Accessed: January 4, 2011).

Steinfeld, H.; Gerber, P.; Wassenaar, T.; Castel, V.; Rosales, M. & de Haan, C. (2006): Livestock's long shadow. Environmental issues and options. Rome: Food and Agriculture Organization of the United Nations (FAO).

Stephens, E. et al. (2010): An economic and technical evaluation of microalgal biofuels, Nature Biotechnology, Vol. 28, No. 2, 126-128.

Sterman, J.D. (2000): Business dynamics: systems thinking and modeling for a complex world. Boston, MA et al.: McGraw-Hill.

Sterman, J.D. (2001): System Dynamics Modeling: Tools for Learning in a Complex World, California Management Review, Vol. 43, No. 4, 8-25.

Sterman, J.D. (2002): All models are wrong: reflections on becoming a systems scientist, System Dynamics Review, Vol. 18, No. 4, 501-531.

Sterman, J.D. (2008): Why Bad Things Happen to Good Technologies. Lecture. Boston, MA: MIT World. http://mitworld.mit.edu/video/553/ (Accessed: May 23, 2009).

Stern, N. (2007): The Economics of Climate Change. The Stern Review. Cambridge, UK: Cambridge University Press.

Stewart, R.F. (1963): A Framework for Business Planning. Stanford, CA: Stanford Research Institute.

Stiglitz, J.E.; Sen, A. & Fitoussi, J.-P. (2009): Report by the Commission on the Measurement of Economic Performance and Social Progress. http://www.stiglitz-sen-fitoussi.fr/documents/rapport_anglais.pdf (Accessed: March 28, 2010).

Strack, R.; Baier, J. & Fahlander, A. (2008): Managing Demographic Risk, Harvard Business Review, Vol. 86, No. 2, 119-128.

Stubbs, W. & Cocklin, C. (2008): Conceptualizing a "Sustainability Business Model," Organization & Environment, Vol. 21, No. 2, 103-127.

Suarez, F.F. & Oliva, R. (2005): Environmental change and organizational transformation, Industrial and Corporate Change, Vol. 14, No. 6, 1017-1041.

Suarez, F.F. & Utterback, J.M. (1995): Dominant Designs and the Survival of Firms, Strategic Management Journal, Vol. 16, No. 6, 415-430.

Sun Tzu (1963): The Art of War. Translated by Samuel B. Griffith. London: Oxford University Press.

Sundaram, A.K. & Inkpen, A.C. (2004): Stakeholder Theory and "The Corporate Objective Revisited": A Reply, Organization Science, Vol. 15, No. 3, 370-371.

SustainAbility (2008): A Survey of SustainAbility's Compass Network: Opinion Research from GlobeScan. Toronto: SustainAbility.

Sweeney, L.B. & Sterman, J.D. (2000): "Bathtub dynamics: Preliminary results of a systems thinking inventory", in: International System Dynamics Conference. Bergen.

Taleb, N.N. (2008): The black swan. The impact of the highly improbable. London: Penguin Books.

Tampi, T. (2009): Poor fuel quality and increased maintenance costs can tax savings from slow steaming, DNV Container Ship Update, No. 3. http://www.dnv.com/industry/maritime/publicationsanddownloads/publications/dnvcontainershipupdate/2009/3-2009/Poorfuelqualityandincreasedmaintenancecostscantaxsavingsfromslowsteaming.asp (Accessed: September 20, 2010).

Tannenbaum, R. & Schmidt, W.H. (1973): How to choose a leadership pattern. Should a manager be democratic or autocratic-or something inbetween?, Harvard Business Review, 162-180.

Tapscott, D.; Ticoll, D. & Lowy, A. (2000): Digital Capital: Harnessing the Power of Business Webs. Boston, MA: Harvard Business School Press.

Taylor, A. (2006): The Birth of the Prius, Fortune Magazine, Vol. 153, No. 4. http://money.cnn.com/magazines/fortune/fortune_archive/2006/03/06/8370702/ (Accessed: March 18, 2011).

Taylor, F.W. (1911): Principles of Scientific Management. New York/London: Harper & Brothers.

TED.com (2009): Shai Agassi's bold plan for electric cars. http://www.ted.com/talks/shai_agassi_on_electric_cars.html (Accessed: September 21, 2010).

Teece, D.J. (2008): Dosi's technological paradigms and trajectories: insights for economics and management, Industrial and Corporate Change, Vol. 17, No. 3, 507-512.

Teece, D.J. (2010): Business Models, Business Strategy and Innovation, Long Range Planning, Vol. 43, No. 2/3, 172-194.

Teece, D.J. & Pisano, G. (1994): The Dynamic Capabilities of Firms: an Introduction, Industrial and Corporate Change, Vol. 3, No. 3, 537-556.

Teece, D.J.; Pisano, G. & Shuen, A. (1997): Dynamic capabilities and strategic management, Strategic Management Journal, Vol. 18, No. 7, 509-533.

The Climate Group (2008): SMART 2020: Enabling the Low Carbon Economy in the Information Age. Global eSustainability Initiative (GeSI).

The Economist (2009): Getting warmer. A special report on climate change and the carbon economy. The Economist.

The Economist (2010): BP and the oil spill: The oil well and the damage done, The Economist. http://www.economist.com/node/16381032 (Accessed: January 24, 2011).

The Economist (2011): Nuclear power: When the steam clears. The Fukushima crisis will slow the growth of nuclear power. Might it reverse it?, The Economist. http://www.economist.com/research/articlesBySubject/displaystory.cfm?subjectid=821240&story_id=18441163 (Accessed: April 12, 2011).

Tilman, D. et al. (2009): Beneficial Biofuels - The Food, Energy, and Environment Trilemma, Science, Vol. 325, No. 5938, 270-271.

Time Magazine (1984): Dividends. End of a Boycott, Time Magazine. http://www.time.com/time/magazine/article/0,9171,954158,00.html (Accessed: October 13, 2010).

Timmers, P. (1998): Business models for electronic markets, Journal on Electronic Markets, Vol. 8, No. 2, 3-8.

TNS Opinion & Social (2006): Risk Issues. European Commission. http://ec.europa.eu/food/food/resources/special-eurobarometer_riskissues20060206_en.pdf (Accessed: October 15, 2010).

TNS Opinion & Social (2008): Attitudes of European citizens towards the environment. European Commission. http://ec.europa.eu/public_opinion/archives/ebs/ebs_322_en.pdf (Accessed: April 20, 2010).

TNS Opinion & Social (2009): Europeans' attitudes towards climate change. European Commission. http://ec.europa.eu/public_opinion/archives/ebs/ebs_322_en.pdf (Accessed: April 20, 2010).

Toyota (2009): 2010 Prius Plug-in Hybrid Debuts at Frankfurt Motor Show, http://www.toyota.com/about/news/product/2009/09/09-1-prius-hybrid.html (Accessed: September 21, 2010).

Toyota (2010a): 2010 Prius Plug-In Hybrid Vehicle, http://www.toyota.com/esq/articles/2010/Prius_Plgin_In_Hybrid.html?type=Topics&title=Advanced%20Technology%20Vehicles&url=/esq/topics/2010/Vehicles.html (Accessed: September 12, 2010).

Toyota (2010b): Toyota Celebrates 10 Years of Prius, http://pressroom.toyota.com/pr/tms/toyota/toyota-celebrates-10-years-of-164612.aspx (Accessed: September 12, 2010).

Toyota Motor Corporation (2008): Worldwide Prius Sales Top 1 Million Mark, http://www.toyota.co.jp/en/news/08/0515.pdf (Accessed: May 26, 2009).

Toyota Motor Sales, U.S.A. (2009): Toyota Model Selector - All Vehicles, Toyota Model Selector. http://www.toyota.com/modelselector/ (Accessed: May 25, 2009).

Treumann, J. (2007): A Brand-New Brew, Time. http://www.time.com/time/magazine/article/0,9171,901070122-1575303,00.html (Accessed: February 7, 2010).

Tuckman, B.W. (1965): Developmental sequence in small groups, Psychological bulletin, Vol. 63, No. 6, 384-399.

Tuckman, B.W. & Jensen, M.A. (1977): Stages of small-group development revisited, Group & Organization Management, Vol. 2, No. 4, 419-427.

Tushman, M.L. & Romanelli, E. (1985): "Organizational evolution: A metamorphosis model of convergence and reorientation", in: Cummings, L.L. & Staw, B.M. (Eds.): Research in organizational behavior. Greenwich, CN: JAI Press, 171-222.

Tushman, M.L.; Newman, W.H. & Romanelli, E. (1986): Convergence and Upheaval: Managing the Unsteady Pace of Organizational Evolution, California Management Review, Vol. 29, No. 1, 29-44.

Tversky, A. & Kahneman, D. (1983): Extensional versus intuitive reasoning: The conjunction fallacy in probability judgment, Psychological Review, Vol. 90, No. 4, 293-315.

Ulrich, P.; Duranton, S. & Koehler, M. (2009): Planes, Trains, and Automobiles. Crossing Paths in European Travel. The Boston Consulting Group.

Umbeck, T. (2009): Musterbrüche in Geschäftsmodellen: ein Bezugsrahmen für innovative Strategie-Konzepte. Wiesbaden: Gabler.

UN (2004): World Population to 2300. New York: United Nations. http://www.un.org/esa/population/publications/longrange2/WorldPop2300final.pdf (Accessed: November 18, 2010).

UNEP (2008): Towards sustainable production and use of resources: Assessing biofuels. Paris, France: United Nations Environment Programme (UNEP).

UNEP (2010a): Annual Report 2009. Seizing the Green Opportunitiy. Nairobi, Kenya: United Nations Environment Programme (UNEP).

UNEP (2010b): The Emissions Gap Report. Are the Copenhagen Accord pledges sufficient to limit global warming to 2° C or 1.5° C? United Nations Environment Programme (UNEP).

UNEP & Mercer (2007): Demystifying Responsible Investment Performance: A review of key academic and broker research on ESG factors. UNEP Finance Initiative Asset Management Working Group and Mercer. http://www.unepfi.org/fileadmin/documents/Demystifying_Responsible_Investment_Performance_01.pdf (Accessed: March 25, 2009).

Unilever (2009a): Unilever & palm oil, http://www.unilever.com/mediacentre/news/unileverandpalmoil.aspx (Accessed: January 16, 2010).
Unilever (2009b): Unilever global company website, http://www.unilever.com/ (Accessed: January 16, 2010).
Unruh, G. & Ettenson, R. (2010a): growing Green, Harvard Business Review, Vol. 88, No. 6, 94-100.
Unruh, G. & Ettenson, R. (2010b): Winning in the Green Frenzy, Harvard Business Review, Vol. 88, No. 11, 110-115.
Urban, G.L.; Carter, T.; Gaskin, S. & Mucha, Z. (1986): Market Share Rewards to Pioneering Brands: An Empirical Analysis and Strategic Implications, Management Science, Vol. 32, No. 6, 645-659.
US EPA (2011): Green Chemistry, http://www.epa.gov/greenchemistry/ (Accessed: January 4, 2011).
USDA (2010): Agricultural Productivity in the United States, http://www.ers.usda.gov/Data/AgProductivity/ (Accessed: October 14, 2010).
Utterback, J.M. & Abernathy, W.J. (1975): A dynamic model of process and product innovation, Omega, Vol. 3, No. 6, 639-656.
Uzzi, B. & Dunlap, S. (2005): How to Build Your Network, Harvard Business Review, Vol. 83, No. 12, 53-60.
del Val, M.P. & Fuentes, C.M. (2003): Resistance to change: a literature review and empirical study, Management Decision, Vol. 41, No. 2, 148-155.
VanderWerf, P.A. & Mahon, J.F. (1997): Meta-Analysis of the Impact of Research Methods on Findings of First-Mover Advantage, Management Science, Vol. 43, No. 11, 1510-1519.
Van de Ven, A.H. (1986): Central Problems in the Management of Innovation, Management Science, Vol. 32, No. 5, 590-607.
Van de Ven, A.H. & Poole, M.S. (1995): Explaining Development and Change in Organizations, Academy of Management Review, Vol. 20, No. 3, 510-540.
Verfaillie, H.A. & Bidwell, R. (2000): Measuring eco-efficiency. A guide to reporting company performance. World Business Council for Sustainable Development (WBCSD).
Vlasic, B. & Bunkley, N. (2008): General Motors, Driven to the Brink, The New York Times. http://www.nytimes.com/2008/10/26/business/26jane.html?pagewanted=all (Accessed: August 17, 2009).
Voelpel, S.; Leibold, M.; Tekie, E. & Von Krogh, G. (2005): Escaping the Red Queen Effect in Competitive Strategy: Sense-testing Business Models, European Management Journal, Vol. 23, No. 1, 37-49.
Volkswagen (2010): Technische Daten > Zahlen & Fakten > Der Golf > Golf > Modelle > Volkswagen Deutschland, http://www.volkswagen.de/vwcms/master_public/virtualmaster/de3/modelle/golf/golf/zahlen___fakten/daten.detail.0.12.html (Accessed: January 4, 2010).
Vranica, S. (2010): Brands Bounce Back, With Some Casualties, The Wall Street Journal. http://online.wsj.com/article/SB10001424052748704652104575493861108689980.html (Accessed: February 17, 2011).
Wack, P. (1985a): Scenarios: shooting the rapids, Harvard Business Review, Vol. 63, No. 6, 139-150.
Wack, P. (1985b): Scenarios: uncharted waters ahead, Harvard Business Review, Vol. 63, No. 5, 73-89.
Wagner, M.; Phu, N.V.; Azomahou, T. & Wehrmeyer, W. (2002): The relationship between the environmental and economic performance of firms: an empirical analysis of the European paper industry, Corporate Social Responsibility and Environmental Management, Vol. 9, No. 3, 133-146.

Wald, M.L. (2009): “Smart” Electric Utility Meters, Intended to Create Savings, Instead Prompt Revolt, The New York Times. http://www.nytimes.com/2009/12/14/us/14meters.html (Accessed: July 6, 2010).

Walley, N. & Whitehead, B. (1994): It's Not Easy Being Green, Harvard Business Review, Vol. 72, No. 3, 46-51.

Wallin, J. (2000): "Operationalizing Competencies", in: 5th Annual International Conference on Competence-Based Management. Helsinki.

Wal-Mart (2007): Sustainability Progress to Date 2007–2008: We're making sustainability our business. Wal-Mart Stores, Inc. http://walmartfacts.com/reports/2006/sustainability/documents/SustainabilityProgressToDate2007-2008.pdf (Accessed: March 25, 2009).

Wal-Mart (2009a): Sustainability Index, http://walmartstores.com/sustainability/9292.aspx (Accessed: November 17, 2010).

Wal-Mart (2009b): Wal-Mart 2009 Global Sustainability report. Bentonville, AR: Wal-Mart Stores, Inc. http://walmartstores.com/sites/sustainabilityreport/2009/ (Accessed: March 25, 2009).

Waltz, E. (2009): Biotech's green gold?, Nature Biotechnology, Vol. 27, 15-18.

Warner, M. (2005): What Is Organic? Powerful Players Want a Say, The New York Times. http://www.nytimes.com/2005/11/01/business/01organic.html (Accessed: January 13, 2011).

WBCSD (1997): Exploring Sustainable Development. WBCSD Global Scenarios 2000 - 2050, Summary Brochure. London: World Business Council for Sustainable Development (WBCSD). http://www.wbcsd.org/includes/getTarget.asp?type=d&id=ODk5MA (Accessed: February 18, 2010).

WBCSD (2000): Eco-efficiency. Creating more value with less impact. Geneva: World Business Council for Sustainable Development (WBCSD).

WCED (1987): Our Common Future: Report of the World Commission on Environment and Development. Oxford/NewYork: Oxford University Press. http://www.un-documents.net/wced-ocf.htm (Accessed: October 1, 2009).

Webb, T. (2010): Few air travellers offset carbon emissions, study finds, The Guardian. http://www.guardian.co.uk/business/2010/aug/30/carbon-emissions-offset-civil-aviation-authority (Accessed: March 13, 2011).

Weber, M. (2002): Wirtschaft und Gesellschaft: Grundriß der Verstehenden Soziologie. Tübingen: Mohr Siebeck, 5th edition.

WEF (2009): The Next Billions: Unleashing Business Potential in Untapped Markets. Geneva: World Economic Forum (WEF) and The Boston Consulting Group (BCG).

Weick, K.E. (1979): The social psychology of organizing. New York, NY: Random House, 2nd edition.

Weick, K.E. (1995): Sensemaking in Organizations. Thousand Oaks et al.: Sage Publications.

Weick, K.E. & Quinn, R.E. (1999): Organizational change and development, Annual Review of Psychology, Vol. 50, No. 1, 361-386.

Weill, P. & Broadbent, M. (1998): Leveraging the new infrastructure. How market leaders capitalize on information technology. Boston, MA: Harvard Business School Press.

Weill, P. & Vitale, M. (2001): Place to Space: Migrating to eBusiness Models. Boston, MA: Harvard Business School Press.

Weiss, J. (2006): The quotable manager: inspiration for business and life. Layton, UT: Gibbs Smith.

Weitzman, M.L. (1998): Why the Far-Distant Future Should Be Discounted at Its Lowest Possible Rate, Journal of Environmental Economics and Management, Vol. 36, No. 3, 201-208.

von Weizsäcker, C.F. (1988): Bewußtseinswandel. Munich: Carl Hanser.

von Weizsäcker, E.U.; Lovins, A.B. & Lovins, L.H. (1995): Faktor vier: doppelter Wohlstand - halbierter Naturverbrauch. Der neue Bericht an den Club of Rome. München: Droemer Knaur.
Welch, J. (2001): Jack - Straight from the Gut. New York, NY: Warner.
Wernerfelt, B. (1984): A Resource-Based View of the Firm, Strategic Management Journal, Vol. 5, No. 2, 171-180.
Wertheimer, M. (1925): Über Gestalttheorie, Philosophische Zeitschrift für Forschung und Aussprache, Vol. 1, 39-60.
Wetlaufer, S. (2001): The Business Case Against Revolution: An Interview with Nestle's Peter Brabeck, Harvard Business Review, Vol. 79, No. 2, 112-121.
WFP (2010): Fighting Hunger Worldwide. Annual Report 2010. World Food Programme (WFP).
WHO (2003): Obesity and overweight. World Health Organization (WHO). http://www.who.int/dietphysicalactivity/publications/facts/obesity/en/ (Accessed: October 14, 2010).
Wikimedia Commons (2006): File:Duck-Rabbit illusion.jpg - Wikimedia Commons, http://commons.wikimedia.org/wiki/File:Duck-Rabbit_illusion.jpg (Accessed: December 14, 2009).
Wikimedia Commons (2007): File:Gestalt5.PNG - Wikimedia Commons, http://commons.wikimedia.org/wiki/File:Gestalt5.PNG (Accessed: December 14, 2009).
Wikipedia (2009a): Gunpowder artillery in the Middle Ages - Wikipedia, the free encyclopedia, http://en.wikipedia.org/wiki/Gunpowder_artillery_in_the_Middle_Ages (Accessed: June 2, 2009).
Wikipedia (2009b): I Ching - Wikipedia, the free encyclopedia, http://en.wikipedia.org/wiki/Yijing (Accessed: July 7, 2009).
Wikipedia (2010a): Green Revolution, http://en.wikipedia.org/wiki/Green_Revolution (Accessed: October 14, 2010).
Wikipedia (2010b): List of system dynamics software, Wikipedia, the free encyclopedia. http://en.wikipedia.org/wiki/List_of_system_dynamics_software (Accessed: February 10, 2010).
Wikipedia (2010c): VW Golf I – Wikipedia, http://de.wikipedia.org/wiki/Golf_I (Accessed: January 4, 2010).
Wikipedia (2010d): World's largest airlines - Wikipedia, the free encyclopedia, http://en.wikipedia.org/wiki/List_of_largest_airlines (Accessed: May 18, 2010).
Wikipedia (2011): Irving Fisher - Wikipedia, the free encyclopedia, http://en.wikipedia.org/wiki/Irving_Fisher#Stock_market_crash_of_1929 (Accessed: November 12, 2011).
Wikiquote (2009): Arthur Schopenhauer - Wikiquote, Arthur Schopenhauer - Wikiquote. http://en.wikiquote.org/wiki/Arthur_Schopenhauer (Accessed: October 1, 2009).
Wikiquote (2011a): Change - Wikiquote, http://en.wikiquote.org/wiki/Change (Accessed: November 15, 2011).
Wikiquote (2011b): Incorrect predictions, http://en.wikiquote.org/wiki/Incorrect_predictions (Accessed: March 25, 2011).
Wilhelm, R. & Baynes, C.F. (1967): The I Ching or Book of Changes. Princeton, NJ: Princeton University Press, 3rd edition.
Willenbrock, H. (2003): Eiskalt abserviert, brand eins Magazin. http://www.brandeins.de/home/inhalt_detail.asp?id=113 (Accessed: August 14, 2009).
Williamson, O.E. (1981): The Economics of Organization: The Transaction Cost Approach, The American Journal of Sociology, Vol. 87, No. 3, 548-577.
Wilson, I. (1994): Strategic planning isn't dead - it changed, Long Range Planning, Vol. 27, No. 4, 12-24.

Winn, M.L. & Angell, L.C. (2000): Towards a Process Model of Corporate Greening, Organization Studies, Vol. 21, No. 6, 1119-1147.

Winter, S.G. & Nelson, R.R. (1982): An Evolutionary Theory of Economic Change. Cambridge, MA: Belknap Press. http://papers.ssrn.com/sol3/papers.cfm?abstract_id =1496211 (Accessed: November 28, 2010).

Wirtz, B.W.; Schilke, O. & Ullrich, S. (2010): Strategic Development of Business Models: Implications of the Web 2.0 for Creating Value on the Internet, Long Range Planning, Vol. 43, No. 2/3, 272-290.

Witt, U. (2008): What is specific about evolutionary economics?, Journal of Evolutionary Economics, Vol. 18, No. 5, 547-575.

Yin, R.K. (2003): Case Study Research. Design and Methods. Thousand Oaks et al.: SAGE Publications, 3rd edition.

Z_punkt (2008): Megatrends. Köln: Z-punkt GmbH The Foresight Company. http://www.z-punkt.de/fileadmin/be_user/D_Publikationen/D_Arbeitspapiere/Die_20_wichtigsten_Megatrends_x.pdf (Accessed: March 24, 2011).

Die Zeit (2011): BMW lanciert neue Elektroauto-Marke, Die Zeit. http://www.zeit.de/auto/2011-02/bmw-elektroauto (Accessed: March 15, 2011).

Zeug, K. (2009): Alternative Energie: Kein Licht für Afrika, Die Zeit. http://www.zeit.de/2009/03/Afrika (Accessed: August 18, 2010).

Zipcar (2010): Green benefits, see why car sharing is sustainable – Zipcar, http://www.zipcar.com/is-it/greenbenefits (Accessed: September 14, 2010).

Zipkin, P. (2001): The Limits of Mass Customization, MIT Sloan Management Review, Vol. 42, No. 3, 81-87.

Zook, C. (2007): Finding Your Next CORE Business. What if you've taken your core as far as it can go?, Harvard Business Review, Vol. 85, No. 4, 66-75.

Zott, C. & Amit, R. (2006): Exploring the fit between business strategy and business model: Implications for firm performance. Working paper.

Zott, C. & Amit, R. (2010): Business Model Design: An Activity System Perspective, Long Range Planning, Vol. 43, No. 2/3, 216-226.

Zucker, L.G. (1987): Institutional Theories of Organization, Annual Review of Sociology, Vol. 13, 443-464.

Zwicky, F. (1969): Discovery, Invention, Research Through the Morphological Approach. New York, NY: Macmillan.

Appendix 1: List of Interviews

Most interviews were conducted via telephone, some also in person. An average interview lasted about an hour. Dates refer to the first conversation, possible later inquiries are omitted.

Corporate representatives:

ABB	Anders Nordström Group Sustainability Advisor Sustainability Affairs 14 September 2010
BASF	Andreas Kicherer Head of Marketing & Sales Sustainability Services 20 May 2010
Bayer	Daniel Steiners Head of Corporate Strategy Projects 14 July 2010
BMW	Martin Arlt Head of Project Management of project i 10 November 2010
BYD	Leevon Tian Europe Launch Manager 24 October 2010
Cognis	Rita Köster Director Global Marketing Home Care / I & I 3 September 2010
Cognis	Viola Möller Corporate Communications Manager 3 September 2010
Cognis	Mareike Klein Corporate Communications Manager 3 September 2010
Daimler	Stefan Pfahl Advanced Vehicle Engineering & Technology Strategy 30 November 2010
Frosta	Felix Ahlers CEO 25 October 2010
General Electric	Carlos Haertel Director GE Global Research – Europe 14 April 2010
IBM	Klaus Buescher Managing Consultant, Strategy & Transformation Management Consulting 4 October 2010
Lufthansa	Martin Koehler Member of the Supervisory Board (also: former Senior Partner at BCG) 18 May 2010
Lufthansa	Joachim Buse Vice President Aviation Biofuel 26 May 2010
Maersk	Soren Stig Nielsen Head of Sustainability, Maersk Line 26 October 2010
Maersk	Erling Johns Nielsen Global Head of Supply Chain Development, Damco 7 September 2010
Nestlé	Claus Conzelmann Head of Safety, Health & Environmental Sustainability 29 September 2010
Osram	Christian Merz Director Sustainability 18 August 2010

Osram
Jochen Berner
Project Manager Off-Grid
25 August 2010

Otto
Andreas Streubig
Division Manager Environmental and Social Policy
10 September 2010

Philips
Nick Kelso
Senior Communications Manager Africa, Philips Lighting
17 September 2010

Procter & Gamble
Len Sauers
Vice President Global Sustainability
25 October 2010

RWE
Marcus Groll
Head of OEM Relations
14 July 2010

SAP
Carsten Richter
EMEA Lead Carbon Management BTS
16 March 2010

SAP
James Farrar
Vice President Sustainability
6 September 2010

Siemens
Ralf Pfitzner
Head Environmental Portfolio & Sustainability Reporting
8 July 2010

Siemens
Sven Scheuble
Vice President Siemens Management Consulting
3 June 2010

Siemens
Gunnar Liehr
Global Head Energy Services & Performance Contracting
30 August 2010

Toyota
Dirk Breuer
Advisor Advanced Technology
10 September 2010

NGO representative:

Greenpeace
Gerd Leipold
Greenpeace International
Executive Director (2001 - 2009)
7 October 2010

Management Consultants:

BCG
Luc de Brabandere
Partner and Managing Director
14 October 2009

BCG
Jens Baier
Partner and Managing Director
1 February 2010

BCG
Tobias Radel
Partner and Managing Director
26 May 2010

BCG
Gerrit Amthor
Principal
27 May 2010

BCG
Frank Klose
Partner and Managing Director
1 June 2010

BCG
Henning von Kottwitz
Principal
12 August 2010

BCG
Jan Justus
Principal
16 September 2010

BCG
Andreas Dinger
Partner and Managing Director
20 September 2010

BCG
Cornelius Anger
Principal
4 October 2010

Appendix 2: Case Study Protocol Excerpt

One of three types of interview templates (A, B, and C) were used to develop questions tailored to the respective firm and interviewee. Template A emphasises the market side and viability-aspects of green business models. Template B focuses on capabilities that are needed for managing Green Business Model Transformations, whereas template C examines the transformation process itself.

For each interview template, main questions are formulated on the first level (i.e., "A.1", "A.2", etc.). On the levels beneath are potential follow-up questions. Questions were only used as loose reference points because interviews geared towards specific issues of firms turned out to be much more insightful than standardised but general questions.

A. Green Business Model Opportunities

A.1. Does environmental sustainability currently offer significant untapped opportunities in the interviewee's industry?

A.1.1. How big are these opportunities?
A.1.2. Do they have the potential to transform your company/industry?
A.1.3. Why are they untapped?
A.1.4. Has the interviewee's company conducted analyses in this field (e.g., megatrends, scenario analysis)

A.2. What would a (hypothetical) green business model in this industry look like?

A.2.1. Which environmental issues would be affected?
A.2.2. What magnitude would the improvements have?

A.3. When comparing this (hypothetical) business model to the current one – which components would need to be changed significantly? (overview question – detailed in A.4 and A.5)

A.3.1. New value propositions?
A.3.2. New target groups?
A.3.3. New key resources?
A.3.4. New key processes?
A.3.5. New financial logic?

A.4. Which non-economic (ecological or social) challenges would need to be resolved in order to realise a green business model?

A.4.1. Are there technical challenges that possibly make it unsafe or unreliable?
A.4.2. Would the business model immediately unfold its environmental potential or is there a transition time (e.g., until customers have learned to use the product)?
A.4.3. Are there problematic systemic effects that make the business model questionable in a large-scale context?
A.4.4. Are there major uncertainties, e.g. with respect to scientific evidence on certain environmental problems?
A.4.5. Are there conditions in the broader business environment (e.g., politics, supply chain, media, etc.) that impede or prevent its realisation?

A.5. What conditions need to change for green business models to be economically viable?

A.5.1. Are radical green business models already economically viable in the industry today?
A.5.1.1. What megatrends are working for (against) it?
A.5.1.2. What are the characteristics of competitive dynamics regarding green business models now and in the future?
A.5.1.3. What are the critical assumptions/elements that influence the overall economics?
A.5.2. Required changes of conditions: Value propositions?
A.5.2.1. Changes in attitude of customers/consumers?
A.5.2.2. Better communication of eco advantages?
A.5.2.3. Elimination of need for "eco-renouncement"?
A.5.3. Required changes of conditions: Target groups?
A.5.3.1. Rise of the next Billion?
A.5.3.2. Establishment of new channels or relationships?
A.5.3.3. Improvement of position or becoming indispensable as a supplier?
A.5.4. Required changes of conditions: Key resources?
A.5.4.1. Green Re-branding?
A.5.4.2. Technological advances, e.g. in clean tech?
A.5.4.3. Development of new infrastructure and assets?
A.5.4.4. New or different kind of business ecosystem (suppliers, partners, etc.)?
A.5.4.5. New internal capabilities? (see interview template B)
A.5.5. Required changes of conditions: Key processes?
A.5.5.1. New primary processes?
A.5.5.2. New support processes?
A.5.5.3. More entrepreneurial risk taking?
A.5.5.4. Use of different steering mechanisms (metrics and or norms)?
A.5.6. Required changes of conditions: Financial logic?
A.5.6.1. New revenue streams?
A.5.6.2. New cost base?
A.5.6.3. Different margin model (e.g., razor-blade)?

A.5.7. Required changes of conditions: Business environment?
A.5.7.1. Change in political conditions?
A.5.7.2. Leadership of respective industry leaders (or at least absence of resistance)?
A.5.7.3. Better-educated public or more differentiated discussion in the media?
A.5.7.4. More constructive approaches by NGOs?

A.6. Could the current business model become fundamentally threatened by environmental issues in the future?

A.6.1. Is there a risk that your whole industry might lose its license-to-operate?
A.6.2. Are other players better positioned to introduce a green business model?
A.6.2.1. Among your competitors?
A.6.2.2. From other industries?
A.6.3. Is there a first-mover advantage for green business models in the industry?
A.6.4. How big is the risk of being a follower?
A.6.5. Are such long-term threats being considered systematically in your company?

B. Capabilities for managing green business model transformations (remark: a particular green business model should serve as a reference for the following)

B.1. Does environmental sustainability currently offer significant untapped opportunities in the interviewee's industry?

B.1.1. How big are these opportunities?
B.1.2. Do they have the potential to transform your company/industry?
B.1.3. Why are they untapped?
B.1.4. Has the interviewee's company conducted analyses in this field (e.g., megatrends, scenario analysis)

B.2. Do mental models of managers within the company impede the realisation of promising green business models?

B.3. What organisational capabilities are most critical for green business model transformations?

B.3.1. Cognitive capabilities?
B.3.1.1. Vision and leadership commitment? What kind (personalised, formalised, etc.)?
B.3.1.2. Systems perspective: ability to understand conditions beyond the org's boundaries?

B.3.1.3. Dealing with uncertainty (e.g., through long-term scenario planning)?
B.3.2. Innovation capabilities?
B.3.2.1. For which types of innovation (product vs. process vs. organisational; radical vs. continuous)?
B.3.2.2. Ability to experiment widely, accept failure?
B.3.2.3. Learning and sharing within the organisation? With the external environment?
B.3.3. General management capabilities?
B.3.3.1. Strategic management (i.e. planning)?
B.3.3.2. Flexibility of norms and metrics (i.e. not discriminating new BMs)?
B.3.3.3. Decentralised accountability?
B.3.3.4. Cross-departmental cooperation?
B.3.3.5. Implementing large-scale transformations (see interview template C)?
B.3.4. Dynamic capabilities: adapting dynamically to changing business environments?
B.3.5. Sustainability-specific capabilities?
B.3.5.1. Using specific sustainability frameworks?
B.3.5.2. Applying tools for measuring and managing performance on sustainability issues?
B.3.5.3. Strong internal expertise on scientific fields of sustainability?

B.4. How important is stakeholder management for green business models?

B.4.1. Which ones are especially critical: Customers? Suppliers? Competitors? Politics? NGOs? Media? The public at large?
B.4.2. How important is learning from stakeholders?
B.4.2.1. From which ones in particular?
B.4.3. How important is influencing stakeholders (stakeholder communications)?
B.4.3.1. Broad stakeholder communication (to media, NGOs, the public at large)?
B.4.3.2. Influencing customer attitudes and needs?
B.4.3.3. Ability to shape regulatory policy?
B.4.4. How important is cooperating with stakeholders?
B.4.4.1. With which ones in particular?

B.5. Are needed capabilities the same for all green business model transformations?

B.5.1. Do they differ by industry?
B.5.2. Do they differ by type of green business model?

C. Transformation process

C.1. How can outdated mental models be changed effectively?

C.1.1. Are there especially suitable tools (like scenario planning)?
C.1.2. For which audiences do they work?
C.1.3. Are they suitable for all industries and firms?

C.2. How can ideas for new, green business models be generated?

C.2.1. Is it more successful to use small groups or wisdom of the crowds (or both)?
C.2.2. How do small groups become creative and productive?
C.2.2.1. What tools/formats are available?
C.2.2.2. Which members should participate in such groups?
C.2.2.3. Which other circumstances affect the success of such groups?
C.2.3. How can the wisdom of the crowds best be exploited?
C.2.3.1. What tools/formats are available?
C.2.3.2. Who should be invited to participate?
C.2.3.3. How can the crowd be directed effectively without curbing creativity?
C.2.4. Are there other effective ways of creating ideas?

C.3. How can ideas be converted into viable business models?

C.3.1. What is the current approach in practice? (art vs. science; ad hoc vs. systematic process)
C.3.2. Are there special organisational units dedicated to this task?
C.3.3. What are common problems of the current approaches and how could they be mitigated?

C.4. How can the implementation of the green business models be planned?

C.4.1. To what extent should it be planned at all?
C.4.2. What planning tools are used currently?
C.4.2.1. Are they suitable for green business models?
C.4.2.2. Where are potential shortcomings?
C.4.3. How flexible should the planning be?
C.4.3.1. How to deal with uncertainties?
C.4.3.2. To what extent are revisions accepted or even encouraged?
C.4.4. Who should be involved in the planning (at what stage)?

C.5. How can the implementation be executed successfully?

C.5.1. Has the company implemented comparable business model transformations in the past?
C.5.1.1. How were they similar?
C.5.1.2. How successful were these implementations and why?
C.5.1.3. What distinguishes transformations of green business models?
C.5.2. How should the implementation be organised
C.5.2.1. Implementation team: size, composition, organisational integration, ...?
C.5.2.2. Governance: e.g., strong central steering vs. decentralised decision making?
C.5.2.3. What level of participation is optimal? Is this level company-specific?
C.5.3. What are relevant organisational hurdles and how can they be overcome?

C.6. How can the implemented business model be managed and monitored effectively?

C.6.1. Are there monitoring systems in place that might be used?
C.6.2. How do requirements differ owing to environmental sustainability aspects?
C.6.3. What important factors are easily neglected?
C.6.4. Are there best practices?

C.7. What change management issues need to be resolved during the various phases?

C.7.1. How can various types of resistance be prevented?
C.7.2. How can existent resistance be dealt with effectively?
C.7.3. Under which circumstances do the above mentioned measures work? When don't they?
C.7.4. Do green business models require special forms of change management?

Appendix 3: Industry Profiles of Relative Environmental Impact

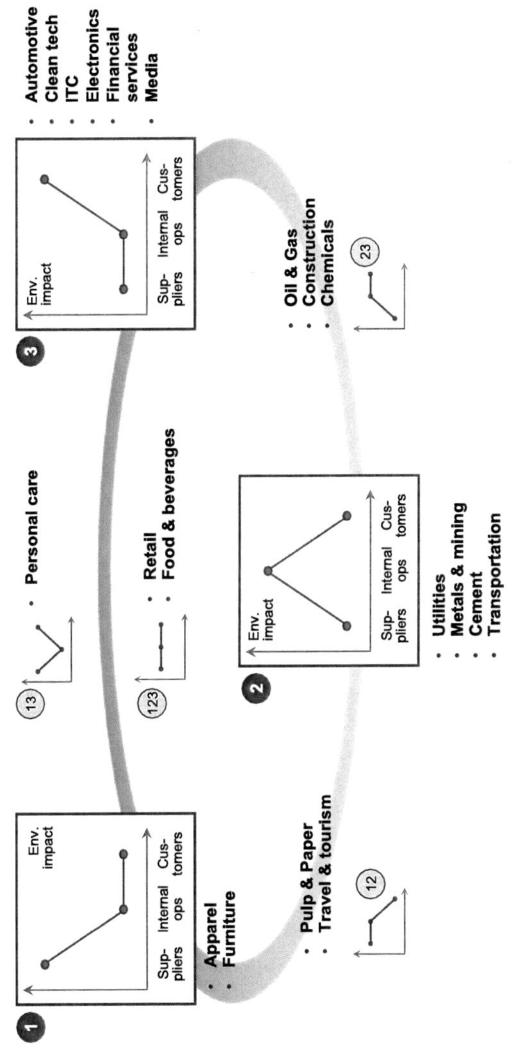